Transforming Automobile Assembly

Springer

Berlin
Heidelberg
New York
Barcelona
Budapest
Hong Kong
London
Milan
Paris
Santa Clara
Singapore
Tokyo

K. Shimokawa · U. Jürgens
T. Fujimoto (Eds.)

Transforming Automobile Assembly

Experience in Automation and Work Organization

With 140 Figures

Springer

Professor Koichi Shimokawa
Hosei University
Faculty of Business
2-17-1 Fujimi, Chiyoda-ku
Tokyo 102
Japan

Dr. Ulrich Jürgens
WZB Wissenschaftszentrum Berlin
für Sozialforschung
Abt. Regulierung von Arbeit
Reichspietschufer 50
10785 Berlin
Germany

Professor Takahiro Fujimoto
University of Tokyo
Faculty of Economics
7-3-1, Hongo, Bunkyo-ku
Tokyo 113
Japan

ISBN-13:978-3-642-64377-4

Cip data applied for

Die Deutsche Bibliothek – Cip-Einheitsaufnahme
Transforming automobile assembly : experience in automation and work organization / K. Shimokawa ... (ed.). - Berlin ; Heidelberg ; New York ; Barcelona ; Budapest ; Hong Kong ; London ; Milan ; Paris ; Santa Clara ; Sigapore ; Tokyo : Springer, 1997
ISBN-13:978-3-642-64377-4 e-ISBN-13:978-3-642-60374-7
DOI:10.1007/978-3-642-60374-7

NE: Shimokawa, Koichi [Hrsg.]

Softcover reprint of the hardcover 1st edition 1997

Typesetting: Data-conversion by M. Schillinger-Dietrich, Berlin
SPIN:10502949 68/3020-5 4 3 2 1 0 - Printed on acid -free paper

Contents

1. Introduction

T. Fujimoto · U. Jürgens · K. Shimokawa

1.1
Main Purpose of this Book

This book focuses on exploring automation and work organization for assembly operations in the world automobile industry. A number of researchers and practitioners from Europe, Japan and the US have contributed papers to this book after attending and exchanging views at two international conferences held specifically for this purpose.

Although the scope of this book may appear somewhat narrow, we believe that the system of assembly automation and work organization is in fact a "microcosm" which characterizes the essence of a broader issue: the dynamic interactions between today's technology, human organization, and their integrated systems performance.

By exploring this research theme, the book will try to understand what kind of system designs exist for assembly operations, how the production technology and human organization are integrated as a total system in each case, what objectives each system tries to accomplish, which obstacles it has faced, what it has achieved so far, how the systems evolved in the past, and where they will go in the future. In other words, the book contains conceptual and/or empirical work for (1) alternative systems, (2) objectives, (3) contexts, (4) actual results, (5) historical evolution, and (6) future prospects in the area of assembly technologies and organization.

The book is designed as a rather loosely organized collection of company reports, case studies, conceptual papers, rather than academically rigorous works. One of the advantages of this type of approach is that academic researchers and industrial practitioners can share interests, problems, hypotheses and test fields in this way. In fact, many of the academic contributors to this book have direct and regular contact with practitioners in automobile and parts manufacturing firms. Practitioners of the automobile manufacturers in Europe and Japan, who participated in the conferences, also contributed papers that describe their efforts in this area. Thus, we have achieved a balanced mixture of industrial and academic contributions to the book, as well as at the conferences.

1.2 International Work Shop for Assembly Automation and Work Organization

Before discussing the subject area of the book, we will briefly explain the international conferences on assembly automation and work organization which formed the basis for this book.

The Japanese Study on Assembly Automation
A team of Japanese researchers, including Shimokawa and Fujimoto, started a domestic study on assembly automation of the Japanese auto makers in 1991. The Japanese study involved virtually all of the Japanese auto companies (11 car and truck makers). A series of questionnaire surveys, as well as plant visits, were conducted by the study group [4,5,14]. Each company selected one most automated and one least automated production line as sample cases for the questionnaires.

The purposes of the study included the following points:

- To compare levels, spans and other aspects of automation (e.g. flexibility, process development, skill development, information systems, etc.) in selected production lines, processes or tasks across manufacturers or factories.
- To identify the strengths and weaknesses of various types of alternative automation systems.
- To explain why a high level of automation has been introduced to some processes (e.g. welding) more successfully than others (e.g. final assembly).
- To identify certain production processes or operations in which large cross-company gaps in levels and spans of automation exist, and to explain how these gaps were generated.
- To compare organization and performance of projects for new process development related to factory automation.

At the beginning, the research agenda of the Japanese study was quite domestic, reflecting the labor shortage problem which dominated the discussions among industrial observers in Japan between 1990 and 1991. However, it soon became clear that international research collaboration would bring about a better understanding of this issue through mutual learning between practitioners and researchers from different countries.

Thus, the Japanese study group approached the Social Science Research Center Berlin (WZB) which had previously just concluded an international comparative study on production modernization in the automobile industry [11]. WZB agreed to host an international conference on this issue.

The Berlin Conference in 1992

At the Berlin meeting, we discussed various issues on assembly automation and work organization, including obstacles facing assembly automation, patterns of work organization in the assembly area, new approaches toward factory automation, and automation concepts for improving productivity, flexibility, and quality of assembly work. For each topic, researchers and practitioners from Europe, the US and Japan made short presentations on their experience and research results. The participants also included members of larger study projects such as GERPISA (Permanent Group for the Study of the Automobile Industry and its Employees) in France and IMVP (International Motor Vehicle Program) in the US. Incidentally, Jürgens, Shimokawa and Fujimoto have all participated in both projects.

The German and other European researchers and practitioners evaluated the current performance of existing assembly automation systems. Jürgens, for example, discussed the strength and weaknesses of the technology-based approach as opposed to the human-oriented approach developed in Sweden and the lean production approach developed in Japan in terms of performance and work attractiveness. The US group, including researchers from MIT's International Motor Vehicle Program, University of Pennsylvania and Harvard, proposed new approaches to deal with variety and contingency in design and manufacturing. The Japanese research team presented the preliminary results of a survey and field studies in Japan. At the end of the meeting, the participants agreed that we should continue international comparative studies and mutual learning on various pre-competitive issues related to assembly automation and assembly work organization.

The Tokyo Conference in 1993

Based on the discussions in Berlin and in order to continue discussions on the issue with broader perspectives and clearer foci, the second workshop was held in November 1993 at Hosei University, located in the suburbs of Tokyo. The conference agenda included the following topics:

1. *Are New Production Systems Emerging?*
   The various speakers discussed the future prospects of the Ford system; the possibility of a "post lean" (or "neo-lean") production system; the limits and prospects of the Volvo approach; the possibilities of integrating technology-oriented; humanization-oriented and lean production oriented systems and the future trajectory of automation in auto manufacturing.
2. *Linking Assembly Automation and Total System Performance*
   Major issues under this topic were questions of linking automation and productivity/quality; of coping with product variety and model changes; the robustness of automation and production systems with regard to contingencies; the need to overcome the problem of alignment accuracy; the importance of product design for automation modularization, simplification, etc.; the prospects of computer-integrated manufacturing and the impacts of automation systems on supplier networks.

3. *Human Aspects of Assembly Automation*
   Among the issues discussed here were the role of automation for improving working conditions (e.g. for senior workers); the implications of assembly automation for work organization, skill development and training programs; the specific conditions at automated factories for quality of worklife (QWL)/total quality control (TQC) and continuous improvements; and finally, the impact of automation on employment and work hour reduction.
4. *International and Inter-firm Collaboration*
   Among the issues discussed here were experience and prospects of automation in the transplants; collaboration between auto makers and equipment makers; and the possibility of enhancing mutual learning on assembly automation among auto firms.

After the workshop, a series of tours to some Japanese assembly plants (e.g. Mitsubishi Mizushima plant, Nissan Kyushu plant, and Toyota Kyushu plant) were arranged for the conference participants.

Key Concept of the Conferences: Loose and Compact Network
One of the purposes of the international conferences in Berlin and Tokyo was to establish international networks between practitioners from different firms, between researchers from different countries and disciplines, and between researchers and practitioners. In other words, the study tried to be international, inter-firm, interdisciplinary, and inter-occupational. In the meantime, a third conference, organized by Volpato and Camuffo, took place in Venice in October 1995.

With regard to academic research, the aim was to create a loose coalition network of researchers. This means that researchers from different institutions or regions were expected to raise their research funds respectively. In this way, each of the local teams maintained a certain level of independence in terms of research objectives and financial support.

The weakness of this type of "weak-tie approach" may be a lack of focus, coherence or leadership for the entire study. However, considering the skills and experience of the researchers involved, we believed that the advantages of the loose network scheme would exceed the disadvantages mentioned above. Also, other projects which follow the "tight" approach, such as MIT's International Motor Vehicle Program, were welcomed into this network, since we believed that the two approaches complemented each other for a better understanding of this multi-faceted industry.

1.3 Motivation for the Book: Why Study Assembly Automation Now?

With the background explained above, let us first illustrate the major motivation for this book. Why did we want to study assembly automation and assembly work organization now? There seem to be at least a few reasons.

Final Assembly is a "Microcosm" of Today's Production System

Although automobile final assembly quantitatively represents only a small fraction of today's manufacturing sectors in industrialized countries, it has been regarded as a "symbol" of mass production systems. Many of the researchers in industrial economics, industrial technologies, labor relations and human organization have analyzed the so-called *Ford system* as the dominant production system of this century, and many of them recognized moving assembly lines to be the core operation of this system[1]. An implicit assumption behind such research has been that key characteristics of the Ford-style assembly system, including high output rates, high productivity, short through-put times, synchronized logistics, interchangeable parts, highly standardized and repetitive work, short cycle times regulated by machines and conveyor lines, can be also observed in many other areas of today's industrial systems.

Although the manufacturing sector represents an ever smaller fraction of national economy, whilst various service sectors have increased their share, and electronics and information-related industries have dramatically increased in importance in the late 20th century, we still believe that additional in-depth studies of the automobile production system will bring about valuable insights not only for this industry, but also for other sectors of the future. We regard final assembly in particular as a "microcosm", where the key problems of today's production technologies and work organization in general are crystallized and can be observed. This is one of the reasons why, over 80 years after the first installation of the moving assembly line, this particular production process still attracts the attention of many researchers.

Final Assembly is the Most Difficult Operation to Automate in the Automobile Manufacturing Process

Another reason for studying assembly automation is the very fact that it is one of the most difficult production processes to automate. For various technological and economic reasons, final assembly (unlike other processes, such as machining, forming, painting, welding, and even subassembly of smaller components) has long been a bottleneck of automation in the auto industry, even after the advent of robotic technologies. Although the transfer of car bodies has been mechanized since the 1910s, direct assembly work itself has seldom been automated. It seems to be a consensus among technical experts, that twenty years from now, a majority of the final assembly tasks will not yet have been automated. In other words, typical final assembly operations of the early twenty-first century will involve robots, semi-automated equipment, as well as many workers closely interacting with one another on the same shop floor. Totally unmanned assembly lines will continue to be an unfulfilled dream of engineers.

---

1 "Regulation" theory, for example, described Fordism as a typical system that sustains social stability and macro-economic growth of post-war capitalist economy.

It follows that, due the difficulties involved in automating final assembly, designing a human-machine interface has been (and will continue to be) a highly complex and subtle task. Neither human organization nor automation technology alone can solve manufacturing problems in this area. So far, no company in the world seems to have developed a clear vision as to how to design this human-machine interface in terms of both higher competitiveness and better quality of work life.

This is one of the reasons why we believe that there are great opportunities for both practitioners and researchers to learn from the experience of various companies in designing human-machine interfaces in the final assembly area. For this purpose, this book will present various case studies on how each of the auto companies, mainly from Europe and Japan, planned and implemented an integrated system of automation and work organization in the assembly area.

There are Alternative Designs for Assembly Automation that can be Mutually Learned Among Firms and Researchers

Another motivation for the present study was that we observed a variety of alternative designs for assembly automation and work organization today. At the level of total production systems, including production technologies and shop floor organization, we have seen a certain variety of firm-specific or region-specific patterns in recent years [3,11]. We have also seen mutual learning or knowledge transfer between the auto firms. As a result, we have observed trends towards both convergence and diversification in the design of automobile production systems.

On the one hand, the so-called lean production system, first developed in Japan, had been at least partially adopted in the US and by European auto makers; which seemed to be part of the reason why some Western auto makers have narrowed the gap against the Japanese in certain competitive indicators in recent years [13]. In the Japanese auto industry, on the other hand, the lean production system of those days was said to be facing various problems. Although the classic lean system may continue to be an effective source of competitiveness, it does not seem to be the all-mighty solution to every situation. Even in Japan, certain negative side effects have become a concern to auto makers who had adopted the traditional lean system, as the era of continuous high growth ends and demographic changes occur. For example, the relatively low popularity of the conventional lean assembly system in the labor market, together with long work hours and chronic labor shortage, became an obstacle for further development and improvements by Japanese auto makers in the 1990s [6,7,9].

The so-called Volvo system (or Uddevalla system in its purest form), an obvious alternative to the Ford system, was also faced with problems in the 1990s: its major test fields, both Uddevalla and the Kalmar plants, were closed by the mid 1990s. Although many researchers advocate that these plant shut-downs were not caused by the inherent limits of Volvoism itself - a claim which is supported by the fact that Uddevalla is now being re-opened, even though in a reduced and modified way -, it is also true that this system, although highly human-oriented, has not proven its capability to compete as effectively as the lean production system in the global market. Other potential solutions, such as neo-craftism and neo-Fordism, do not seem to have demonstrated superior performance.

In summary, as of the mid 1990s, the world auto industry has not yet reached a consensus as to the next dominant design for automobile assembly. To be sure, we have observed a trend of convergence towards the lean production system to some extent [13], but this has not resulted in the emergence of "one optimal way" which dominates in the post-Fordism era. We have even observed proliferation of many variants, as "hybridization" of different industry models occurs among competing/cooperating firms [3,9,11, etc]. Convergence and diversification have been happening simultaneously.

In this situation, we see that there is ample opportunity for mutual learning between the European, US and Japanese auto industries and their production systems. We are not sure whether the coexistence of various hybrid systems is long-lasting, or whether it merely reflects a transition from the Ford system to another dominant production system. In any case, however, we believe that both empirical and conceptual studies of assembly automation and work organization will facilitate such mutual learning processes.

1.4 Factors Affecting the Choice of Assembly Systems

1.4.1 Objectives of Assembly Systems

There are various factors which affect a firm's decisions regarding assembly automation systems and assembly work organization. First, objectives of the manufacturing firms, as well as priority among them, will affect the choice of their assembly systems. Competitiveness is obviously the most important criterion for evaluating assembly systems for many companies. Ironically, a higher assembly automation ratio, no matter how it is defined, did not contribute to improvements in firms' overall competitiveness in the 1970s and 1980s. It could even be observed that some of the new assembly lines with relatively high automation ratios failed to function as well as certain non-automated (but well managed) assembly plants; this was due to various unexpected side effects, such as down times and insufficient flexibility. In the 1990s, competitiveness will continue to be an important criterion for decisions concerning assembly systems.

However, competitiveness is not the only criterion for evaluating and selecting assembly systems. Improvements in quality of work life or employee satisfaction are other important aspects. Better working conditions, smaller physical workload, less danger, more participation, more collaboration in the work group, more meaningful work, as well as broader task assignment, all contribute to improvements in this criteria. Some firms, which voluntarily accept employee satisfaction as one of the company goals and as part of its social responsibility, will emphasize this aspect. Other firms, facing serious labor shortages and/or high turn-over ratio, may also be forced to emphasize the employee satisfaction aspect of the systems in order to remain competitive. Advocates of Volvoism, socio-technical work designs, and ergonomics will naturally give this criterion high priority.

Environmental friendliness has become another important criterion in recent years. Although final assembly is a relatively "clean" process as far as air and water pollution is concerned, it may play a pivotal role with regard to the recycling of motor vehicles. In other words, how to incorporate "disassemblability" to product designs, as well as how to design disassembly processes themselves could be a critical issue in the coming decades.

There may also be other motivating factors for firms to automate assembly operations. For example, some engineers may give high priority to the adoption of advanced production technologies per se, such as highly intelligent robots. Others may try to use such technologies in order to shift political power on the shop floor from workers to management. The track records of such assembly systems, driven by high technology for the sake of technology or power shifts on the shop floor, have usually been unimpressive. Such attempts tended to result in the erosion of overall competitiveness or deteriorated labor relations.

In any case, final assembly is the area where the trade-off among different objectives (e.g. customer satisfaction, employee satisfaction and environmental friendliness) is difficult to make: it is difficult to automate; it is difficult to attract workers; it's impact on customer satisfaction through cost and quality improvements is high. How each firm balances these objectives will affect the basic design of assembly automation and work organization in profound ways.

1.4.2 Context

Not only objectives, but also various contextual factors will affect the choice of assembly automation and work organization. First, the market environments affect the choice of assembly systems. Diversification of user needs may trigger product proliferation which in turn calls for flexibility on the side of assembly processes. High growth of the market may enable the firms to invest heavily in high-technology automation with relatively low risks of fixed cost burden, while a slow down in growth will force companies to be more cautious about investing in expensive high-tech automation equipment.

The nature of inter-firm relationships among auto makers (e.g. competition, cooperation and conflicts) will also affect choice. Intense international competition may accelerate diffusion of certain assembly technologies which have proven their competitive advantages, while competition may make firms more conservative in regard to adopting high-technology approaches, which often involve risks of erosion in cost competitiveness. Inter-firm cooperation, on the other hand, may facilitate the diffusion of certain assembly technologies among them.

Labor market situations will also affect the choice of assembly systems. Labor shortages, local or macroeconomic, will accelerate assembly automation in at least two ways: a) by attempts to reduce labor demand per assembly line; and b) by attempts to make the work place more attractive to the workers. In recent years, the second way has become particularly important in Japan.

However, when the unemployment ratio is relatively high, there will be pressure from labor unions against automation of labor-intensive operations, such as final assembly. In this situation, both management and labor have to evaluate the net effect of assembly automation on decreased employment and increased competitiveness.

Product and process technologies also affect the choice of automation technologies. As for product technologies, improvements in *design for assembly* (DFA) will accelerate companies' attempts to automate assembly. Compared with many other products, such as electronic products and even some auto components, the production of the total motor vehicle has been inherently difficult to automate, without deteriorating product quality and/or raising assembly costs. However, to the extent that future automobile design incorporates DFA factors (e.g. reduction of the number of discrete parts per vehicle, higher number of common parts between product variations, product variety reduction, unification of bolting directions, modularization of product designs, changes in product layout and assembly sequence for simpler motions of robot arms, better technologies for parts connection, etc.), we may see a dramatic increase in assembly automation ratio in the future.

To sum up, each firm's choice of assembly systems will be affected by its objectives and strategies, as well as by context and environment. To the extent that different firms or different regions deal with different objectives and constraints, the assembly systems that we observe will also differ significantly. However, it will still be useful for each firm to learn from the experience made by other firms.

1.5 Main Perspectives of the Book

As mentioned earlier, the purpose of the book is to explore trends and approaches of production system design for assembly operations, focusing on Europe and Japan. In the last decades, companies in Japan and Europe seem to have followed a different trajectory. Our main interest is related to the question of divergence versus convergence without, however, neglecting company differences in the world regions. We have two conceptual perspectives which helped us organize this book: the total system perspective and the evolutionary perspective.

1.5.1 Total System Perspective: Alternatives and Actual Results

It is our belief that the future assembly automation system must be discussed in the context of a transformation of the automobile production system as a whole, which includes automation technologies, human organizations, logistics and process configurations as integrated subsystems.

In this regard, the 1980s - 90s has been a period of transition in which the traditional Ford (i.e. American mass-production) system faced a competitive challenge from the "lean" (i.e. Toyota-style) production system [16]. As the limits of the conventional Ford production system became obvious, various alternative production systems were proposed and tried out. At this point in time, we are not certain whether the next period will be characterized by the coexistence of alternative production systems, or by a convergence towards the next dominant system.

1) *Neo-Ford System*: The introduction of advanced automation technologies and computer networks into more or less traditional mass production paradigms was one direction which companies, such as GM, VW and FIAT, pursued in the 1970s and 80s. The technology approach changed traditional process layout and work organization significantly with ambivalent consequences for performance as well as quality of work. Work structures became polarized: automation "islands" were created, while surrounding manual operations remained mostly conventionally organized. Qualification requirements in the high-tech area became polarized with jobs for highly skilled specialists, on the one side, and simple feeder jobs on the other. Personnel development and training tended to focus on the specialists and neglected rank and file workers. In order to cope with machine break downs, large buffer areas became necessary. Jobs were improved ergonomically, however, as many strenuous jobs (heavy loading, overhead work) were often the primary candidates for automation.
   The technology approach failed to demonstrate superior international competitiveness in the assembly area. However, there seems to be a significant difference within the neo-Ford approach in terms of who controls the high-tech equipment. In the US and to a certain extent in the Italian version too, automated equipment was assumed to be controlled by engineers, while direct workers were simply de-skilled or eliminated. In the German version, by contrast, it was assumed that a type of skilled worker, the *Anlagenführer* (equipment controller or system regulator; cf. Kuhlmann and Schumann in this book) controlled the automated equipment - while other "residual" work could be assigned to unskilled workers. Thus, the high-tech approach led to quite different jobs in work organization solutions [cf. 11].
2) *Uddevalla System*: Another proposal for an alternative to the conventional Ford system was the one known as *Volvoism*, which attempted to make assembly work more attractive by essentially abolishing moving assembly lines and returning to a modern version of stationary production. The Volvo system led to a new "factory-design" to support group work. The assembly line was abandoned and so was machine-paced work. A new type of craftsmanship for assembly work was developed, and work was planned for holistic job contents of several-hour job cycles. The system required extensive training. Work groups were given partial autonomy with regard to the planning and execution of their work. Groups were also autonomous, i.e. they could regulate their own affairs.

The Volvo system, while being reasonably successful in attracting and satisfying workers, has not demonstrated its competitiveness in terms of productivity. Its main experiment sites, Volvo's Kalmar and Uddevalla plants, were both shut down in the early 1990s, although this may not mean the failure of Volvoism itself (cf. the chapters by Ellegård and Berggren in this book). Some elements of this system are still being tested in various factories, and, as of recently, Uddevalla is being re-opened to produce sports cars in the frame work of Autonova, the new joint venture between Volvo and TRW, the British sports car maker.

3) *Neo-Craft System*: Craft systems of production still play an important role in prototype shops and in some plants for small volume production of customized cars. Honda's plant in Tochigi for the assembly of its NSX sports car up to recently served as an example (cf. chapter 3.5 by Tanase, Matsuo and Shimokawa in this book). Although such experiments were seen as "craft renaissance" or anti-theses to work alienation in modern assembly lines by some observers, its productivity was less than one-tenth of the "lean" production line. Thus, the application of neo-craft systems has been limited to special luxury models and customized cars. As Tanase et al. show, however, they can serve quite successfully as a test bed for new concepts and training lab in a mass production environment.

4) *Lean System*: The so-called *lean production system* (i.e., the Toyota-style system) attracted the widespread attention of Western auto makers during the 1980s and early 90s because of its competitiveness in productivity, manufacturing quality and flexibility [1,16]. The Japan-oriented lean production system is characterized by process design aimed at low buffers and no-errors in production. Teams are the basic unit of work organization as well as of improvement activities. Task profiles are broad and the personnel development system aims at creating generalists with multiple skills. Training takes place mainly on the job and is closely related to Kaizen activities. In contrast to the Volvo system and to the work situation in the high-tech assembly areas of automated plants, there is no time sovereignty for assembly workers and the assembly line remains the back bone of work organization.

   Although the lean system can be regarded as a derivative of Fordism in many senses, it also has unique features in managing human resources, supplier networks, material flows, inventories, as well as productivity and quality. In fact, many American and European auto makers introduced a part of the system to catch up with more competitive Japanese firms during the 1980s.

   However, the labor shortage and recession in Japan during the early 1990s revealed some weaknesses in the existing lean production system in attracting the domestic work force and handling fluctuations in total production volume [6,7]. It is now obvious that the existing version of the lean system (i.e. lean-on-growth system) needs reform in the long run. Direct assembly line work has been a particularly problematic area; this is due to the fact that it is essentially an extension of the Ford-type moving assembly line and has hence carried over its inherent problem of lack of attractiveness - despite the fact that the lean system added mechanisms for flexibility, self maintenance, self inspection and continuous improvement.

Thus, none of the existing alternatives to the traditional mass production system have clearly demonstrated long-term advantages over the others - neither in terms of competitiveness nor of employee satisfaction. However, as of the early 1990s, there is no sign that any other totally new approach to automobile production will suddenly emerge as a dominant production paradigm. Instead, many of today's auto companies in the world seem to be seeking better solutions by fusing elements of existing alternatives. Thus, in the foreseeable future, auto companies are likely to rely on "hybrid" production systems rather than entirely new production concepts. In other words, the twenty-first century will be marked by various hybridization experiments by the world's auto makers.

In any case, it is our opinion that discussions on the future form of assembly automation need to take the future of the automobile production system as a whole into account, as the former is one of the subsystems of the latter.

1.5.2 Evolutionary Perspective

Another assumption in the current paper is that history also plays an important role. In other words, the choice of future assembly automation system for each company needs to take into account its evolutionary path of organizational learning and dynamic capability building [8,15]. Facing challenges from its environment (e.g. product markets, labor markets, competitors, etc.), it tries to acquire new capabilities, add these to existing managerial resources, and create a new set of core capabilities. Although some elements of the old system may be abandoned or modified as new capabilities are acquired, other elements will remain in the new blend, making the capability-building process cumulative. Whether the old and new elements collide or fuse with each other may affect the subsequent performance of the total system [10].

This evolutionary view does not imply that there is only one deterministic trajectory or sequence of capability building. The evolutionary paths may be region-specific or even firm-specific. At the same time, it is our opinion that the overall trend for future assembly automation will be that of *convergence* and *mutual learning* across regions and firms on a basic level of assembly automation strategies. In other words, we believe that future assembly automation systems will emerge as a result of the *hybridization* [cf. 2] of different automation strategies, rather than competition for survival among the pure strategies. Also, this very convergence may create *diversity* of the future assembly automation systems, as each company will have to build a hybrid system based on its own unique capability base.

1.6
Organization of the Book

We have organized the contributions to this book into three main sections. The first section of this introduction deals with *concepts and histories*.

The contributions in this section provide introductions into the theme of assembly automation from different view points and disciplinary backgrounds. It begins with the historical review from a European perspective by Hsieh, Schmahls and Seliger from the Berlin Technical University. Whitney from M.I.T. analyzes the physics and economics of robot deployment in assembly operations. Tidd from the University of London's Management School discusses different patterns of robot deployment. He shows that from the background of different engineering traditions and production systems, the deployment of robots and even the technical choice of robots differs fundamentally when comparing the UK and Japan. Fujimoto from Tokyo University discusses basic problems of definition by asking the question, "What do you mean by Automation Ratio?" In fact, companies use different definitions and this leads to considerable confusion in survey research when degrees of automation are reported.

The second section of the book presents the *diversity of approaches*. This section contains ten case studies on the present state and future vision of Japanese and European companies with regard to the issue of automation and work organization of their assembly operations. Most of these case studies were written by company representatives in charge of production engineering in their respective companies. Mishima of Mitsubishi Motor Corporation's office of production engineering focuses on the issue of the "human fit" of technology and of Kaizen activities to assure quality conformance and process yield in the automation areas. Niimi and Matsudaira from the Production Control and the Manufacturing Engineering Division of the Toyota Motor Corporation have set a completely different focus. They highlight the issue of ergonomic improvements and work structuring in order to increase the quality of work and overcome the labor problems which had surfaced at the end of the 1980s during the height of the bubble boom and, as Niimi and Matsudaira point out, which have not lost their strategic relevance, even under recession conditions in the mid 1990s. The reasons behind Honda's excursion to a "neo-craft sytem" for its NSX sports car and the role this plant has played for organizational learning within Honda is explained by Tanase from Honda's engineering affiliate Matsuo from Tokyo University and Shimokawa from Hosei University. The contribution by Naitoh, Yamamoto, Kodama and Honda from Production Engineering at the Nissan Motor Corporation concludes the case studies from the Japanese side. Naitoh et al. present the Intelligent Body Assembly System (IBAS), the company's new technology for framing the steel panels of the car body. The IBAS responds to the requirements of increased flexibility for process equipment for different car types and models.

Approaches taken by European companies are presented in the following five case studies. Enderle, the former Board Member at Adam Opel, in charge of production operations, describes the case of the new production system implemented by Opel at the Eisenach plant which is currently regarded as the most efficient

European assembly plant. Wilhelm, currently the Technical Director of Volkswagen's Assembly Plant in Brussels, explains Volkswagen's platform and modular concepts and their affects on the assembly process. Decoster from Renault's Vehicle Engineering Department and Freyssenet, researcher at the French National Research Center, discuss Renault's automation strategy which involves the introduction of teams taking responsibility for production and maintenance tasks at the automated equipment. Camuffo and Volpato from the University of Venice discuss the different approaches towards assembly automation which were taken up at Fiat since the early 1970s as a process of dynamic capability building and organizational learning. Finally, Ellegård from the University of Gothenburg describes the different approaches discussed during the planning process for the layout at Volvo's Uddevalla car assembly plant as a process of developing the reflective production system.

The fourth section of the book deals with *issues and dynamics* in the area of assembly automation and work organization. The issues discussed in altogether eleven chapters range from quality of work and ergonomic requirements, new forms of work organization, process design for ease of assembly to the question of recycling and disassambly.

Fujimoto discusses the different strategies by Japanese manufacturers in response to the labor programs at the end of the 1980s. These strategies range from automation to approaches, which aim at ergonomical improvements, to more meaningful work tasks on the work group level. MacDuffie from the Wharton School of Business and Pil from the Katz School of Business report on international trends in assembly plant automation and work organization based on their experience from the international assembly plant studies carried out by them in the context of the International Motor Vehicle Program. They observed a steady replacement of fixed automation by flexible automation and a development towards low-cost initiatives.

Jürgens from WZB deals with the trend towards re-establishing the classic assembly line design principle in European plants - a trend which seems to mark a roll back of socio-technical systems and the humanization of work principles, which, in the past, had sought a process design with extended job cycles and work decoupled from the rigid pace of the line.

The two chapters by Springer from Mercedes Benz's Department for Work Organization and Labor Politics and by Kuhlmann and Schumann from the Institute of Sociological Research in Göttingen discuss the introduction and the effects of group work in assembly operations, a measure which played a central role in the restructuring effort by German car manufacturers in the early 1990's. The two chapters by Freyssenet and Ellegård discuss a common question - although in different companies and in different work structures: how can workers' experience and knowledge be used for solving programs and process optimization. Freyssenet makes a plea for new automation concepts which keep the inner functioning of the machinery "readable and intelligible" for the workers, thus moving away from Taylorist principles, i.e. the division of knowledge from work. Ellegård describes a system of worker-generated improvements, or Kaizen, at the Uddevalla plant. It is a worker-based system assisted by personal computers, which Ellegård helped to develop. Her report demonstrates that the Uddevalla system did not just follow the "logic" of craft systems, relying on the skill of individual work men. The strength and merits of the Uddevalla system also form the core of the next chapter by Berggren from the National Work Life Institute in Stockholm. He compares the deliberate "low-tech" Uddevalla approach with the high-tech approach pursued by Japanese car makers in the generation of new plants which were established by the end of the 1980s.

The following two chapters also deal with issues which are related to the socio-technical systems tradition in Europe. Mishina from the Japan Advanced Institute of Science in Technology (JAIST) reports from a survey on workers perceptions regarding stationary work structures with work performed at automated guided vehicles (AGVs). Such systems, which were introduced by many European firms in the 1970s and 1980s, are currently being shelved in many cases. As viewed by the workers, the AGV-based process layout did not seem to be a decisive factor in explaining work satisfaction either. The chapter by Benders and Dankbaar, Nijmegen University Business School, deals with the organizational change in the Dutch automotive industry under the impact of Japanese manufacturing concepts. NedCar's new assembly line is discussed as an example of the merging of lean production and socio-technical systems traditions.

Last, but not least, in the section on issues and dynamics, the question of recycling and disassembly is raised by Seliger, Hentschel and Kriwet from the Berlin Technical University. The authors discuss current recycling technology, disassembly techniques and the principles of product design supportive for disassembly and recycling.

Many people have contributed to make this book possible. At this point, we would like to thank them all for their cooperation.

Our special thanks go to the conference participants and to the authors, especially from the practitioners' side. We know that it was not always easy for them to find time in their busy agendas for writing papers and answering our queries.

We express our thanks to the Ministry of Education in Japan for their research funding and the Japan Automobile Manufacturers Association (JAMA) for their financial support.

We are indebted to WZB and Hosei University for hosting our conferences and all the individuals involved who helped to make these meetings enjoyable and productive events. Furthermore, we would like to express our gratitude to the managers and staff of the plants we were able to visit and to the members of JTTAS for their perfect organization of the auto plant tour throughout Japan.

Furthermore, we wish to express our thanks to our technical staff in Berlin, namely Volkhard Röseler for his effort and technical know-how, as well as Ralph Wittgrebe and Helen Dalton-Stein, our English editors, for their tireless support. And last - but not least - we would like to thank Sigi Leslie; without her patience and invaluable assistance we would have been lost - more than once.

1.7 References

1 Abernathy W J, Clark K B, Kantrow A M (1983) Industrial Renaissance. Basic Books, New York

2 Berggren C (1993) Volvo Uddevalla - A Dream Plant for Dealers? Working Paper. Royal Institute of Technology, Department of Work Science

3 Boyer R, Freyssenet M (1995) The Emergence of New Industrial Models. Actes du for GERPISA No 15 December, Paris: 75-142

4 Fujimoto T (1992) Why Do Japanese Auto Companies Automate Assembly Operations? Presented at the Berlin Workshop on Assembly Automation. November. Research Institute for the Japanese Economy Discussion Paper 92-F-15, University of Tokyo

5 Fujimoto T (1992) What Do You Mean by Automation Ratio? Presented at the Berlin Workshop on Assembly Automation. November. Research Institute for the Japanese Economy Discussion Paper 92-F-16, University of Tokyo

6 Fujimoto T (1993) At a Crossroads. Look Japan. September: 14-15

7 Fujimoto T (1994) The Limits of Lean Production. Politik und Gesellschaft. Friedrich-Ebert-Stiftung, Germany, January: 40-46

8 Fujimoto T (1994) Reinterpreting the Resource-Capability View of the Firm: A Case of the Development-Production Systems of the Japanese Auto Makers. Paper to be presented to Prince Bertil Symposium, Stockholm, June

9 Fujimoto T, Takeishi A (1994) Jidosha Sangyo 21 seiki he no Scenario (The Automobile Industry: The Scenario toward the 21st Century). Seisansei Shuppan

10 Fujimoto T, Tidd J (1993) The UK and Japanese Auto Industry: Adoption and Adaptation of Fordism. A Paper Presented at the Conference on Entrepreneurial Activities and Enterprise Systems, University of Tokyo Research Institute for the Japanese Economy, Gotenba City, January. Japanese Translation: Keizaigaku Ronshu (The Journal of Economics) 59,2 and 3, The Society of Economics, University of Tokyo

11 Jürgens U, Malsch T, Dohse K (1993) Breaking from Taylorism. Changing Forms of Work in the Automobile Industry, Cambridge/New York Oakleigh, Cambridge University Press

12 Jürgens U, Dohse K, Malsch T (1986) New Production Concepts in West German Car Plants. In: Tolliday S, Zeitlin J (eds) The Automobile Industry and Its Workers: Between Fordism and Flexibility. Polity Press, Cambridge: 258-281

13 MacDuffie J P (forthcoming) International Trends in Work Organization in the Auto Industry: National-Level versus Company-Level Perspectives. In: Wever K, Turner L (eds) The Comparative Political Economy of Industrial Relations, IRRA 1995 Research Volume

14 Shimokawa K (1992) Japanese Production System and the Factory Automation. Discussion Paper for the Berlin Workshop on Assembly Automation. November 1992

15 Teece D J, Pisano G, Shuen A (1992) Dynamic Capabilities and Strategic Management. Revised, June. University of California at Berkeley Working Paper

16 Womack J P, Jones D T, Roos D: The Machine That Changed the World. New York: Rawson Associates

2
Concepts and Histories

2.1 Assembly Automation in Europe – Past Experience and Future Trends

L.-H. Hsieh · T. Schmahls · G. Seliger

2.1.1 Automobile Factory: Production and Assembly

Traditionally, different ideas as to what makes up an assembly plant existed in the U.S.A. and in Europe. In the U.S.A., two types of automobile factories have developed: Production Plants and Assembly Plants. In a production plant, all the manufacturing operations necessary for the production of automobiles are carried out. These operations include, among others, the casting, machining and assembly of the machined parts to functional units, i.e. engines, gearboxes and axles, as well as the so-called *assembly operations*, including the stamping of the sheet metal, welding the pressed sheets to a carbody, painting and final assembly. A plant which performs only assembly operations is an assembly plant.

Until 1960, production plants were predominant in the U.S.A. In the River Rouge plant in Detroit, Ford even integrated the steel working and the attaching parts production of trim components. A higher rate of insourced manufacturing operations has never been implemented in any other automobile factory. Concentrating all automobile manufacturing processes in one single plant did not prove to be a competitive approach, neither in terms of overall productivity, nor in terms of organizational effectiveness. In the new plants of the nineteen-sixties, manufacturing and assembly operations were performed in different plants, but the distinction between these two types of plants still prevails.

In Europe, separate plants have been traditionally set up for manufacturing parts and for assembly operations. The assembly plants are further divided into plants with or without a press shop. In Europe, plants with a press shop are regarded as production plants, whilst plants without a press shop are called assembly plants.

The different ideas as to what makes up an automobile factory corresponded with the different role of independent supplier companies. In the U.S.A., the growth of a strong supplier industry was hindered by the dominant role of Ford. Henry Ford's claim was that nobody could produce parts and cars as efficiently as the Ford Motor Company with its modern production techniques and work organization. Europe's car manufacturers, in contrast, needed strong supplier companies for their development and companies like Continental, Bosch etc. hence became an integral part of the automobile production system in Europe.

This article describes, chronologically, the innovations made in the press shop, body shop, paint shop and final assembly.

Fig. 2.1.1. Assembly of the flywheel magneto for the *T* Model at Ford

2.1.2 A Chronology of Innovations

1900: Craft Production

At the beginning of the 20th century, craft production was typical of the automobile industry throughout the world. Before being assembled, parts had to be machined and matched. This work took time and effort. Matching large parts sometimes required two or more craftsmen. Due to these working conditions, automobiles were assembled in teams. These teams consisted of highly qualified craftsmen who machined the parts on machine tools located close to the assembly area. The craftsmen were very often self employed, but worked in close cooperation with the automobile company.

1913: Mass Production (Ford, U.S.A.)

Henry Ford set up an efficient automobile production in his *Highland Park* plant. He achieved this by integrating F. W. Taylor's ideas: i.e. by using standardized parts, the conveyor belt system and by limiting production to a single car type, the *T* model. Product, process and organizational innovations were introduced almost at the same time. Within one year, these ideas were gradually introduced to all areas of the plant. At first, innovations were introduced in the foundry, and subsequently in the assembly of the flywheel magneto and the body. The assembly time could thus be drastically reduced from 12 to 1.5 hours.

The introduction of these techniques was not always successful. The *Brennabor* automobile company in Germany was one of the first to introduce them. As the product was not suited for mass-production, the company, however, was not able to take advantage of the Fordist system. It did not survive the "Black Friday" crisis of 1929.

Due to the high labour-saving affects, the Fordist and Taylorist ideas were transferred to plants all over the world and strongly influenced the way in which work was organized in these factories. The efficiency and effectiveness hence became the benchmark for all subsequent approaches towards new production concepts in car manufacturing.

1947: Integral Body Design Frame

The integral body-frame concept was developed by Opel for the Olympia before World War II (time of production: 1935 - 40). After the war, this design feature was successively adopted by all automobile companies. Low engine efficiency, shortage of steel and a demand by customers for more comfort in these post-war years formed the necessity for more light-weight automobiles. As long as other suitable materials - such as aluminium and fibre-reinforced plastics - were not available, the integral body-and-frame design was the only reasonable solution to these requirements.

External forces and torque were absorbed by the unitized body and transmitted to the platform, wheels and drive system. Due to this particular restraint, the freedom of body design became relatively limited. As a consequence of the new product architecture almost all special bodies were eliminated and many independent body makers had to give up.

For automobiles with a separate frame, external forces and torque can only be absorbed by the frame. Figure 2.1.2. shows the assembly of the "Beetle" body onto a central-pipe frame with a welded platform, to which the external forces were fully transmitted.

1960: Mechanization of the Paint Shop

After the introduction of the conveyor belt in the body shop and the assembly shop, the paint shop remained the missing link in the mechanization of a continuous transportation system for car bodies in automobile plants. The necessary process time of up to three days had made mechanization of the paint shop seem unnecessary.

The development of annealing synthetic-resin varnishes, as a substitute for pyroxilin paints, reduced the process time to just a few hours. The buffing process, a strenuous physical activity, could also be omitted. Most of the maintenance efforts for the painting by customers was also largely reduced.

In order to maintain the quality of the body painting, the transport of the car bodies in the factories was mechanized, thus reducing surface scratches resulting from manual handling. A reduction in the number of workers in the paint shop and an improvement in the working environment were thus achieved.

Fig. 2.1.2. The "wedding" between the chassis and body of a Volkswagen *Beetle*

1960: Product-Specific, Multi-Spot-Welding Machine
The demand for the efficient production of larger product quantities and the shortage of workers on the labour market triggered off the development of product-specific, multi-spot welding machines for the body shop.

A reduction in lead time became possible by using multi-spot welding machines. The availability of the machines was 5 - 10% lower than that of manual welding guns. The increased productivity of the mechanized solution compensated this disadvantage. Prerequisites for the implementation of these machines were control units, which temporarily delayed the setting of the spots in order to avoid overload conditions due to the simultaneous setting of spots. Control units with different time-lag relays were one of the solutions.

The development of the synchronous control unit was followed by the asynchronous control unit, because asynchronous control avoided the setting of spots at the same time. There was no flexibility of multi-welders in terms of different types, variants and changing-over possibilities. Automobiles produced in smaller numbers still had to be welded manually.

1965: Higher Positioning Precision of the Transport Systems
The shortage of workers in the labour market in the early nineteen-sixties led to an increased introduction of mechanization in the automobile industry. One focal point of the rationalization wave was the body shop.

Repeatability of the joining positions was a prerequisite for mechanized assembly operations. The higher the degree of mechanization, the higher the re-

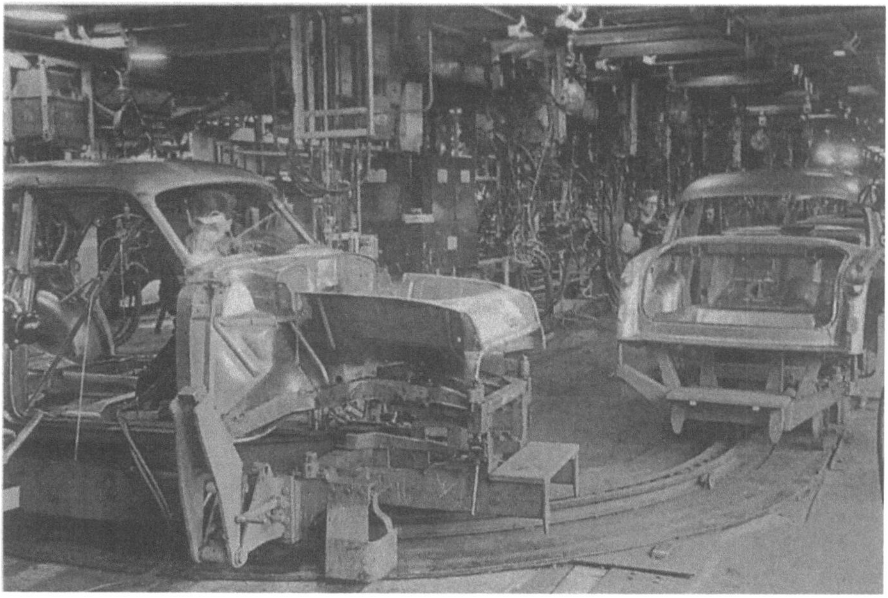

Fig. 2.1.3. Rail-guided transport system in a body shop at Opel

quirements for positioning accuracy. Figure 2.1.3. shows an early example of a rail-guided transport system used by Opel. This transport system could only be used for transport operations between manual assembly stations. This system, however, was continuously perfected. As an alternative, rail-less transport systems, independently controlled by micro-chips, were developed.

1970: Microelectronic-Controlled Handling Devices

Freely programmable control units for handling devices opened up new perspectives for rationalization, whilst the economic effects of further mechanization were exhausted due to lack of technical flexibility.

The installation of industrial robots was difficult because of technical problems, such as the handling weight of an industrial robot - i.e. the weight of the welding gun and the welding transformer. In order to cope with the increasing complexity of the assembly process, programmable controllers for industrial robots had to be developed.

In the case of final assembly operations, the requirements were even more stringent. As long as continuous path control systems were not available industrial robots could not be introduced in this area. Organizational demarcations were another obstacle. User demands were often insufficiently considered in the process of designing these devices which took place in separate production planning departments.

Fig. 2.1.4. Framing of the Ford *Capri I* side panels

1974: Teamwork at Volvo / Kalmar

The decline in product quality severely hit manufacturers, especially manufacturers of high-value automobiles. Quality problems were due to the short cycle times - resulting from an excessive division of labour -, as well as from the manifestations of workers' dissatisfaction with working conditions specifically in assembly plants.

In 1974, Volvo opened its Kalmar plant where final assembly systems with short cycle times were transformed into systems with longer cycle times and a different kind of work organization. Final assembly operations were now performed by 20 working teams. One team consisted of 15 - 20 workers. Each team assembled parts in a cycle time of 3 - 4 minutes while car bodies were stationary. This design has been seen as exemplary for humanized work for a long time since.

Buffer areas in between the work stations allowed for "time-sovereignty" in performing the work task. But the central shop floor control of the assembly and material planning operations could only decouple the machine cycle to a certain degree. The bodies were moved forward at the end of the cycle time, irrespective of whether the operation was finished or not.

A new transport system was necessary in order to implement a new operation method for final assembly. The *automatic guided vehicle* (AGV) was hence developed for the Kalmar plant. These AGVs were powered by batteries and could be controlled individually.

Fig. 2.1.5. Robogate at Fiat

At the beginning, improved assembly quality, lower absenteeism amongst workers and a reduction in assembly time were observed. After a couple of years, however, these advantages had disappeared.

1975: The Fully Mechanized Body Shop
In the mid-seventies, the degree of mechanization in body shop operations was close to 100%. The high capital investment for this development could only be justified by the mass production of automobiles on production lines. This production quantity could only be reached by A and B class automobiles. The flexibility of these production lines was largely limited and could only cover derivative types which differ insignificantly from the base body.

The machines in these production lines were to a large extent special machines which were custom-designed and built to order. Figure 2.1.4. shows the framing of the side panels for the Ford Capri I. There was no flexibility for other car bodies, and even for the successor type, flexibility was very limited.

Workers were released from heavy physical work. The assembly quality of the automobiles increased, but the availability of the production line decreased as a result of its technical sophistication.

1975: Flexible Automation in a Body Shop (FIAT, Robogate)
Flexible automation theoretically enables the body shop either to produce a mix of different car models, or aim at using the equipment over several product life cycles. Before its implementation, it was believed that this kind of technology

would not pay off economically. However, the hope of achieving an advantage over competitors induced more and more companies to move in this direction.

At FIAT, labour problems motivated the decision to take the risk. At the beginning of the nineteen-seventies, FIAT faced a serious shortage of workers for physically strenuous working places. At the same time, unions had begun to organize, and wildcat strikes were a daily event.

Automation of labour-intensive work areas, by means of flexible, automated machines, was declared to be the solution to this problem. In order to achieve a sufficient degree of flexibility with regard to the type of cars and the derivative types, a body shop was planned with robots for most of the welding operations, especially for the framing of panels. Both the positioning and clamping of panels were performed by means of clamping frames. Changing the car model merely required changing the clamping frames. The material transport in the body shop was carried out by AGVs, so that a high availability of the Robogates could be ensured (Fig. 2.1.5.).

The availability of these Robogates was lower than expected, because of an inadequate design of the parts for automated assembly processes. But with the transition from the Ritmo/Regata to the successor models Tipo/Tempra, most of these problems had been eliminated.

1975: Automated "Wedding" of Chassis and Body

Rapidly rising labour costs and the highly automated body shop provided the motivation for an automated final assembly. Almost at the same time as the introduction of the Robogate, Fiat developed the automated "wedding" of drive train and body for the "131" model in the Mirafiori plant.

Frame and chassis, designed for automated assembly, made the wedding possible in one step. The introduction of a horizontal separation plane between drive train and body - in order to achieve a defined joining direction - was a substantial prerequisite for automation.

Components - such as axles, drive shaft, exhaust, engine, etc. - were assembled to a subassembly. In three sequential system assembly stations, this sub-assembly was automatically connected to the body. This process is generally known as the "wedding". In the first station, the chassis is placed onto the sub-assembly, guided by 30 centring pins and is subsequently indexed. With 28 lead screws, the body and drive train are joined together in the second station. The third station places the assembled subassembly on the conventional conveyor belt, and finally transports it to the final assembly section. Upstream and downstream buffers decoupled this system from the line flow.

This solution had a strong influence on the automation of final assembly operations in future developments, even though it was never economically justified.

1978: Semi-Flexible Transfer Lines in the Body Shop

The automobile industry's answer to the market demand for customized cars was the introduction of the concept of installing different derivatives (saloon, coupé, fastback, station wagon) on the same platform. The traditional transfer lines did not offer the set-up flexibility necessary for such a concept, therefore, flexible equipment had to be installed on the production line. On a semi-flexible transfer line, industrial robots were linked with multi-spot welding machines, so that

Fig. 2.1.6. Rolling-axis transport system at BMW

large series production of an automobile type with its different derivatives was technically and economically feasible. Although the technical availability of such a system might be lower than that of a traditional, specific machine system, the resultant flexibility nevertheless justified the investment.

1978: Improvement of the Assembly Quality by Automation and Better Working Conditions

Increasing customer demands for product quality, along with an increase in the number of assembly operations, led to the introduction of automated final assembly islands at Daimler Benz. One example was the automated assembly of underbody parts, such as heavy axles or parts which required two workers to cooperate.

A transport system was introduced which enabled the turning of the whole car body. This tilting mechanism eliminated the overhead working position, and hence improved the quality of the manual assembly of safety-critical parts, such as the brake line. This type of rolling mechanism has since been adopted by other automobile manufacturers as standard equipment. Figure 2.1.6. shows the transport system in different tilting positions at BMW.

1980: Cathodic Dip Replaces Anaphoretic Dip

Improvements in engine technology in the seventies resulted in differences in the "life expectancy" of engines and car bodies. The engine usually outlived the car body by several years. The transition from anaphoretic dip to cathodic dip in the paint shop, multi-layer painting and the conservation of hollow spaces together led to a reduction in this durability gap.

In order to avoid surface oxidation, panels were covered with an oil film in the body shop. This film, however, also inhibited an adequate adhesion of the coat, so that the bodies had to undergo careful mechanical and chemical cleaning before coating. In order to achieve a high quality coat, this form of pretreatment became a crucial process in the seventies .

The dip coat was the first coat capable of protecting panels from mechanical damage. In the case of anaphoretically dipped bodies, this coat did not form a durable panel cover. The change in electric polarity of the body at the dip permitted a durable cover of the panels. Dipping under a specified temperature ensured better penetration of paint into hollow spaces.

1983: Assembly Shop Number 54 (Wolfsburg, Volkswagen)

In order to cope with the boom in compact-class cars, one possible solution was the reduction in the cycle time. This could be achieved by increasing the number of workers along the assembly line. However, employing more workers at an assembly line with 100 working stations on a length of 580 m could lead to unacceptable working conditions. In some working places, this would mean up to five workers on an area only 5.80 m long. The assembly shop number 54 was an attempt to solve this problem: it did this by increasing the automation ratio in the final assembly shop.

Interior trim and flexible attachment parts had, up to now, been fitted manually. The assembly of the interior trim was followed by the automated assembly of attaching parts (usual for that time). Most of the parts were subassemblies, such as fuel pipes, batteries, brake lines, fuel tanks, engines, exhaust pipes, rear axles, spare tires and tires.

At that time, assembly shop number 54 featured the highest automation ratio world-wide. The automation ratio immediately increased from 5% to 25% in just one single step. Product innovation was a substantial prerequisite for this process innovation. All the parts and subassemblies had to be designed and evaluated with regard to their suitability for automated assembly. The measures thus implemented included assembly with linear joining movements, reduction in the number of flexible parts and development of new joining technologies, e.g. by means of clips. However, the short cycle time, which made it more difficult to recover losses due to short break-downs, reduced the economic effectiveness of the automation area.

1984: Emerging Glue Technology

Gluing - as an alternative to joining - permitted the substitution of welding operations in the body shop and led to the aerodynamic design of the transition between windows and body.

Heat generated during the welding process may lead to small, local wrinkles on the panels. This small unevenness of outer panels must be ground smooth. The gluing of panels made this operation superfluous, but now, improved chemical preparation before the joining process became a vital necessity. Gluing was hence only used for a few operations in the automobile industry.

Substituting the rubber seal, the gluing of the windshield glass led to a smooth transition between windshield and body. The glue not only served as a joining, but also as a sealing compound. Due to the vast strength of glue joints, the windshield glass also performed a supporting function in the body. One of the results was the enlargement of the windshield.

The substitution of the rubber seal made automated windshield glass assembly possible. The assembly of large, flexible parts has not been investigated sufficiently with regard to the implementation of a feasible, automated solution. A more wide-spread introduction and application of the gluing technology is limited because of the problems connected with the recycling of glued parts.

1985: The Cockpit Subassembly

The modular design concept adopted by most companies resulted in the separation of subassembly lines from main assembly lines. This made the assembly line shorter and more reliable.

The dashboard as a basic part - integrated by the fascia, the shaft and the foot pedals to form a subassembly - permitted the uncoupling of these operations in a separate assembly line. This modular assembly scheme was adopted by Opel when the first Omega model was introduced.

In the subassembly line, cockpits were assembled on AGVs. Good accessibility of all joining positions was ensured, and the workers were relieved of the error-prone work area of the foot pedals inside the car body, a task which also involved strenuous physical work in the past.

The subassembly line was information-linked to the main assembly line. Synchronized information controlled the delivery of subassemblies in accordance with the requirements of the main assembly line (Production Information System, PIS). A buffer, which allowed sequencing, compensated peak loads on any line.

The automated assembly work was performed by a gantry robot in three working cycles. Fixtures and cockpit subassembly were designed in such a manner that they could be used or assembled automatically or manually. Figure 2.1.7. shows the manual assembly of the cockpit at Opel's Eisenach plant.

1987: Just in Time (JIT)

Inventories, i.e. work in progress, were identified to be the largest source of waste in automobile production. In order to reduce storage costs, the production-synchronized delivery of parts to the assembly line was introduced in the nineteen-eighties.

In accordance with this philosophy, the delivery of a large range of parts, and expensive parts such as tanks, bumpers and seats, was taken into consideration in the planning and control of the assembly process. A direct and integrated flow of information between suppliers and assembly plant helped reduce part inventories for the assembly area to just two hours. When the information flow is too slow, parts then have to be delivered in sequential order. This can lead to an increase in inventories of up to an average of two days.

Most deliveries are, in fact, still performed without a tight information network, so that, in reality, the roads serve as storage facilities for inventories.

Fig. 2.1.7. Manual assembly of the module cockpit at the Eisenach plant, Opel

1988: Flexible Automation in the Body Shop with Clamping Robots

At the end of the eighties, VW implemented a new, flexible body shop in order to cope with increasing flexibility, cost and quality requirements.

The concept meant that only flexible devices were used for joining and handling operations.

In order to assemble underbody and framing, robots were used for the spot welding and the clamping of the parts. In order to reach a high degree of flexibility for the production flow and preparatory operations, the arrangement of the machines in a production line was substituted by automatic welding centres, supplied by AGVs.

At the Emden plant, 20 of these welding centres, called flexible honeycombs or boxes, were established to produce 1,200 automobiles per day. A flexible honeycomb consisted of a precision frame, clamping robots, welding robots, lifter and welding guns, as well as tool magazines. In order to achieve the car geometry, the sub-assemblies are clamped and spot-welded in 11 geometry boxes by means of 344 industrial robots. Up to 32 clamping robots and six welding robots can be installed in one box. Respotting of the bodies then follows in nine respotting boxes.

The tolerance requirements for the panels and subassemblies in the honeycomb posed problems for the flexible automated assembly of car bodies. Since the implementation of the body shop, there has been no change in body type, nor has a successor model been introduced, so that up to now, no experience has been gained with regard to the possible flexibility of the system. The concept of using

Fig. 2.1.8. Standard press line and one-large-piece press at Wolfsburg, Volkswagen

industrial robots in order to clamp panels was also adopted by Nissan, and was then perfected to what is known as the *Intelligent Body Assembly System* (IBAS).

1988: One-large-piece presses

The one-large-piece press, with a quick die change (QDC), reduced the number of parts required and also led to reduced storage requirements. In contrast, press lines linked by mechanical pick-and-place manipulators or industrial robots required approximately four hours for a die change. Panels hence had to be pressed in one lot to meet the demand for several days and in order to achieve a high utilization rate of the press. Figure 2.1.8. shows a conventional press line and a new one-large-piece transfer press in Wolfsburg.

The dies for the one-large-piece presses were prepared outside the presses. This time was nearly the same as the preparation time at press lines. The difference, however, is that in the case of the one-large-piece press, the die is prepared parallel with the operation of the press. The shutdown time for the die change is reduced to 6 - 10 min. The size of the batch job can be reduced to the just a single day's demand, or even less.

The integration of all forming steps in one press resulted in a very high geometric accuracy of the panels. This led to the development of the one-piece side panel in which several parts are integrated into one, thus eliminating several assembly operations in the body shop.

Fig 2.1.9. Automated assembly of the vehicle ceiling at Fiat's Monte Cassino plant

1988: Automation in the Assembly Shop (Fiat / Monte Cassino)
In accordance with the company's automation strategy, Fiat established a highly automated assembly shop at the Monte Cassino plant, where the Tipo and Tempra models were built. Nearly 45% of assembly operations was automated, and information was integrated in a highly computerized shop-floor control system. Automation of final assembly operation and implementation of Computer Integrated Manufacturing (CIM) were seen to be a solution towards increasing the quality of the assembled product.

Subassembly lines for modules were introduced. These modules were then assembled in the main assembly line. High product variations, which affected the continuous flow in the main assembly line, were banned from the sub-assembly lines. Synchronous assembly in the main and subassembly lines reduced the length of the main assembly line tremendously.

In order to improve the accessibility of joining positions by automated machines in the car body (e.g. assembly of the vehicle ceiling), the doors were removed at the beginning and reassembled at the end of the assembly shop process. The doors were fully assembled to a module in a sub-assembly line. Figure 2.1.9. shows the automated assembly of the vehicle ceiling.

Thanks to the preparatory operations, final assembly of 15 modules was performed by 200 automated screwing operations. Automated assembly was adopted mainly for doors, cockpits, windows, bumpers and front ends.

Several problems occurred during the implementation phase. Assembly of the front end had to be performed manually. The robot-integrated assembly of the front end could not be performed with sufficient reliability. It took four years after

Fig. 2.1.10. Complete assembly by teamwork at Volvo's Uddevalla plant

production start up until a shop floor control system for the automated door assembly could be installed.

1989: Qualified Teamwork (Volvo)
Building on the experience from the Kalmar plant, Volvo tried to develop teamwork-based assembly even further. In order to increase workers' motivation and achieve a good assembly quality, cars were fully assembled by teams. Conventional sequential assembly was replaced by the parallel assembly of cars. This concept was implemented by Volvo at its plant in Uddevalla.

The teams consisted of 8 - 10 workers and each team had to assemble one car completely within two hours. The painted bodies were transported on a body carrier, which was moved by AGVs to the assembly area of the team. Parts and subassemblies were provided on seven material support units. Figure 2.1.10. shows a body and a material support unit at the Uddevalla plant

The transport of all the parts and subassemblies took place in a highly automated area. Between 3,000 and 4,000 parts, most of them body-specific, had to be made available for one car. The subassembly of several components was integrated in this preparation area.

At this plant, Volvo never did achieve a degree of productivity comparable to that of the Dutch plant. Volvo closed the Uddevalla and Kalmar plants at the end of 1992.

1992: "Lean" Automobile Plants in Europe

As an answer to the challenges set by Japan's enormous lead in productivity, Opel built a new plant in Eisenach, Germany. The experience GM had made in its joint venture with Toyota at Fremont in California (NUMMI) influenced the design and planning of this plant.

Innovation was mainly in the field of work organization: workers were organized in teams of 6 - 8 members. The body and assembly process was set up in such a manner that the team approach was reflected by the plant layout. As for the process equipment, Opel used proven solutions, such as the Robogate (Opel Bochum and Zaragoza), and module assembly for the cockpit (Opel Bochum and Rüsselsheim). Three-shift work was introduced in order to increase productivity. Shifts overlapped by 30 minutes, so that workers were able to talk to one another about special work requirements.

The standardized operation sheet forms the basis for teamwork. In this sheet, workers and engineers determine the sequence of operations at their work stations. The standardization of the work sequences is the same for the two car models, the *Corsa* and the *Astra*, produced in this plant, so that the results of Kaizen can be transferred and generalized.

The SEAT plant at Martorell and the Volkswagen plant at Mosel were other new plants where lean production was introduced. In these cases, the implementation of the Toyota Production System was only decided upon at a later planning stage, however. Therefore, the Eisenach plant is presently the only plant in Europe which was planned under this aspect from the very beginning. The productivity of this plant is very high, a fact which also results from the reduction in the number of variants of the *Corsa* and *Astra* models compared to the degree of variation found in the other Opel plants producing these models.

1992: Water-Based Coating

Legal requirements under the German *Air Quality Guideline* (TA Luft) and the *Water Quality Guideline* (TA Wasser) helped to speed up the development and introduction of hydrophilic coats. In Japan, these coats were developed in order to achieve an increased lustre of the final coating.

In these hydrophilic coats, harmful chemical solvents are replaced by pure water. This required the development of a different coating process, however. The rationale for the process innovation is the narrow climate window, where water-based coats can be processed. The use of modern microelectronics and software solutions allows the temperature in spray booths to be controlled between + 20°C and 28°C at a relative humidity of between 60% - 70%. In the years that followed, all European automobile companies built new with paint shops using water-based coats.

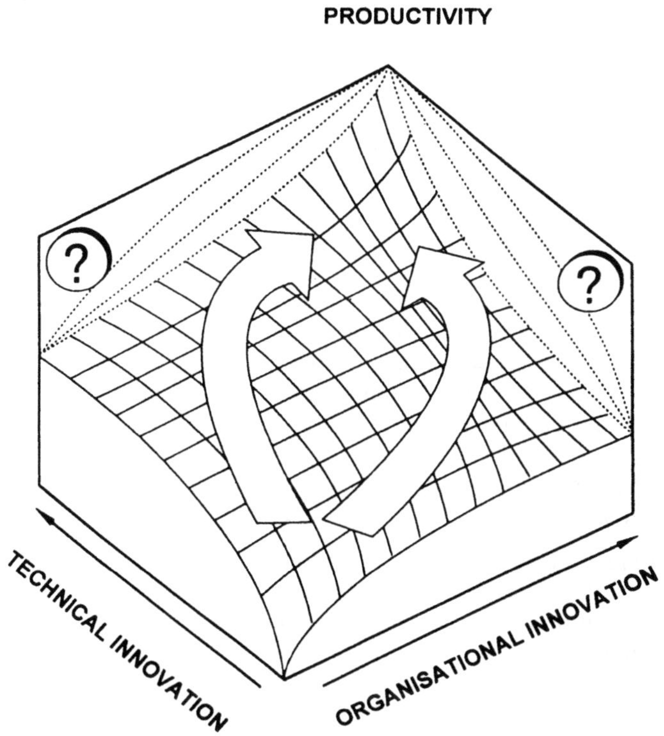

Fig. 2.1.11. Impact of the progress due to technical and organizational innovation on productivity

2.1.3 Innovation Paths

The analysis of the trajectory of innovations in the preceding chapter shows that each manufacturer focused on just one innovation path for a certain period of time. Other innovation paths were given lower priority and merely adjusted to the main innovation path. The potential for rationalization in these other innovation paths was not fully utilized. In the past, this approach was viable because economic efficiency was, in most cases, the dominant objective.

In this decade in Europe and Japan, two additional objectives, i.e. protecting the environment and making working life more humane, have become increasingly important. Optimization is now a multi-goal issue. It is only by simultaneously focusing on the three innovations paths - i.e. product, process and organization - that efficient rationalization, environmental protection and humanization can be achieved (fig. 2.1.11.).

2.1.4 Conclusions

At present, European automobile companies, hoping to gain a leading edge over their competitors, focus mainly on the implementation of the Toyota Production System. The way in which innovations are pursued today is the same as in the past: focus on just one single innovation path which is currently centred on production organization. This seems indeed the appropriate choice, as an enormous backlog needs to be made up in this respect.

In the 1980s, it was the Japanese automobile companies who developed and implemented large scale automation projects for the body and assembly shops. This was partially motivated by the problems in recruiting factory workers due to the *3D's*. In the 1990s, however, Japanese manufacturers are concentrating on concepts of product design and human-oriented automation, based on the established organization.

Drawing from the lessons of past developments, we suggest the following hypotheses for further discussion:

1. At the moment, there are no technical limits to automation which, in theory, could not be overcome by the research and development now underway.
2. Additional improvements in product functions and quality can be achieved by further investment in automation.
3. User skills and knowledge for mastering the complex functions and structures of the flexible technical systems limit the further advancement of automation. Modern information techniques, including implementation of computer-aided systems, can further enhance the transparency of the complex systems and processes.
4. Technical systems will be increasingly established on a plant-wide and company-wide basis. Developments in computer-integrated systems permit the internal and external integration of all flexible automated machines. The introduction of appropriate CAQ-systems for customers and suppliers is one example.
5. Flexible, automated production systems are developed for applications on the shop floor and are integrated into a total system architecture. This development is not yet complete. Further integration, aimed at mastering even the most complex systems, can be expected.
6. Control systems, simulation systems, artificial intelligence, database methods are the next elements to be integrated. The main objective of the technical development is therefore to improve the communication and integration capability of hardware and software.
7. The communication between information systems will be further improved by means of new input devices. The new communication ports for speed recognition and the sensomotoric ports between man and computer are examples of this.
8. Problems related to the synchronization of information and material flow in the production process are leading to the development of mobile data units. These data units carry all of the important product, organization and technology data, documenting present production status, machine data and quality.

9. Product life cycle considerations require computer-integrated recycling systems, where data regarding the materials used, information concerning disassembly processes and other recycling information can be stored. The data can be used to control machines or to support planners in their decisions. A closed information loop can be established by integrating the recycling-related information into design and planning systems.

2.2 Basic Trends in the Physics and Economics of Automated Fabrication and Assembly Operations

D. E. Whitney

The economics of automation in fabrication and assembly operations cannot be separated from some basic physics. This is especially true of assembly where the only way to compete economically with people is to be very fast. Speed requires high accelerations related to starting and stopping, since assembly motions are basically of a back and forth nature. Accelerations immediately bring forces, masses, lengths, moments of inertia and similar properties of physical objects into play, forcing us to contend with the connection between economics and physics.

2.2.1 Fabrication

Fabrication processes include shaping objects by bending, deforming or cutting. In the main, people cannot do these operations without tools, and cannot do them at economical speeds unless the parts are very small, such as parts of watches. For these reasons, most fabrication must be done by machines, and often very large ones. Fabrication operations are said to add value to raw material, since they give individual parts their required shape and internal material properties. Such operations typically take a long time; operation cycles can last from several minutes up to as much as an hour. During these cycles, several different operations may be carried out on the same part, such as drilling several holes, or drilling and threading holes.

The machines typically do not move very fast but, instead, exert a lot of power by means of low speeds and large forces. Distances moved during an operation are typically the same size as the size of the piece being worked. The in/out time (the time to remove a part and replace it with a new one), or the time to change a tool, is typically very small compared to the operation cycle time, perhaps 1% to 10%. For all these reasons, a large, slow, powerful machine can be quite economical, since it adds value to a part by doing things that cannot be done any other way except by another machine.

An economically competing machine would have more accuracy or higher operating speed for the same or lower acquisition and operating cost. Both trends are occurring.

New applications with these properties (high power and high mass at low speed and possibly high force) include numerically controlled cutting and bending machines, and robot welding machines.

2.2.2 Assembly

Assembly is basically different from fabrication in the sense that for the most part people can do assembly operations. Thus any assembly machine or robot must compete directly with people on a cost and speed basis. Only in special cases must machines be used for assembly: the required precision or cleanliness is beyond the abilities of people, or there is outright danger to people, such as radioactive or chemical contamination. This means that most robots must either be very fast or very low cost.

Assembly operations are typically fast and must be said to add only a little value. Typical operations last from just a few seconds to no more than a minute. Each assembly operation typically adds only one part to an assembly, except when the "part" is a subassembly of many parts assembled elsewhere. An assembly robot arm moves from a part supply to the assembly, back to another part supply (if more than one operation is being done), back to the assembly, and so on. The robot must therefore accelerate and decelerate at high rates. Typical distances moved are in the order of 10 to 100 times the size of a typical part. In pure dimensional terms, the power required to achieve such motions is in the order of

$$\pi = \frac{ML^2}{T^3} \qquad (2.2.1.)$$

where π is the power, M is the mass being moved, L is the distance it is moved and T is the duration time of the move. L is also typically the robot's overall size, equivalently its reach or the distance from the robot's center of motion to the tip of the extended wrist.

Equation (2.2.1.) means that the required power rises quickly as either the time decreases, the reach or distance moved increases, or the accelerated mass rises (including the workpiece, the end effector or gripper, and part of the robot's own mass). Larger parts have larger size and mass, increasing both M and L. The power must be supplied by motors, some of which must also be moved since they are in the hand, wrist, or elbow. Since motors also have mass, the problem is compounded.

This combination of economic and physical facts leads to a number of results. First of all, there is not much "mass capacity" left over for carrying parts once the mass of the robot itself is handled by the motors. Thus, higher speed robots tend to carry only small parts. Secondly, very fast operation times mean that in/out and tool change times loom very large, and are often as long as the operations themselves (e.g. 1 to 3 seconds). Therefore, economics force the robot system designer to have the robot do many operations at once before an in/out or a tool change occurs. Thus, either one tool must be able to grip and insert a series of dissimilar parts, or an array of similar assemblies must be within the robot's reach. (Remember, serving more assemblies increases L.) Otherwise, the robot must carry several tools (no time to change tools, remember), but this only adds more mass at the largest distance L from the robot's center of motion, thus adding to the power required or reducing the speed.

A major conclusion is that part mass is basically negligible in high-speed situations, with the mass of a set of tools on the wrist being much larger. Therefore, tool mass determines overall robot power requirements, not part mass. For all practical purposes, this means small part mass, or, equivalently, small parts. Required power determines motor size, and motors are the main cost component of robots.

When tool or part mass must be large, then accelerations will have to be low, meaning that motions will be slow and operation times long. Such robots cannot compete with people on a speed basis. Instead, they must compete by carrying a heavy load (such as a spot welding gun). Economically, the trend toward larger and slower robots cannot be pursued very far, except in the case of construction cranes. In such cases, people cannot compete on a load-carrying basis.

So we return to the case of the fast, small robot that carries several tools and assembles several parts at once. An interesting design of such a robot is the SCARA, which is arranged so that all its motorized axes except one are vertical, causing the arm to move in a horizontal plane. The outermost motion axis (providing vertical assembly action) is the only exception to this rule among axes that provide the main motions. Therefore, the motors that drive the planar motions do not have to support the weight of the arm against gravity (which otherwise would require them to provide effectively 980 cm/s^2 acceleration before moving the robot at all). These motors can exert all their power in useful accelerations, permitting smaller motors for the same motion speeds, and lowering the mass of the robot without reducing its speed.

The advent of the SCARA robot eliminated the myth that robots must have 6 degrees of freedom in order to do useful assembly. (People have from 7 to 30 degrees of freedom in one arm, depending on how one counts.) But the restricted motions of the SCARA, basically XYZ, required designers to seriously confront design for assembly for the first time. By now, even very complex items such as Polaroid cameras and Sony CamCorders can be assembled by robots working from above the assembly. However, very dexterous end effectors, almost robots in themselves, are required.

For the most part, fixed automation machines assemble small items, typically less than 10 cm cube. Machines to assemble larger items would be very big (see below) and too expensive. Large items are usually not made in large enough numbers to justify the expense of these machines, which must last a long time due to the difficulty of "reprogramming" them.[1] (Again, see below.) The assembly motions of these machines are very small and deliberate. Since programmable motions are not employed, high power cam and lever drives can be used instead of electric motors, providing more than enough power for the accelerations

---

[1] They are not impossible to reprogram, which is done by removing all the end effectors, cam actuators, and part feeders. The resulting foundation and main power drive are reused. New cam drives and actuators can often be designed quickly using CAD, but part feeders are difficult to design. Sony's SCARA robot systems use a novel part feeding method that can be designed relatively quickly compared to conventional means.

needed, which are not large since L is small. Typical of products made by fixed automation are cigarettes, ball point pens, spray bottle mechanisms, and so on.

2.2.3 Simplified Economic Models

Even simplified economic models can capture the essence of the above discussion. Equation (2.2.1.) showed how important task operation time T is to required motor power. In the models below it can be seen that task time is again the pivotal variable.

Below are simplified cost models for manual assembly, fixed automation assembly, and flexible automation assembly. Fixed automation is a system of workheads in a line in which each workhead does exactly one operation. Flexible automation and manual assembly comprise "workstations" that can change tools and do several operations on one assembly.

The cost of labor includes wages and benefits. The number of people needed is inversely related to the operation time they can achieve, which varies inversely with the mass of the parts. The cost of fixed automation is essentially proportional to the number of operations that must be done which, at a minimum, is the number of parts that must be assembled. Machines for assembly are typically 10 to 100 times the size of the assembly itself, and overall cost depends directly on machine size. The number of robots needed is, as with people, inverse to the operation time achievable. Robot cost is roughly proportional to the size and number of motors. A useful proxy is robot size. Thus larger, more powerful, faster robots that can carry heavier parts or more tools will cost more. In addition, a set of robots needs tools or end effectors in a quantity proportional to the number of parts to be assembled.

These basic facts are captured in the equations in table 2.2.1. that follow, except that robot, fixed workstation, and robot end effector costs are represented as parameters rather than as functions of part size. To provide a way to compare the fixed costs of machines and the variable costs of people, the investments are amortized by a capital recovery factor f_a. The equations capture the resulting cost to assemble one unit comprising N parts.

Figure 2.2.1. utilizes these equations to graph the cost per unit as a function of annual production volume, using the equations in table 2.2.1.

The assumptions are:
T = 5 sec for both people and robots (typical for early 1980s)
Labor cost = $15/hr (again typical for early 1980s)
N = 10 parts/unit
fa = 0.38 (roughly 25% annual rate of return over 5 years)

S\$ = \$40000 per robot, \$50000 per fixed station
\$/tool = \$5000
w = 0.25 worker/station

Figure 2.2.1. shows that in the early 1980's, neither robots nor fixed automation were really competitive with manual labor, unless production volume was over 1,000,000 units per year. Such volumes are rare, even in the auto industry. Such volumes are typically achieved only in consumer electronics and with small items, such as pens and cigarettes. (fig. 2.2.3.) Since designs change frequently in these products, fixed automation is not really economic because it is difficult to change over to a new model unless that model is quite similar to the old one.

Table 2.2.1. Simplified economic models of assembly alternatives

a) Manual automation:

$C_{unit\ Manual}$	$= T\ L_H\ N / 3600$	
where	T	= operation time per part, sec
	L_H	= labor cost, \$ / hour
	N	= number of parts in one unit
	3600	= number of seconds / hour

b) Fixed automation:

$C_{united\ Fexed}$	$= f_a\ N\ S\$ / Q$	
where	Q	= annual production volume, units / year
	f_a	= fraction of machine cost paid for per year
	S\$	= cost of one station in the machine

c) Flexible automation:

$C_{unit\ Flex}$	$= (f_a\ 1 / Q) + (L\$ / Q)$	
where	1	= total investment in machines and tools
	L\$	= annual cost of workers associated with the system
1	= (# Machines * \$ / Machines) + (# Tools * \$ / Tools)	
# Machines	= T N Q / 2000 * 3600	
# Tools	= N	
L\$	$= w\ T\ N\ Q\ L_H / 2000 * 3600$	
where	w	= number of workers / station
	2000	= number of hours per year per shift

Combining the above yields:

$C_{unit\ Flex} = (f_a\ \$ / \text{Machine}\ T\ N) / (2000 * 3600) + (f_a\ \$ / \text{Tool}\ N) / Q + (w\ T\ N\ L_H) / 3600$

To show the dramatic effect of robot speed on relative economics, figure 2.2.2. shows the same comparisons as figure 2.2.1. except that robot operation time is now 2 seconds instead of 5. This speed increase occurred in the early 1980s with the advent of SCARA robots. The large increase in required power came from a new generation of direct drive high torque low speed electric motors.

Figure 2.2.2. indicates that robots might be economical at volumes as low as 200,000 units per year under the assumption of greater speed. This opens a wide area of new applications, but these are still in the consumer electronics area where model change and intricate assembly requirements lead to expensive dexterous end effectors and rapid changeover of both end effectors and assembly control programs. Application areas in general remain limited, as can be seen in figure 2.2.3. which schematically compares production volume with product size.

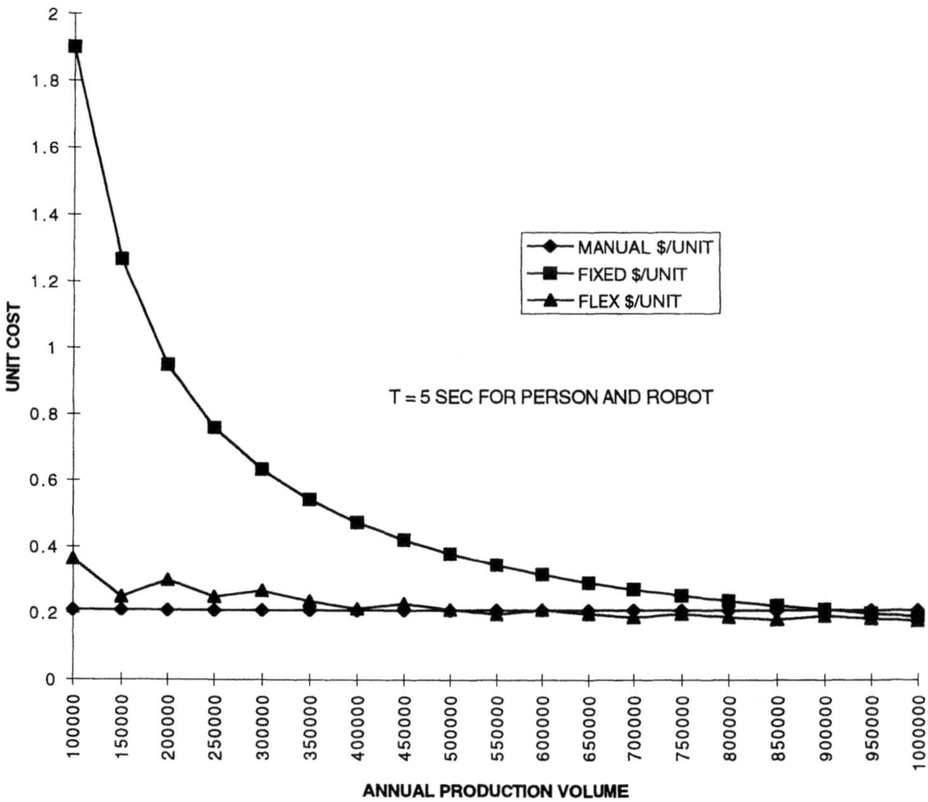

Fig. 2.2.1. Simplified cost models for three kinds of assembly automation: robot and human operation time is 5 seconds for person and robot

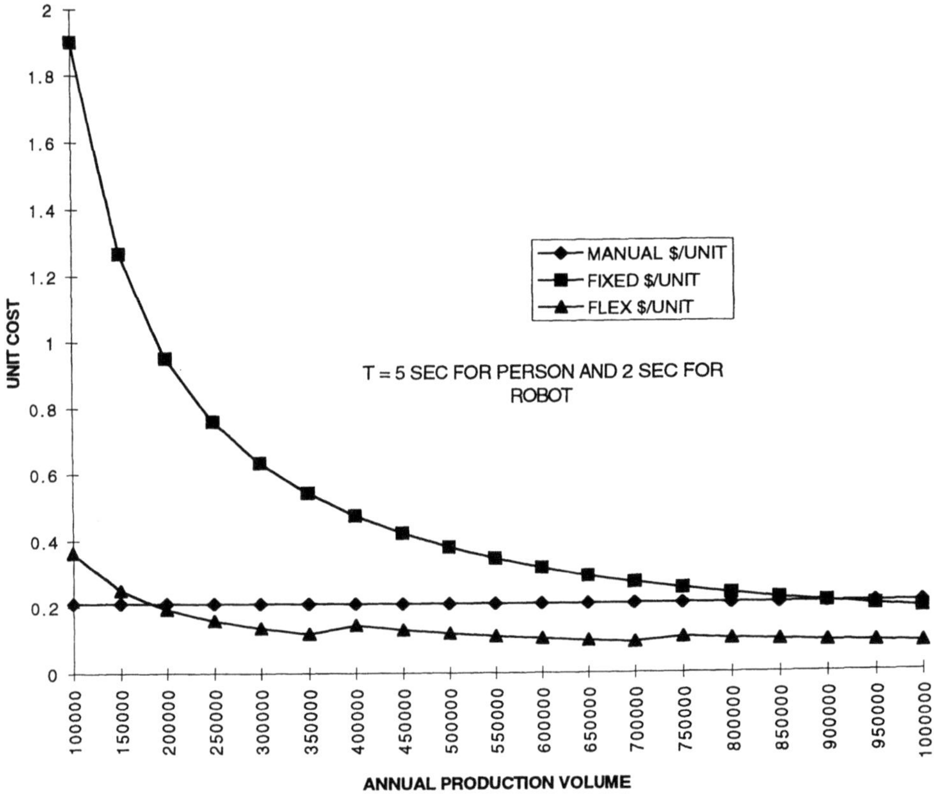

Fig. 2.2.2. Comparison of unit assembly costs when robot operation time is 5 sec. for person and 2 sec.for robot

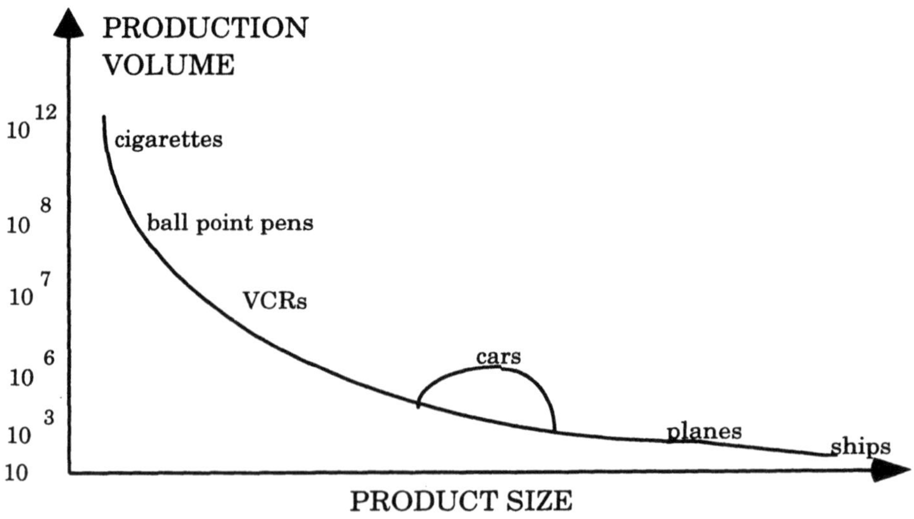

Fig. 2.2.3. Schematic of the relationship between product size and annual production volume

2.2.4 Summary

Choices in automation are governed by a combination of economics and physics. Economics make investments feasible or infeasible, while physics make them possible or impossible. Table 2.2.2. summarizes the combinations.[2]

Table 2.2.2. A rough cut at what is possible physically, what is feasible economically for fabrication and assembly automation

		Parts and Work:	
		Small	Large
Process or Machines:	Slow	Possible Infeasible for assembly	Possible Infeasible except in some fabrication applications
	Fast	Possible Feasible for some assembly applications	Impossible Infeasible

2.2.5 References

1 Nevins J L, Whitney D E (eds) (1989) Concurrent Design of Products and Processes. McGraw-Hill, New York

[2] More detail on the issues raised in this chapter may be found in [1].

2.3
Key Characteristics of Assembly Automation Systems

J. Tidd

2.3.1
Automation, Complexity and Flexibility

Studies of the automobile industry suggest that organisational and human resource factors have a greater impact on manufacturing productivity and quality than levels of automation [1]. However, previous research may have under estimated the significance of assembly automation for three reasons.

First, a relatively crude measure of the level of automation is typically used in studies of the automobile industry, for example, levels of automation are commonly expressed as a percentage of operations which have been automated. This is not adequate, and in most cases it is necessary to differentiate between different types of automation, and the automation of different tasks. Second, most studies have avoided the issue of product complexity by defining a standard set of assembly operations and comparing levels of automation of these operations. However, it may be useful to examine the affect of product and task complexity on the type and level of automation. Third, the primary focus of most studies has been on factors affecting productivity and quality, rather than flexibility.

This chapter reviews the main characteristics of assembly automation and examines the relationship between different types of assembly automation, task complexity and production flexibility, using differences between trends in the UK and Japan to illustrate the options and trade-offs which exist. The first section provides a brief summary of the development of assembly automation, specifically the growing application of robots to assembly tasks. The second section reviews the main technological developments responsible for the application of robotics to assembly automation. The third section identifies the main factors currently limiting the widespread application of assembly automation, specifically task complexity and flexibility requirements. The final section contrasts the experiences of users of robotic assembly in the UK and Japan to illustrate the wide options available to users.

2.3.2
Growth of Robotic Assembly

Until the 1990s, automation in the automobile industry was largely confined to machining operations, that is component, transmission, and engine fabrication and the resultant increase in rigidity had to some extent been offset by flexible, labour-intensive assembly. Assembly has historically presented an 'automation bottleneck': typically all parts fabrication, up to 95% of all body welds, most wash, phosphate and paint processes have been automated in the automobile industry, but chassis and final assembly remain labour-intensive, and even in state-of-the-art plants only 25-30% of such tasks are automated. Consequently, assembly now accounts for about two-thirds of all direct labour employed, and over a third of total manufacturing time. However, robotics is now being applied to component assembly, and the automation of final assembly is now receiving attention in plants worldwide.

For the assembly of high-volume, standardised, but relatively simple products, dedicated, purpose-built cam and pneumatically driven assembly machines have been in use since the 1940s. In an effort to overcome some of the limitations of such machines, cheaper and more flexible modular designs were introduced in the 1960s, followed by programmable controls in the late 1970s. However, because of their high cost and limited capability such assembly machines are only suitable in very specific circumstances: relatively simple products ideally consisting of less than ten components; high quality components; few product styles, ideally just one; few design changes throughout the products life; and high production volumes.

Because of these limitations only around 5% of all assembly work was automated by the 1980s, and most of this automation was based on dedicated, special-purpose assembly machines. The main users of this technology were the general machinery and electrical engineering industries. More than a third of the assembly machines used in the general machinery industry were programmable, and almost a fifth of those used in the electrical engineering industry (including electrical automotive components). In contrast, less than a tenth of the assembly machines used in the automobile sector in the 1980s were programmable. This suggests that the flexibility of assembly machines was less of a limiting factor than their inability to perform complex assembly tasks.

During the 1980s, robots began to be applied to assembly work. Robots had the potential to perform more complex tasks and offered greater programmability. Industrial robots had become increasingly task-specific, moving away from the original concept of 'steel collar' worker, or 'universal automation'. In terms of generic tasks robots can be divided into three broad groups, which also approximate to their chronological development:

1. *Handling robots,* where the workpiece is handled by the robot, for example, loading machine tools or injection moulding;
2. *Process robots*, where the tool is gripped by the robot, for example, welding or paint spraying;
3. *Assembly robots*, where the robots are used in the assembly of parts into components or complete products.

In almost every industrialised country the automobile industry was the first and has subsequently become the dominant user of both handling and process robots, particularly for welding and painting. However, since the widespread diffusion of assembly robots this dominance has decreased: in 1980, the automobile industry accounted for 34% and 38% of all robots in use in the UK and Japan respectively,

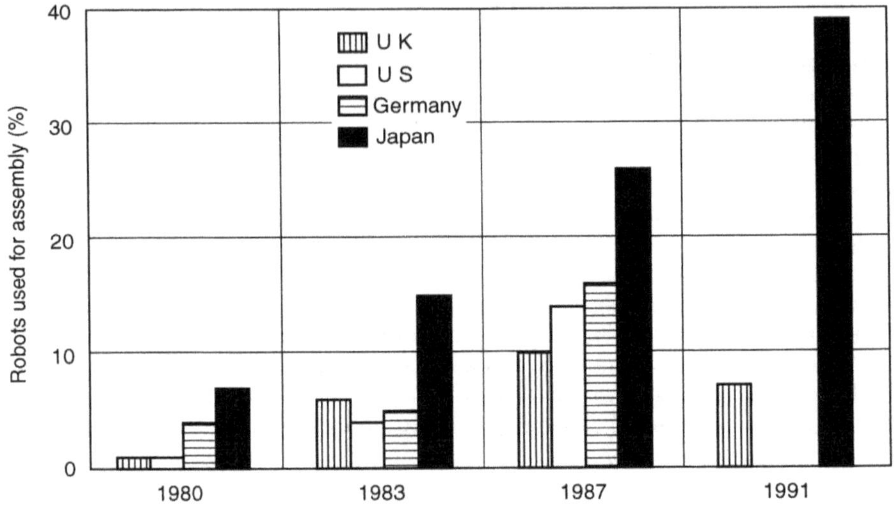

Fig. 2.3.1. Growth of robotic assembly, 1980-91. Source: Tidd [12]

but by 1987 the industry's share had fallen to 30% in the UK, and just 17% in Japan; in the latter case, the electronics industry had become the major user with almost a quarter of all robots.

The first commercial assembly robots were introduced during the late 1970s, the first being DEA's Pragma and Olivetti's Sigma in Italy, followed by the PUMA developed by Unimation in conjunction with GM in the US and launched in 1978. Japan's first major contribution to robotic assembly was the SCARA family of robots launched in 1981. Since then the growth of robotic assembly has exceeded all other robot applications in almost every industrialised country (fig. 2.3.1.). Britain and Japan are clearly at opposite ends of the spectrum in this respect as there are still only 500 assembly robots currently in use in Britain, compared to over 25,000 in Japan; Japan also leads the world in terms of assembly robot density, ie. the number of assembly robots per 10,000 assembly workers, in the case of the UK by more than an order of magnitude, 224 compared to just 21. In terms of the proportion of all industrial robots used for assembly, the US and Europe now appear to be very close, but the aggregate statistics hide considerable national disparities: Germany and Italy lead Europe, each with around 30%, whilst the UK currently uses less than 10% for assembly.

To some extent the observed international disparity is due to differences in industrial structure: the electrical industry is normally the first and largest user of robotic assembly, usually for the insertion of odd-form components, but increasingly for more complex electro-mechanical assemblies such as VCR mechanisms; the use of robotic assembly for larger, heavier, and more complex mechanical products is a more recent development, and the automobile industry is pioneering such applications. The use of robotic assembly in Japan where the technology is currently most widespread suggests that although the electrical industry (but this includes electrical automotive components) is still the main user, the automobile industry is catching up fast. On one hand, there is evidence of technologies developed primarily for the electrical industry being 'scaled up' for use in the automobile industry, for example, the use of larger SCARA-type robots to assemble mechanical components; and on the other, the use of sensors to improve the positional accuracy of larger hydraulic robots for use in final assembly.

2.3.3
Robotics Technology

A robotic system typically consists of: a mechanical structure; actuators or motors; end-effector; sensor system; control system; and its programming language. In addition, a robotic assembly system will incorporate some form of parts feeding and orientation system, and a transfer device. Therefore, a large range of configuration options exists:

1. *Assembly Robots* - basic configuration, eg. Cartesian (ie. where all axes are linear, like a conventional gantry crane), Jointed-Arm (where all joints are rotary and normally vertical, like a restricted human arm), SCARA (a simpler, horizontally jointed-arm robot), or Cylindrical (two linear axes, one rotary), the number of axes or powered joints, in addition to more specific make/model decisions based on cost/payload/repeatability/speed requirements;
2. *Sensors* - whether or not to use, robot or fixture mounted, type, eg, proximity, force, tactile, vision;
3. *Component feeders* - conventional vibratory or track, multi-part feeders,pallets of identical parts, pallets of mixed parts, programmable feeders, sensor-based feeders, random parts bins;
4. *Gripper/Tool design* - dedicated tools, multiple tools on turret wrist, universal socket and tool changes, flexible gripper;
5. *Transfer system* - manual, indexed line or rotary, free-transfer, gantry, automated guide vehicles (AGV).

In addition to these hardware options, two distinct system configurations exist, although in practice hybrid examples are also common: Cell or parallel: analogous to manual bench assembly where a complete assembly is built at each station; Line or series: analogous to the classic assembly line where just one or two parts are inserted at each station. The choice of system configuration is significant because:

1. Robots used in line assembly systems will normally each insert fewer parts than those used in cells, and therefore will not need to be as technically sophisticated;

2. Line assembly cycle times are generally shorter than the cell equivalent, and therefore parts are presented to each robot more frequently, which reduces costs as parts feeders are generally much faster than the robots;
3. With good product design, and appropriate sequencing of tasks, line systems should not require time-consuming gripper changes;
4. The special-purpose part of the system is smaller per assembly with line systems.

In the automobile industry there are essentially two layouts for final assembly: the long line or short line. The long line layout takes into account the fact that current assembly robots are not as space-efficient as workers. The short line layout allows for this by reducing the main line by increasing the number of subassembly lines. The short line approach simplifies the assembly process, but requires modular product design. In most cases it is easier to automate these subassembly operations because the components are smaller, lighter and less complex.

The main technological developments responsible for the increased availability of assembly robots are the use of electric motors and microprocessor control. First-generation assembly robots are powered by electric motors for improved repeatability than the earlier hydraulic robots, but this also limits the maximum payload of such robots, which is typically less than a few kilogrammes, and the size of the work envelope, which is a function of the robots reach. The development of lightweight arms, constructed from composite materials, is no longer as popular as it was a few years ago, although new arm configurations continue to be developed for assembly work.

Direct-drive robots, which directly couple the drive motors to the arm joints, are currently receiving a great deal of attention worldwide, and one direct-drive machine has been commercially available for several years now. In principle, the direct-drive robot is a much simpler device than its conventional counterpart: the drive transmission of conventional industrial robots consists of various gear trains, leadscrews, belts and linkages, all of which limit both the speed and accuracy of such machines. However, the elimination of a drive train introduces a new set of technical problems for robot designers to solve: the absence of gearing means that more powerful motors are needed, increasing the weight of joints and arms; and the absence of mechanical stiffness and damping inherent in conventional drive systems means that these must instead be provided by the control algorithms.

The application of sensors to improve the repeatability of the more powerful hydraulic robots may overcome the payload and reach limitations of existing electric-drive assembly robots, but this will require faster and cheaper sensing and image processing capabilities. Whether or not a robot makes use of sensory information is often used to distinguish between so-called 'first' and 'second' generation machines. The main drawback of first generation machines is that they cannot obtain information concerning their work environment and therefore, require well-structured environments involving expensive and inflexible fixtures and parts-orientation devices; in addition, they must also be programmed on-line, ie, taught a particular sequence of movements on the shop floor.

Thus in a typical first-generation robotic assembly system fixtures, grippers, and parts-feeders may account for between a half and two-thirds of total system costs. Although second-generation robots are more expensive, in principle at least, the cost of such fixtures should be significantly reduced, and they promise the ability to be programmed off-line, promoting greater CIM. However, significant technical obstacles still exist:

1. Sensors are still relatively expensive and require considerable computational power to translate signals into appropriate action;
2. Information from sensors cannot readily be fed into robot or process control systems; there is still a proliferation of interface standards, compliance with Manufacturing Automation Protocol (MAP) is still the exception rather than the rule, and in any case, MAP does not deal with software compatibility;
3. On-line programming languages are still used extensively, but are unsuitable for CIM; off-line languages currently under development cannot exploit CAD databases in real-time, and are unable to incorporate sensor input or the effect of tool compliance;
4. The absolute positional accuracy of existing robots still results in deviations between the geometric model used off-line and actual conditions on the shop floor.

Consequently, in practice current sensor-based, second-generation robotic applications tend to be expensive and application-specific, rather than cheaper and more flexible than their first-generation counterparts. For example, the use of machine vision usually requires the use of special light sources, high optical-contrast between parts and environment, accurate positioning of cameras and parts, and dedicated programming. Flexibility is therefore sacrificed to ensure system robustness and reliability.

The first commercial examples of machine vision and direct-drive were launched over ten years ago, but to date, these technologies have failed to produce the predicted 'revolution in robotics'. Development continues, but for most industrial applications these technologies are still too expensive, and in many cases technically impractical. Practical off-line robot programming is probably even further away.

In the final assembly of automobiles, the tracking of moving conveyors represents a major technological challenge. Stationary robots require complex and expensive sensing and control systems in order to track a moving subassembly. Moving robots, synchronised with the conveyor, are also used, but require a great deal of space and may endanger workers. The most flexible approach uses automatically guided vehicles in place of a conveyor, but again this is expensive. However, the most common solution is some form of shuttle transfer system in which the subassembly is temporarily decoupled from the main moving assembly conveyor to an automated assembly area. This approach allows less sophisticated and less expensive robots to be used for assembly.

Thus the choice of robotic assembly system - type of robots, sensors, grippers, parts feeders, and transfer system - is often constrained by the work environment. The complexity of the assembly task and required flexibility will also help determine the type of system adopted. Two of the main advantages robotic assembly has over earlier forms of assembly automation are that it is capable of performing more complex or difficult tasks, extending the range of assembly tasks that can be automated; and that it can potentially easily be reconfigured to perform different tasks, changing the mode of assembly automation, allowing smaller batch production.

2.3.3.1 Task Complexity

Task complexity remains one of the key factors affecting the feasibility of robotic assembly, and as noted earlier was one of the main barriers to the widespread adoption of special-purpose automation. To some extent the complexity of a given task is fixed, being determined by factors such as geometry, materials and component design and weight. However, design for assembly guidelines and evaluation procedures for automated assembly have long been available for small, light assembly and software-based packages have been developed for more general use.

Numerous measures of system complexity have been developed [2], but the more practical measures tend to be industry-specific. For example, in the automobile industry complexity is some function of the model mix and the variety of component parts [3]. For the purpose of assessing the potential for automation, the complexity or difficulty of a particular assembly task can be evaluated by assessing various elements of the task:

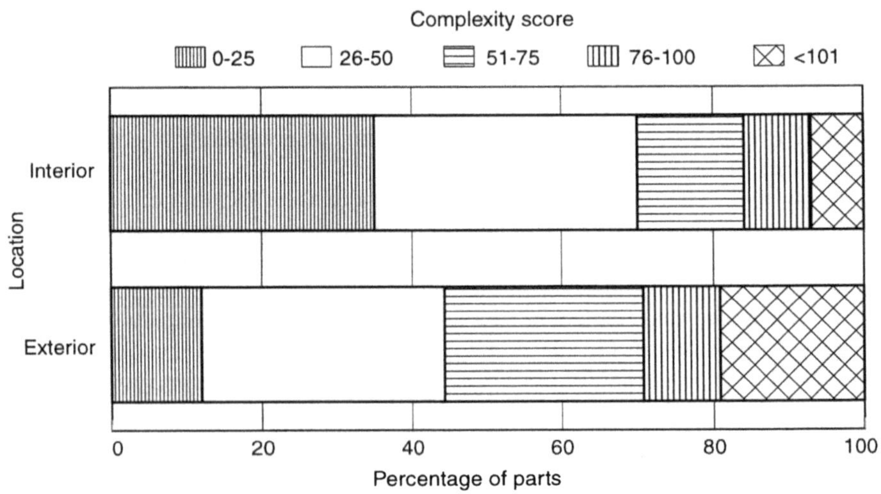

Fig. 2.3.2. Complexity of final assembly in the automobile industry. Source: Arai [11]

1. Physical properties of the component, for example, its size, weight, and rigidity;
2. Positional accuracy necessary, determined, for example, by functional requirements, component design and tolerance;
3. Required location of the part, for example, whether inside the automobile, outside or underneath;
4. Method of fastening, for example whether interference fit, by bolt, screw or adhesive.

Using these different dimensions, it is possible to develop a measure of the complexity of an assembly task by combining the scores for these different factors (fig. 2.3.2.). In most cases the complexity of an assembly task can be reduced by careful attention to design for assembly, but in practice there will be some compromise between design for assembly, process design and product performance and cost.

2.3.3.2
System Flexibility

There is little agreement on the definition, costs, benefits or best ways to achieve flexibility [4]. Clearly, flexibility is a multi-dimensional concept which may exist at different levels: business, functional and component level. The flexibility required from the production system will depend on product strategy, competitors behaviour, demand and other environmental factors. A common weakness of many studies of flexibility is the assumption that greater flexibility is always beneficial, whereas in practice, there may still be a trade-off between flexibility, efficiency and quality [5].

There has been little theoretical work on the strategic or organisational aspects of flexibility, or empirical studies of the link between flexibility in a firm's operations and a firm's strategy, competitive environment and performance. Suarez *et al* [6] propose a strategic framework based on three interrelated sets of variables, and hypothesize that firms that achieve the best fit between the need and type of flexibility will consistently perform better:

1. Four different types of flexibility - mix, volume, new products and delivery time;
2. Five factors that affect the need for flexibility - product strategy, consumer demand, competitor behaviour, product life cycle, and product characteristics;
3. Seven factors that affect implementation - production technology, production management techniques, product development process, worker skills and training, labour policies, relationship with suppliers and distributors, and accounting and information systems.

At the plant level several types of flexibility will be important:

1. *Mix flexibility* - the ability to manufacture a given set of product variants, measured by the changeover time/batch size;

2. *New product flexibility* - the ability to respond to engineering changes or to manufacture a new set of products economically and quickly, measured by the time and cost necessary to change fixtures and tooling;
3. *Volume flexibility* - The ability to manufacture a set of products profitably at different production volumes, measured by the minimum level of profitable production and the time and cost of changing the size of the system.

In practice, quantitative evaluation of all types of flexibility is difficult, as in most cases trade-offs between the different forms of flexibility will be necessary. In addition, unlike more conventional measures of performance, such as productivity or quality, flexibility is essentially a potential, making any rigorous optimisation calculation impossible. Therefore, it may be better to first assess the types of flexibility likely to be needed in a particular plant, and then audit the plants ability to respond to these demands. Research suggests that mix and new product flexibility are the most important types of flexibility in many cases [7], therefore, the number of product styles the system was originally designed to accommodate, together with the anticipated life of the system/product range, are useful indicators of flexibility. For example, if a plant of given capacity increases the number of product variants or models assembled, the plant must, on average, produce a smaller volume of each product variant or model. Therefore, dividing the capacity of a plant by the number of different variants or models provides a crude indication of the flexibility.

Product diversity and customer options have only limited impact in the body shop, as each style has its own unique stamped component and fixture; in the paint shop, the main effect is the increased use of materials as the systems must be purged between colour changeovers. However, in trim and chassis shops customer options place serious constraints on vehicle sequencing because of labour content differences; if all options are considered typically 80% of production is unique in these areas. Both 'fixed' and 'variable' constraints exist, for example, capacity limitations and option content are to a large extent constant, whereas material and component availability are variable, especially in the case of mul-tiple, distant sourcing.

Most modern auto plants operate mixed-model assembly lines. However, where the complexity of the assembly task differs significantly between models this may cause problems for automated assembly systems. Two approaches are common in such circumstances. First, avoid the problem altogether by automating only high volume, high content models. Second, by absorbing differences in the subassembly lines.

2.3.4 Trends in the UK and Japan

The specific type of assembly automation a firm adopts will depend on a number of variables, including the relative cost of labour and capital, proximity of system

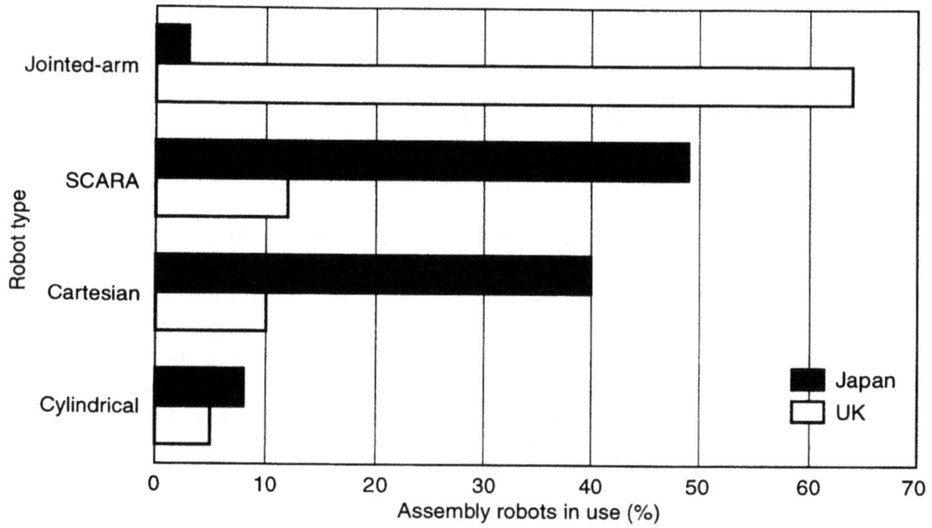

Fig. 2.3.3. Type of assembly robot - UK versus Japan. Source: Tidd [12]

suppliers, product design and skills of the workforce. In this section we will concentrate on two broad factors which appear to have the most significant affect on the broad automation concept adopted by a company: its product strategy and work organisation [8].

Product strategy has a profound affect on the level and type of flexibility a firm will require. For example, it is possible to position a firm on a product variety/life cycle matrix [9]. A firm competing in the variety-intensive region will demand different forms of flexibility to a firm competing in the region characterised by low variety, but short life cycles. The most dynamic region is characterised by both high variety and short life cycles, and will demand the greatest production flexibility. Work organisation has a significant independent affect on flexibility, but also affects the type of assembly automation adopted and its efficiency. This section contrasts the experiences of users of robotic assembly systems in the UK and Japan to illustrate the links between the type of system adopted, work organisation and product strategy.

Firstly, the type of robotics assembly systems used in the UK and Japan are different. In the UK, the most common type of robotic assembly system in use is the robotic assembly cell, whereas in Japan the robotic assembly line is more common, although many are used in 'mixed-mode', that is were each robotic station performs several tasks. Consequently, the British installations tend to employ technically more sophisticated systems than their counterparts in Japan: the most common assembly robots used in Japan are the relatively simple SCARA and

Cartesian type, typically with just four or five axes; in the UK, more versatile jointed-arm robots are favoured, typically with six axes (fig. 2.3.3.). In addition, most of the systems in the UK utilise multi-purpose grippers or involve tool changes, and almost all use some form of sensors. Surveys of potential users in the UK confirm that users in the automobile industry, more than those in any other industry, want cheaper but improved sensors and greater machine intelligence.

The robotic assembly of light and small components such as gauges, pumps, lamps, and so on is now relatively commonplace, and considerable progress has been made in the automation of engine and transmission assembly. However, because of the size, weight, and elasticity of most parts used, difficulty of robot access, and the proliferation of customer options, the automation of final assembly is only now receiving serious attention. To date, the most common applications of robotic assembly in trim and final assembly are: windscreen/backlight insertion; wheel mounting; battery and spare wheel installation; headliner installation; cockpit/facia installation; seat placement. The automation of door removal and replacement, and boot/rear hatch installation are currently being developed. There is a general trend toward the use of complete sub-assemblies on a modular basis, and all successful robotic assembly applications require the product to be specifically designed for automated assembly. In Japan, the latest automobile final assembly lines have much higher levels of automation. One of the major motivations appears to be the labour shortage, rather than to improve competitiveness.

Secondly, work organisation is different in the two countries. In the UK operators often receive little training; communication between design, manufacturing and sales is poor; and relations with suppliers are weak. As a result, machines are made 'idiot-proof', products are not designed for ease of manufacture, and components are of uncertain quality. Sophisticated technology is employed in an effort to overcome these organisational shortcomings. Consequently, users have found the technologies difficult to justify financially, or have subsequently experienced difficulties. A few British firms have introduced new positions such as 'operator-setter' and 'operator-technician' in an effort to break down the tradition demarcation lines and improve labour flexibility, but this is still uncommon. In contrast, users in Japan begin with the advantage of a highly-trained, multi-skilled work force, good communication between different functions and close relationships with suppliers. This has allowed Japanese manufacturers to adopt less complex technology, and to focus on improving flexibility. It is common for a worker to operate a number of different machines at the same time, and to be responsible for quality control and routine maintenance. In short, users of robotic assembly in the UK have been constrained by organisational shortcomings, whereas users in Japan have leveraged their advantages in work organisation to adopt less sophisticated, but more flexible automation systems.

Thirdly, the product strategies of users in the UK and Japan are very different. In the UK, robotic assembly is most commonly used to assemble complex products, with few variants and relatively long life cycles (fig. 2.3.4. and 2.3.5.). In most cases robotic assembly has replaced manual assembly, so overall flexibility

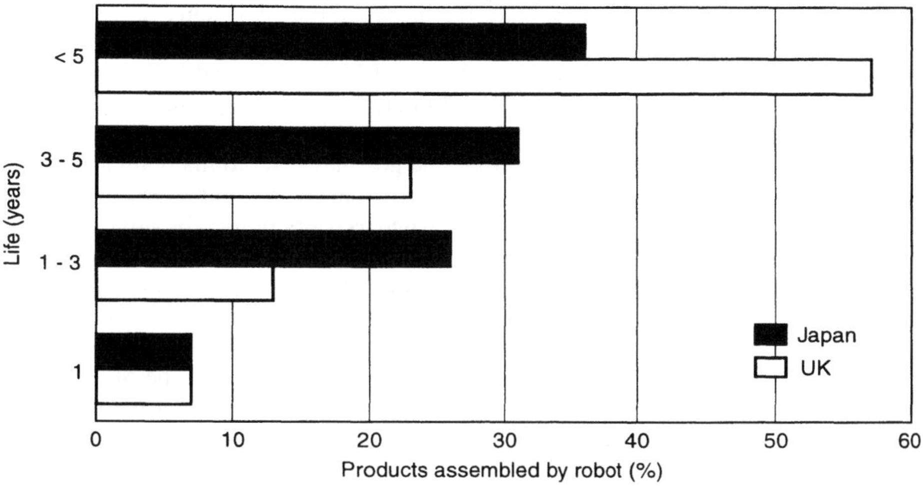

Fig. 2.3.4. Assembly product life - UK versus Japan. Source: Tidd [12]

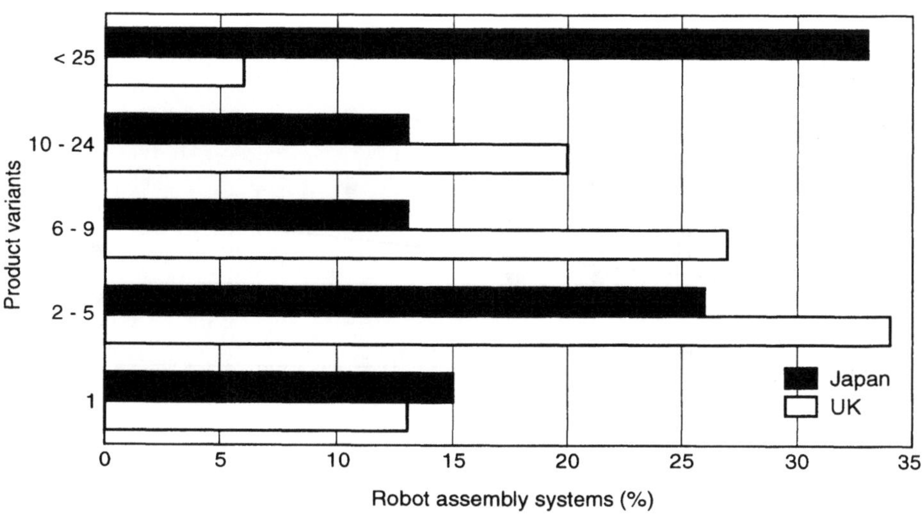

Fig. 2.3.5. Assembly product variety - UK versus Japan. Source: Tidd [12]

has been reduced. The main rationale for the adoption of robotic assembly has been cost reduction and improved product consistency. In Japan, robots are used to as-semble a wider range of product variants, and life cycles tend to be shorter. In many cases robotic assembly has replaced dedicated, special-purpose machinery, the aim being to improve flexibility. More recently, the main motive in Japan has been the shortage of labour, and therefore, the goal has been to maintain flexibility whilst automating these tasks. In short, it appears that users in Japan are more fully exploiting the flexibility of the technology, whereas users in the UK are exploiting the ability of robots to perform more complex tasks.

Finally, there appears to be some relationship between task complexity and the required sophistication of assembly automation. As might be expected, the greater the complexity of the assembly task, the more sophisticated the assembly automation system (fig. 2.3.6.). The relationship between flexibility and type of assembly automation is less clear. There is no significant correlation between the technical sophistication of the system adopted and manufacturing flexibility (fig. 2.3.7.).[1] There are two possible interpretations. Either users are not fully exploiting the potential flexibility of the systems, or the level of technological sophistication has little affect on flexibility. In the latter case, factors such as good product design, component standardisation and quality, well-structured robot work

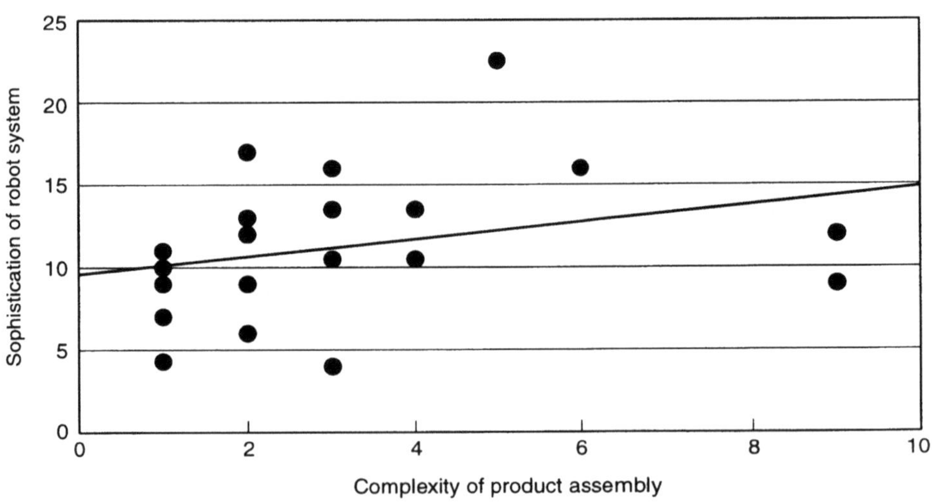

Fig. 2.3.6. Product complexity versus robot sophistication. Source: Tidd [12]

[1] The Spearman rank correlation between the sophistication of the technology and the complexity of the assembly task was 0.51, which was significant at the 0.10 level. The correlation between the sophistication of the technological sophistication and manufacturing flexibility was 0.26, which was not significant at the 0.10 level.

environments, work organisation, and production management systems may be more important.

2.3.5 Future Potential of Robotic Assembly

Recent surveys of Japanese automakers suggest that the two most important barriers to assembly automation are related to *flexibility* - excessive parts variety, insufficient flexibility of equipment, or *complexity* - physical property of parts and difficulty in parts' orientation [10]. However, financial pressures are likely to force automobile manufacturers to increase their product life cycles, reduce product variety, increase parts commonality across models and design for manufacture. In short, it is likely that there will be a simultaneous reduction in complexity and flexibility requirements. This, combined with the shortage of manual labour, is likely to improve prospects for assembly automation in Japan. At the same time, manufacturers in Europe and the US are beginning to place less emphasis on advanced manufacturing technology, and more emphasis on new working practices, and are beginning to benefit from aspects of 'lean' production. This switch in manufacturing philosophies suggests that Japanese, European and British manufacturers may still have much to learn from each other.

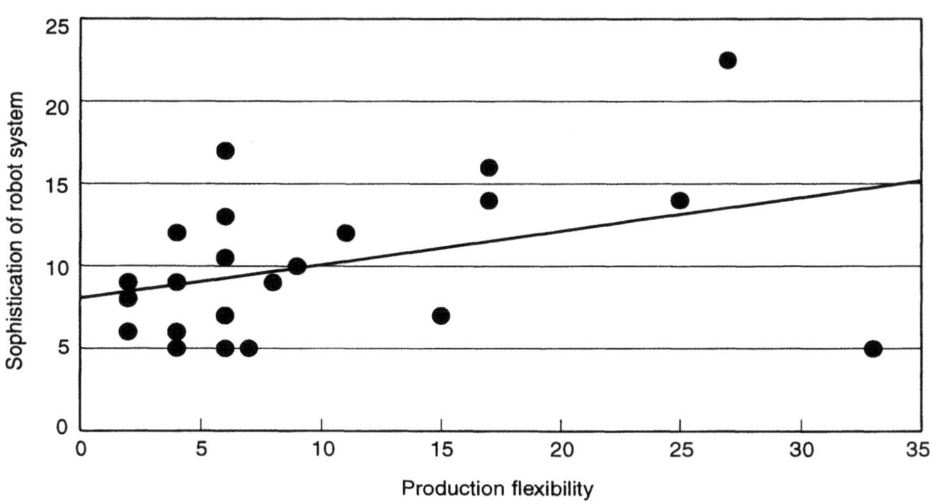

Fig. 2.3.7. Product flexibility versus robot sophistication. Source: Tidd [12]

2.3.6 References

1 Womack J, Jones D, Roos D (1991) The Machine that Changed the World. Rawson Associates, New York
2 Singh K (1993) The Concept and Implications of Technological Complexity for Organizations. Working Paper, School of Business, Michigan
3 MacDuffie J P, Sethuraman K, Fisher M L (1993) Product Variety and Manufacturing Performance. IMVP Research Briefing Meeting, Auto 21, Cape Cod
4 Gupta Y P, Somers T M (1992) The Measurement of Manufacturing Flexibility. European Journal of Operational Research Vol 60: 166-182
5 Gaimon C, Singhal V (1992) Flexibility and the Choice of Manufacturing Facilities under Short Product Life Cycles. European Journal of Operational Research Vol 60: 211-223
6 Suarez F F, Cusumano M A, Fine C H (1991) Flexibility and Performance: A Literature Critique and Strategic Framework. Working Paper, Sloan School of Management, MIT, Cambridge Mass
7 Dixon J R (1992) Measuring Manufacturing Flexibility: An Empirical Investigation. European Journal of Operational Research Vol 60: 131-143
8 Tidd J (1993) Manufacturing Strategy, Technology and Organization. In: Storey J (ed) New Wave Manufacturing Strategies, Paul Chapman Publising, London
9 Uzumeri M V, Sanderson S W (1992) Model Variety, Technical Change and Manufacuring Flexibility. Paper presented to the Academy of Management Meeting, August
10 Fujimoto T (1992) Why Do Japanese Companies Automate Assembly Operations: A Survey in the Auto Industry. Berlin Workshop on Assembly Automation
11 Arai T (1988) A Model of Automisation on a Final Assembly Line. International Conference on Assembly Automation: Japan vs Europe Forum, IFS, Bedford: 25-36
12 Tidd J (1991) Flexible Manufacturing Technologies and International Competitiveness. Pinter, London

2.4 What Do You Mean by Automation Ratio? Definitions by the Japanese Auto Makers[1]

T. Fujimoto

2.4.1 Research Questions

It has often been pointed out that definitions of automation ratio or degree of automation differ from company to company and from process to process. In one of the European plants known for its advanced automation, for example, the engineers at the company claimed that the automation ratio of their final assembly line was about 50%. An engineer at a Japanese company, who visited the same plant recently, admitted that the assembly line in question was one of the most advanced in the world, but reported that the automation ratio was estimated to be around 10%. Another engineer from another Japanese auto maker estimated it to be about 15%. Of course, no one is telling a lie, and no one is making a bluff in this case. This seems to be a simple matter of difference in definition.

Automation ratios are often reported in trade journals and other publications. Auto companies often disclose automation ratio numbers when they announce the start-up of their new factories. And yet, production engineers from Japanese auto companies do not seem to take these numbers seriously. One reason for this disregard by experts may be that automation is a complex phenomenon that cannot be summarized by a single number called automation ratio anyway. Whilst this argument may be valid, it still seems to be desirable to have compatible definitions of the automation ratio. After all, the automation ratio - whatever it may be - is the most commonly used indicator when patterns of automation are compared among multiple factories. Once the numbers are disclosed, non-experts tend to compare them across plants and companies and draw their conclusions, without taking the different definitions into account. It is hence important for researchers and practitioners to, at least, be aware of the potential differences in the definitions and, if possible, to use a common set of definitions when describing and analyzing automation across companies.

---

[1] This paper is based on [1,2]. A survey for this paper was conducted by the Committee for Research on Optimal Automation Systems in the Automobile Industry, which is chaired by Professor Koichi Shimokawa of Hosei University and is organized by Japan Technology Transfer Association (JTTAS). Professor Hisanaga Amikura of Chiba University and Mr. Takashi Matsuo, Doctoral Candidate at Tokyo University, were particularly instrumental in compiling the data. Mr. Seigo Onishi and the staff of JTTAS facilitated distribution and collection of the questionnaire. The author is grateful to the respondents of the survey, as well as the above people.

In reality, disseminating a common definition is not an easy task. Each company has its own history of using a certain definition of the automation ratio for internal purposes, so that it tends to stick to the existing set of definitions in order to retain a consistency for time series analyses within the company. It may be possible for each company to develop a dual system, one for internal use and another for external use, just like management accounting and financial accounting; but companies tend to be reluctant to do so because the double standard may create confusion on the factory floor. Nonetheless, it seems to be helpful to report the difference in definitions, which auto companies are actually using, as a reminder of this problem. This brief paper describes and analyzes different types of definitions of the automation ratio which the eleven Japanese auto makers have been using since 1992 in the following nine production processes: stamping, body welding, painting, final assembly, engine assembly, engine machining, engine casting, engine forging, and plastic injection molding. After discussing the different types of automation ratios, the paper will present actual data on how often each type of definition is used by the companies.

2.4.2 Types of Automation Ratios

As discussed above, automation ratios can differ from company to company and from process to process. Theoretically, however, we can set up some categories for the operational definitions of automation. Suppose that *workers* and *machines* (equipment) transform *materials* at each *step of the production process*. The automation ratio can be measured by any of the four process elements shown above (fig. 2.4.1.).

1. *Machine-based definition*: Ratio of automated machines or machine-hours.
2. *Worker-based definition*: Ratio of workers or person-hours saved by automation.
3. *Material-based definition*: Percentage of materials transformed by automated equipment. Ratio of automatically welded spots, ratio of automatically painted areas of a body, ratio of automatically assembled parts, etc.
4. *Process-step-based definition*: Ratio of process steps or work stations which are recognized as automated. The criteria of automation may be determined by the companies in question.

Again, which definition is to be used depends on the types of processes, companies, timing, etc.

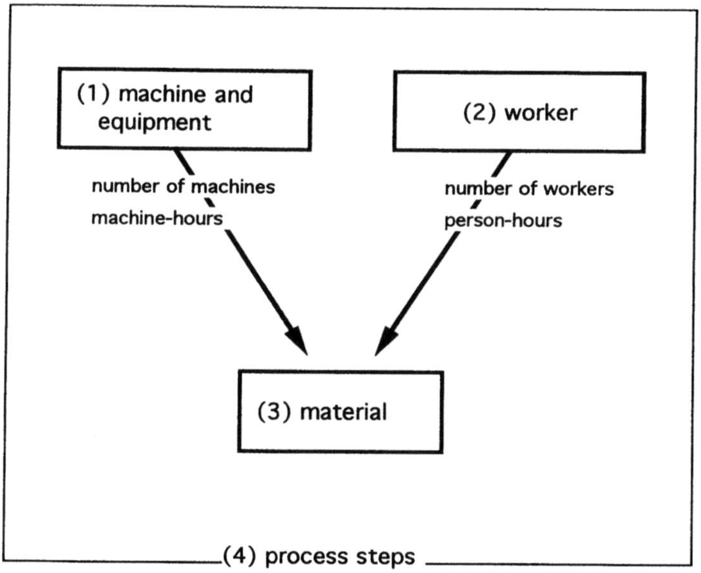

Fig. 2.4.1. A framwork on definition of automation ratio

2.4.3
Outline of the Survey of Automation Ratios

The survey was conducted from late 1991 until early 1992, by the Committee on Optimal Automation Systems in the Automobile Industry at JTTAS (Chaired by Professor Shimokawa of Hosei University). All of the 11 auto assembly companies in Japan participated in the survey, 9 of which were car makers and 4 were large and medium truck makers (two companies made both types). The questionnaire asked the individual engineers of each company to define the automation ratio for each of the 9 production processes. The survey also contained questions regarding automation ratios by process types, based on each company's own definition.

Let us first look at the average automation ratios, while disregarding the definition problem. Average automation ratios and their standard deviations (11 samples) for 9 major production processes are shown in figures 2.4.2. (for *A lines*, or the most automated lines from each company) and 2.4.3. (for *B lines*, or the least automated lines from each company).

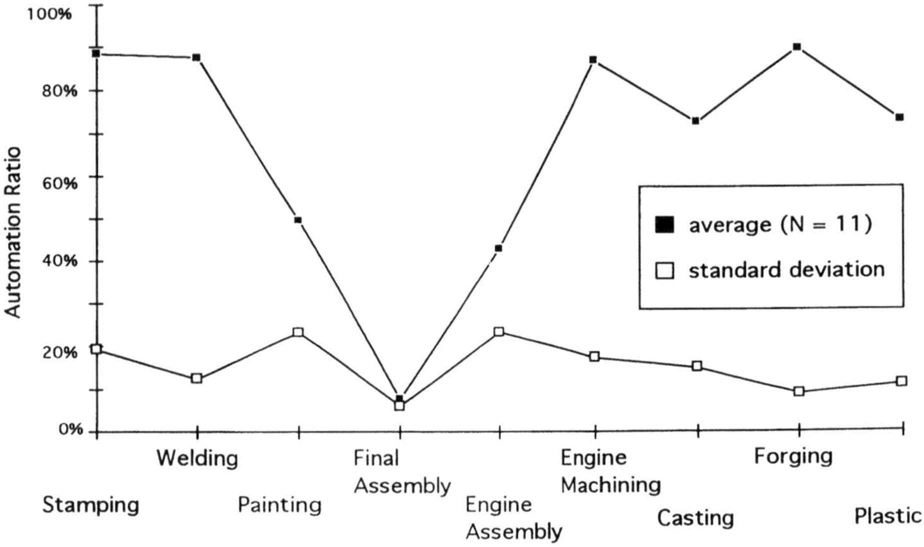

Fig. 2.4.2. Automation ratio of the most automated lines : *A Lines* (2 companies didn't answer)

In the case of the most automated production lines in the companies (i.e., A lines; fig. 2.4.2.), automation ratios are high (around 90%) in stamping, welding, engine machining and engine forging, followed by casting and plastic molding. This applies more or less to all the companies studied, as the relatively low standard deviations indicate. By contrast, the degree of automation is very low in final assembly areas, even in the A lines (10%). The data roughly ranges from 0% to 20%, although, again, we have to be careful about the difference in definitions of the automation ratio. Painting and engine assembly processes are found in the middle field, with relatively high standard deviations. This seems to indicate that some companies are currently automating painting and engine assembly processes fairly rapidly, which creates differences between companies.

In the case of the least automated lines (i.e., B lines, fig. 2.4.3.), automation ratios are naturally lower throughout, whilst standard deviations tend to be higher, indicating significant inter-firm differences in automating older production processes. As in the case of the A lines, the automation ratio of the final assembly is the lowest among all the processes. Both average and standard deviations are high in engine machining, which reflects the fact that the firms are divided into two groups: those which have aggressively automated engine machining operations, even in the B lines, and those who have not. A similar pattern is observed in stamping.

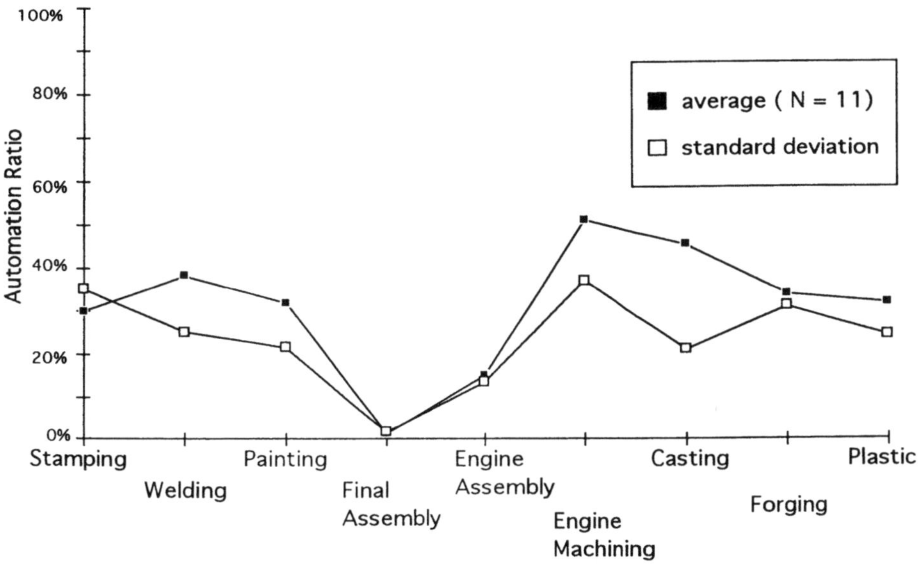

Fig. 2.4.3. Automation ratio of the least automated lines: *B Lines* (2 companies didn't answer)

In summary, there are significant gaps between the most and least automated lines for each production process. Also, some Japanese companies tend to automate both the A and B lines, whilst others automate their A lines only. This seems to indicate two different strategies: company-wide automation and focused automation.

The accuracy of the above analysis may be limited, however, by the possibility of different companies using different definitions of automation ratios at different processes in the first place. We hence have to examine how each firm defines automation ratios.

2.4.4 Results: Different Definitions of Automation Ratios by Different Firms

The result of the survey on the definitions of automation ratios is shown in table 2.4.1. As expected, the types of definitions differ, depending on the company and the process. There were some processes where one definition was used by almost all of the Japanese companies (e.g. welding), while there were other processes where a consensus virtually did not exist (e.g. painting, final assembly). The most frequently used definition for each process was as follows:

Table 2.4.1. Distribution of definitions of automation ratio

definition type \ process		stamping	welding	painting	final assembly	engine assembly	engine machining	casting	forging	plastic molding
machine -based	number								1	
	hours	1								1
worker -based	number	2		4	1			1	1	3
	hours		1	2	6	5	1	1	1	
material -based			9	3	3	1				
process -step -based	process -steps	5		2	1	2	4	4	5	3
	overall evaluation						1			
others; n.a.		2			1	2	3	3	2	2
automation ratio of the most advanced lines by the above definitions (average of 9 car makers)		94%	91%	51%	10%	55%	93%	79%	90%	73%

Stamping: Number of automated steps (process-based)
Welding: Number of spots welded automatically (material-based)
Painting: Number of workers saved (worker-based)
Final Assembly: Number of person-hours saved by automation (worker-based)
Engine Assembly: Number of person-hours saved by automation (worker-based)
Engine Machining: Number of automated steps (process-based)
Casting: Number of automated steps (process-based)
Forging: Number of automated steps (process-based)
Plastic: Number of automated steps, or number of workers saved

As table 2.4.1. shows, worker-based definitions tend to be used for relatively labor-intensive processes (e.g. assembly), whilst process-step-based definitions are often chosen in machine-paced operations, such as stamping, machining and casting. In total, however, there seems to be little consensus for definitions of the automation ratio.

This appears to apply, in particular, to painting and final assembly. In painting, Japanese companies defined automation ratio differently, e.g. in terms of the number of workers, person-hours, areas of finish coating, and number of automated process steps. In final assembly, the definitions used by the Japanese included the person-hours saved by mechanization, ratio of bolts screwed automatically, the ratio of parts automatically assembled and the ratio of automated work stations. Therefore, we must be particularly careful when we interpret the automation ratios of these processes which the companies have disclosed.

In the case of process-step-based definitions, a potential problem was how to draw a line between automated steps and non-automated ones. In defining automated stamping processes, for instance, one Japanese company only regarded a stamping line as automated when there was no operator, whilst another company excluded palletizing operations at the end of press lines from the automation criteria. Therefore, a tandem stamping line which has fully automated transfer mechanisms between the machines, but which does not have an automatic palletizer, may be judged as "automated" by the latter company, but it may not be so at the former company.

2.4.5
Implications: Multiple Indicators for Automation Ratio

This brief paper has pointed out that automation ratio definitions, or the degree of automation, tended to differ across companies and processes in the case of the Japanese auto industry. What follows from this finding might be a proposal to standardize the definitions for companies and countries, or even for processes. In Germany [3], the "degree of automation" is defined by the German Industrial Standard (DIN) #19233 as "the percentage of mechanized or automated functions of the total number of functions." The report by the International Motor Vehicle Program (1990) used the "number of direct steps" as a uniform definition of the automation ratio in its international comparative study of assembly plants. In Japan, however, there is neither an industry-wide standard (e.g. JIS), nor a common understanding in the engineering community as to what the automation ratio is. As international comparative studies of automation practices have become increasingly common in the 1990s, we may need some kind of definitions which can be applied both internationally and across different companies.

However, it would be unrealistic to propose a formal and totally uniform set of definitions for companies and countries. Each company may already have accumulated automation data based on certain firm-specific definitions for the purpose of internal control. Some countries (e.g. Germany) already have industry standards, whilst others (e.g. Japan) do not. In this situation, any effort to disseminate a uniform definition would face friction and resistance.

What seems to be more realistic for auto manufacturers is an informal exchange of knowledge among them, rather than a formal agreement on a standard for multiple indicators for the automation ratio. If each company measures the degree of automation by more than one indicator, for instance, it would become easier for practitioners and researchers to perform meaningful benchmarking studies between competitors.

The adoption of multiple indicators would be desirable, not only because it facilitates meaningful inter-firm comparisons, but also because automation itself is a multi-faceted phenomenon which is hard to capture by just one indicator. In spot welding, for instance, the ratio of automatically welded spots has been widely used, as table 2.4.1. indicates. However, now that the automation ratio by this definition is nearly 100% in many of the recent welding lines, where workers are mostly involved only indirectly, the definition focusing on direct jobs does not seem to effectively measure the real degree of automation. The definition based on the spots was apparently selected because of its clarity and reliability, but it cannot capture important aspects of welding automation, such as indirect jobs, sealing operations, arc welding, finishing, etc. Thus, it seems to be desirable to have at least two definitions of welding automation: the existing one based on the number of spots and a new one based, for example, on person-hours saved.

It would not be necessary for each company to measure the automation ratio by all the definitions listed in table 2.4.1. As the table indicates, for instance, just two definitions for each process seem to cover most of the current practices. For example:

Stamping:
1) The ratio between the number of automated press lines (i.e. with automatic transfer mechanisms, palletizers, moving bolsters, etc.) and the total number of press lines (including transfer press machines);
2) the ratio between workers or person-hours saved by automation and those necessary for a totally non-automated process.

Welding:
1) The ratio between spots automatically welded and the total number of spots welded in-house;
2) the ratio between workers or person-hours saved by automation and those necessary for a totally non-automated process.

Painting:
1) The ratio between the area that was automatically painted and the total area painted;
2) the ratio between workers or person-hours saved by automation and those necessary for a totally non-automated process.

Final Assembly:
1) The number of parts assembled automatically in the main line (excluding bolts and fasteners) in comparison to the total number of parts assembled;
2) the ratio between workers or person-hours saved by automation and those necessary for a totally non-automated process.

Engine Assembly, Machining, Casting, Forging, Plastic Molding:
1) The number of totally automated work stations in comparison to the total number of stations in the line;
2) the ratio between workers or person-hours saved by automation and those necessary for a totally non-automated process.

This means, if each company measures the degree of automation by a few (instead of just one) indicators, it may be possible to implement a meaningful benchmarking analysis without destroying the existing measuring system.

To sum up, when inter-firm and international comparisons of automation practices become increasingly important to practitioners and researchers, auto companies might benefit from a mutual understanding of each other's definition of automation, as well as from the adoption of a multiple-indicator system for measuring the degree of automation. Again, an informal exchange of knowledge seems to be much more important than formal agreements between auto companies.

2.4.6 References

1 Fujimoto T (1992) What Do You Mean by Assembly Automation? Definitions by the Japanese Auto Makers. Toyko University Faculty of Economics Discussion Paper 92-F-16

2 Fujimoto T, Matsuo T (1992) Note on the First Findings of the Assembly Automation Study, Presented at International Workshop on Assembly Automation and Work Organization in the Automobile Industry, Berlin, November 20-21

3 Jürgens U, Dohse K, Malsch T (1986) New Production Concepts in West German Car Plants. In: Tolliday S, Zeitlin J (ed) The Automobile Industry and Its Workers: Between Fordism and Flexibility. Polity Press, Cambridge: 258-281

4 Womack J P, Jones D T, Roos D (1990) The Machine That Changed the World. Rawson Associates, New York

3

Diversity of Approaches

3.1 Present State and Future Vision of Vehicle Assembly Automation in Mitsubishi Motors Corporation

Y. Mishima

3.1.1 Introduction

The operations at the end of the auto assembly line - i.e. trim and final assembly - are the most labor-intensive stages of car manufacturing. This is primarily because the introduction of automation in diverse and very complex assembly operations is often hindered by limits of automation technology, effective return on investment, availability of plant space, etc. On the other hand, however, automation of trim and final assembly operations has been considered to be a very important issue, especially with regard to a better working environment, shorter working hours and problems of an aging workforce. The basic concept of assembly automation of trim and final assembly operations, and the present state of automation at the Mitsubishi Motors Corporation passenger car plant are introduced in the following.

3.1.2 Automation of Assembly Operations

3.1.2.1 Concept

The basic concept for organizing trim and final assembly operations is to seek and establish compatibility between requirements due to the human nature of operators assigned to the line and improvement of productivity of such operations. The basic philosophy of this concept is to eliminate "*muri, mura* and *muda*" in the work process. These are shown in Table 3.1.1.: *muri* is the physical and mental burden facing operators, *mura* is the lack of uniformity in product quality as a result of unevenness in the work process, and *muda* is unnecessary operator movements and excessive consumption of resources. Elimination of *muri, mura* and *muda* immediately addresses the connection between the human nature of the people working under specific conditions and the issue of productivity improvement. In other words, the basic principle must be to create conditions in which people can carry out their work free from avoidable burdens. Daily *kaizen (change for the better)* in order to eliminate *muri, mura* and *muda*, means respect

Table 3.1.1. Basic concept

1)	*muri*	pushing beyond capacity
2)	*mura*	unevenness of operations
3)	*muda*	waste of resources

for human nature, reduction of physical labor, mechanization and automation and eventually productivity improvements.

3.1.2.2 Progress of Automation of Assembly Operations

The following is an example where mechanization and automation were implemented by eliminating *muri*, *mura* and *muda*. To achieve such an effect, it is necessary to select an operation where *kaizen*, reduction of physical labor, mechanization and automation are of high priority. In order to ensure the appropriate selection of such an operation, every task must be broken down into work elements, such as identification and selection of parts, setting, assembling, etc., and the time required as well as strain and stress related to each of these elements must be analyzed. Furthermore, it is necessary to analyze capital investment, achievable benefits, the technical feasibility etc. before an operation is finally selected for automation. Up to now, no conclusive selection method has been established in this regard, and every single manufacturer has its own method. Before a final decision is made at Mitsubishi Motors Corporation, the task elements regarded as strenuous by the individual workers are registered and the capital investment, achievable result and technical feasibility are comprehensively evaluated in each case. The result of such an evaluation may be that there will be only a reduction of human labor or just a kaizen with regard to the work environment instead of introducing process automation. In the following, we will discuss the case of automating the tire fitting operation.

Tires belong to the heaviest items to be handled in final assembly and the worker's posture during the fitting operation tends to be very strenuous. The first kaizen step was to eliminate the physical work of lifting the tires to the necessary height and to equip each nut runner with a torque checking function. This eliminated the strenuous effort required for the manual torque and nut checks. In addition to the above, 1) the nut runner was equipped with a self-return mech-anism which avoids unnecessary effort, 2) an indicator lamp was fitted which supports the selection of tires, and 3) the placement of the tire and the positioning of the nut runner were completely automated, thereby avoiding further unnecessary, cumbersome activities. Altogether, the automation of the tire fitting operation was the result of a gradual *kaizen* process. The method used here for reducing physical labor and the technical know-how acquired on the basis of this automation approach were further developed and reflected in the process design of other later production facilities.

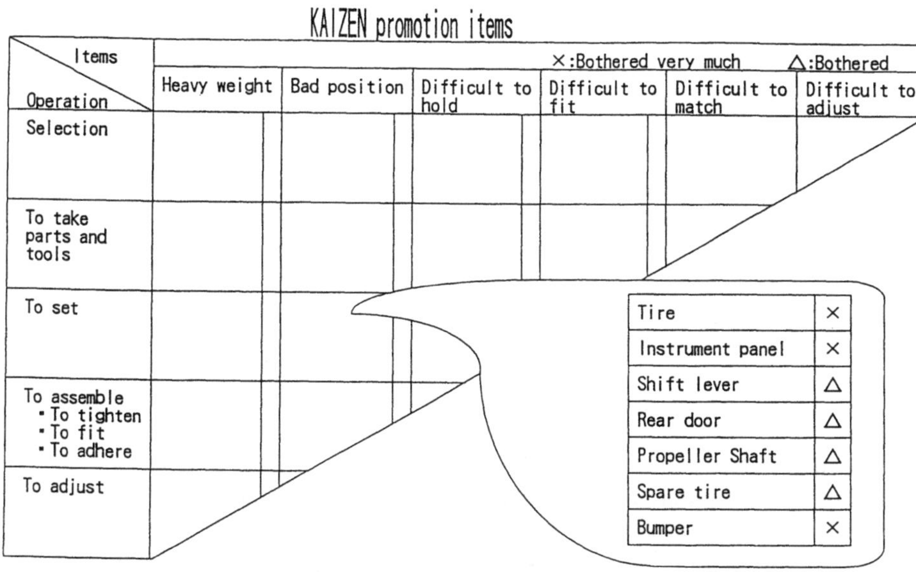

Fig. 3.1.1. Analysis of assembly workload - investigation chart of assembly load

3.1.2.3
Key to Success of Automation

The most essential elements for the successful introduction of automation to final assembly operations are compiled in Table 3.1.2.

Table 3.1.2. Key to successful automation of trim and final assembly operations

1)	Reduction of Deviation
2)	Teamwork, Education, Training of employees
3)	Engineering technology

The first item is the elimination of deviation from product specifications. Deviation from quality specification can occur with regard to the vehicle body as well as with the parts to be fitted. Vehicle bodies are generally built on the welding line of the vehicle manufacturer's plant, and therefore quality deviations must be controlled by *kaizen* activities to both welding operations and plant facilities. As most of the parts are sourced from suppliers, close cooperation with suppliers is necessary in order to achieve uniformity in the quality of finished parts by solving problems of design and parts production through kaizen. The method of controlling quality deviation is essential know-how for any car manufacturer. The second item is education, training and team work specifically with regard to main-

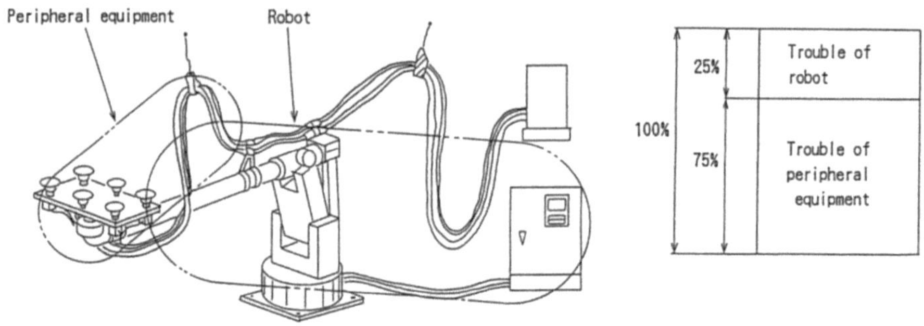

Fig. 3.1.2. Problems - contents of trouble

tenance tasks. No matter how well an automation facility is designed, breakdowns or system failures are inevitable. It is important, therefore, to determine how quickly such a failure is repaired or how fast normal conditions are restored, how soon such a failure is detected in advance or how preventive maintenance is carried out before such a failure actually occurs. For this purpose, it is important to explore the problems and necessary measures with regard to the automated facility at a very early stage of the process planning. This can only be achieved with the full cooperation of those people who will eventually operate and maintain the facility. It is also important to organize and establish close team work among those involved in the education and training of operators and in the development of the maintenance method to be implemented once the plant is completed and commissioned. Customarily, the teaching of robots and their maintenance has been carried out to a certain extent by the operators assigned to such facilities at the Mitsubishi Motor Corporation plants. These are the people who closely monitor the operating conditions of facilities day by day, and, therefore, are in the best position to see and detect wrong or defective conditions first, and to take immediate remedial action as long as it is of a minor nature. A major failure should be repaired by maintenance experts who have the appropriate expertise. But even in such a case, close cooperation of the operators concerned is mandatory in order to find the real cause of failure and to develop a method for its future prevention. Thus, the issue is not just the capability to set up a facility equipped with modern technology, but also the capability to build a plant which is easy to operate, less prone to failure, and which offers easy access for maintenance. In other words, this means the capability to design and build a facility where emphasis is placed on the central role of the people who are to operate and maintain it. Fig. 3.1.2. shows the rate of failure due to trouble with robots and trouble with peripheral equipment. It should be noted that the majority of failures is due to trouble with peripheral equipment and not with robots, and therefore peripheral equipment is of increasing importance. Mitsubishi Motors Corporation has long experience developing and producing such in-house peripheral equipment which is of a low cost, easy to operate and less prone to failure. Problems and new requirements learnt from operating such equipment were recorded and this information serves as feedback when similar equipment is designed in order

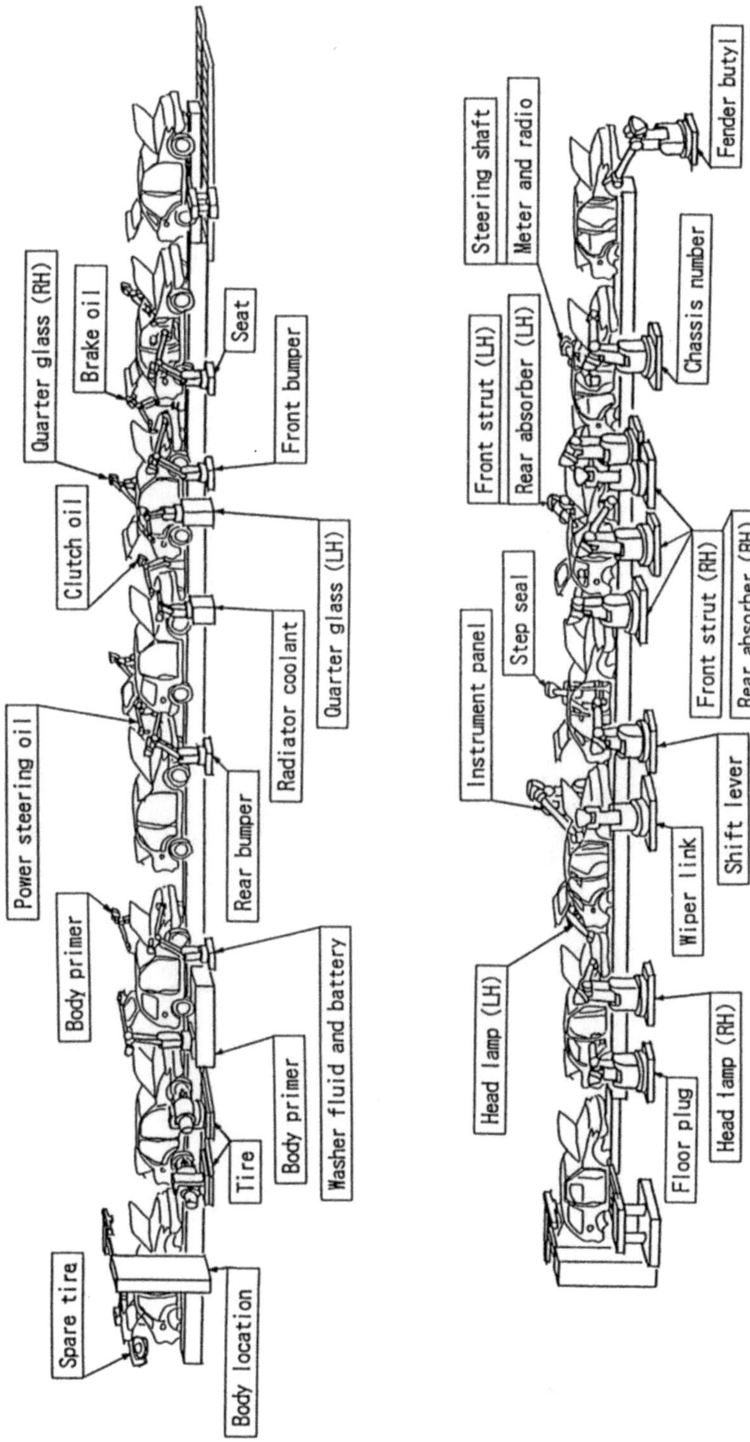

Fig. 3.1.3. Outline of automated final assembly

Fig. 3.1.4. Tire fitting operation

Fig. 3.1.5. Bumper fitting operation

Fig. 3.1.6. Glass fitting operation

to achieve even further improvements with regard to easy operation and failure reduction. Operators are striving for *kaizen* on such new equipment on a daily basis.

3.1.3 Practical Examples

A summary of automation in the area of trim and final assembly at the Mitsubishi Motors Corporation Mizushima Motor Vehicle Works is shown in table 3.1.3. The outline of the final assembly line with the automated assembly operations is shown in figure 3.1.3. Details of the tire, bumper and glass fitting operations are shown in figures 3.1.4., 3.1.5. and 3.1.6. respectively.

Table 3.1.3. Summary of automated operations

	Automated assembly Operation in the trimming line	Automated assembly Operation in the final assembly line
No. of automated operations	5	5
No. of robots	17	16
No. of the automated operation items	26	23

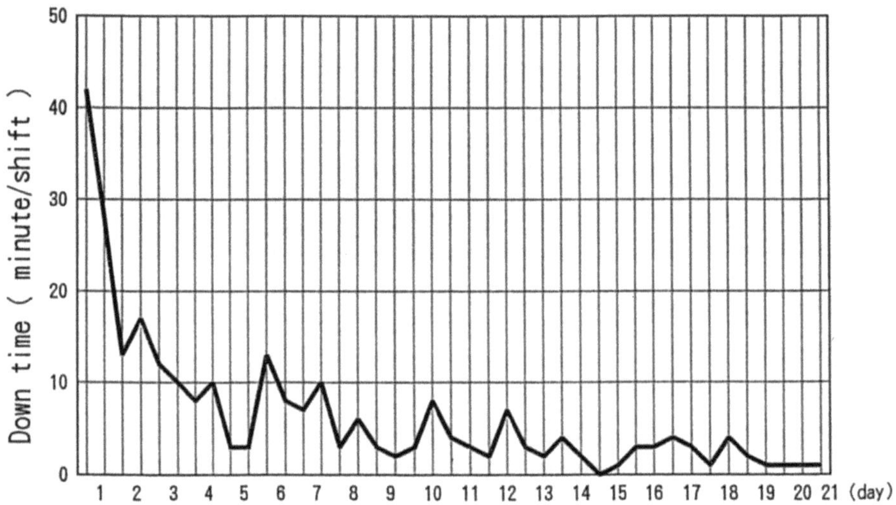

Fig. 3.1.7. Downtime

Figure 3.1.7 shows the improvements which were made with regard to equipment availability after the introduction of automation. Downtime per 9-hour shift initially totaled about 40 minutes, but it has now been reduced to a mere 90 seconds per shift, and production is running very steadily. This achievement was the result of all the remedial actions taken after exploring the causes of troubles, and of minor *kaizen*-based improvements which were introduced continuously.

3.1.4 Conclusions

The demand for automation and mechanization of vehicle assembly operations will increase in order to cope with even more sophisticated and complex conditions in car manufacturing. Driving forces behind it are the increased diversification of car models and variations which are introduced in order to meet cus-tomers' demands, the need to increase product performance while reducing the weight of the vehicles, the pressure resulting from the soaring value of the Japanese currency and the present trend towards a further reduction in working hours. Under these circumstances, it is imperative that car manufacturers provide assembly facilities which are flexible enough to permit variations of models and varying production volumes. To achieve this, they must carry out incessant research and development work. However, automation of assembly operations is a means only, the real aim should be the continuation of untiring efforts to eliminate *muri*, *mura* and *muda*. Doing a natural job always in a natural way is the key to successful automation. We are convinced that this can only be achieved through close cooperation of not only the people who produce the cars, but also of those who are working at the supply companies and, last but not least, of those who are involved in the development of cars. We will continue building assembly lines along this principle and with a view to the human nature of the people who work there.

3.2
Development of a new Vehicle Assembly Line at Toyota: Worker-oriented, Autonomous, new Assembly System

A. Niimi · Y. Matsudaira

3.2.1
Development Background

In the past, our company has built epoch-making assembly lines based on the *Toyota Production System*. After the mid-1980s, however, with the diversification of users' needs, automobile structures became increasingly intricate and assembly work became more complex. Consequently, the rise in productivity began to slow down. Moreover, a change in people's sense of value for labor caused the so-called "dislike of the manufacturing industry" phenomenon and the decreasing birthrate led to a reduction in the number of young workers and resulted in an aging workforce on assembly lines.

Meanwhile, automobile manufacturers in Europe and America were facing problems such an increasing rate of absenteeism and the "blue Monday" syndrome. They tried out *group work without a conveyor system* or *large-scale automation systems* in order to solve these problems. Both systems, however, were not necessarily successful; the former system often failed to reconcile productivity requirements and quality with workers' morale, and the latter failed to reach the anticipated level of automation, due to the limits on automation - both technically and commercially -, or flexibly cope with a model change.

Therefore, we were faced with the task of building an assembly line based on a new concept to solve these problems.

3.2.2
Development Progress and Target

In the late 1980s, the above-stated problems facing assembly processes began to emerge and we started to examine them on a company-wide scale in order to find a solution. We examined a system to improve workers' morale while securing productivity. After management staff had experienced actual work in the assembly line, they tried out a system of optimum work processing oriented towards securing product quality, and carried out repeated research projects and experiments on the automation of assembly. Finally, we established our vision of the future assembly line.

We decided to incorporate this vision in the assembly line in the Toyota Motor Kyushu Miyata Plant, erected on Kyushu island, our first automobile manufacturing plant to be built here. The main idea behind the new assembly line is that this new assembly line will continue to be improved and will continually evolve in response to changes in the social environment and the *growth of workers* in terms of skill and attitude towards Kaizen there. With this aim in mind, we forwarded the development of the new assembly line on condition that the results achieved here should then be applied to existing assembly lines elsewhere in the company.

3.2.3 Description of the Development

Automobile assembly work consists of a vast variety of processes: parts identification, preparation, conveyance, positioning, fastening and checking. Actually, a large part of assembly work cannot be completed without people and their abilities; techniques for handling soft parts haven not yet been fully developed. Moreover, in contrast to people, machinery is seldom equipped with the various judging functions performed by workers intuitively (whether the right parts are picked, whether the parts' positions are correct, whether they can be inserted and so on), and which are vital for assembly work.

In a word, human beings play a major role in assembly lines, and it is important that people with a strong motivation to work are responsible for assembling automobiles and that they manufacture products (automobiles) of high quality with concerted efforts.

Accordingly, our basic concept for building a new assembly line is that the new assembly line should be an attractive work place for workers. To realize this, our concept is built around the following four elements:

1) Increase worker motivation.
2) Design processes that can be performed by all.
3) Employ automation that people want to work with.
4) Make a comfortable work environment.

The following describes the specific development of each element.

3.2.3.1 Increase Worker Motivation

The tasks involved in assembly processes are broken down and divided among many workers. Due to the diversification and high-grade specifications for automobiles in recent years, the variety and number of parts to be assembled have increased and assembly work has often become a complex miscellany of incoherent tasks where the meaning and purpose of the work being carried out is unclear

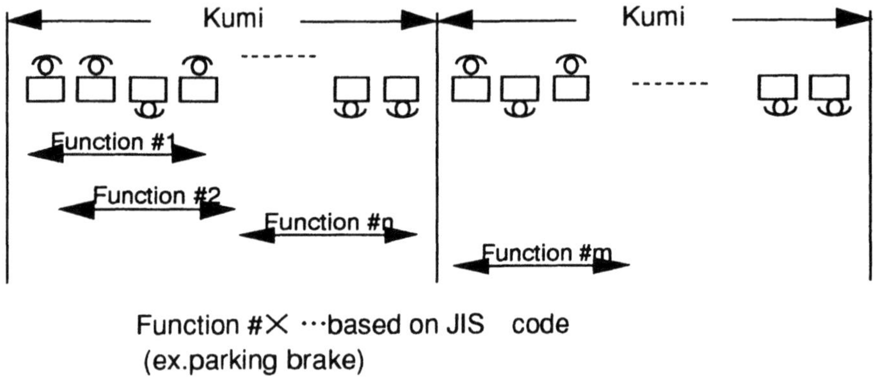

Fig. 3.2.1. Correlation between the *kumi* and the function

to the worker. In addition, the responsibility for quality assurance has become vague and the significance of each individual task has been almost lost.

The conventional assembly line consisted of three to five lines, each of which was a continuous conveyor belt with a length of up to 300 meters. Each line had three or four operation groups (*kumi*) i.e. where 60 to 80 people are working and trying to keep pace with one another.

In such assembly processes, where the independent management of the operation groups was restricted, it was feared that workers' morale would be lowered. Therefore, it became necessary to design processes which would enable workers to obtain a sense of achievement and improvement with their work. The key concepts for this are described in the following.

Manage the Line by Emphasizing Group Activities

According to social psychology, it is said that 20 to 30 is the appropriate number for the size of a group. With a group of this size, human ability can be maximized and displayed in the group activities in the production lines. In our line management organization, the number of workers in each *kumi* is equivalent to this group. Therefore, in this development, the kumi is regarded as the core organization for managing the line.

Clarify the Meaning and Goal of the Work and Complete the Work Within the Kumi

For human beings, high recognition of their job performance is an important incentive and cultivates their abilities. Entrusting them with more responsibility for their job improves their will to work and develops their abilities. Similarly, a prerequisite for designing an assembly process is that the individual work content can be completed fully and accurately. In such a work process, the content and

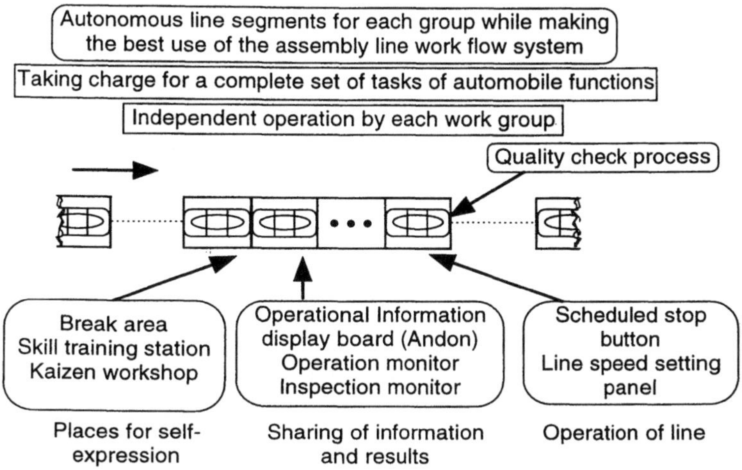

Fig. 3.2.2. Structure of an autonomous - complete process

purpose of the work can be understood easily and the results soon become obvious, e.g. in the quality of the products. Implementing this kind of work process can help workers acquire the necessary skills in a short time and leads to a better understanding of the overall range of skills which have to be mastered.

Autonomous-complete process was a basic rule for composing the processes based on this concept and was developed in the following steps:

1) Classify the assembly work according to the automobile functions and specify the purpose and necessary skills for this assembly work.
2) Prescribe standard assembly sequence in accordance with the above classification and determine *coherent work*, conforming with the automobile functions.
3) Assign line management organization (kumi) to the *coherent work* and develop appropriate process structure to complete the work.

Figure 3.2.1. illustrates the correlation between the kumi and the classification of the automobile functions based on the Japanese Industrial Standards (JIS)code.

Line Structure which Enables Autonomous Management by the Kumi

To make the kumi autonomous, the line is divided for each kumi and proper buffers are set up both in the front and rear of each line. In addition, various displays and management tools to support the kumi operations are available for each kumi (fig. 3.2.2.)

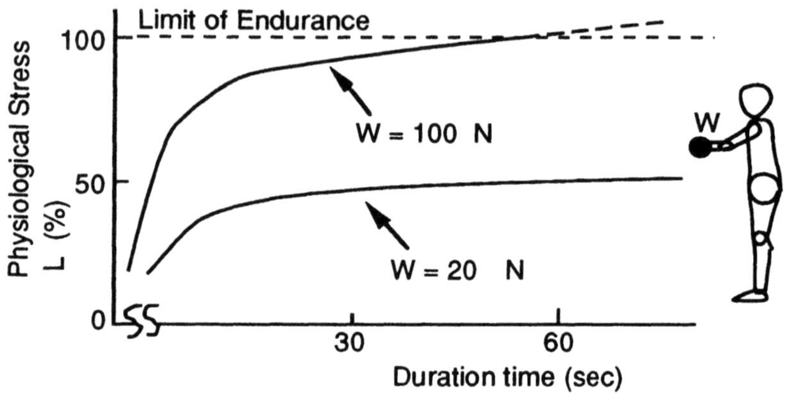

Fig. 3.2.3. Duration time vs. physiological stress

3.2.3.2
Design Processes that Can Be Performed by All

In order to enable older workers or women to work in the assembly lines, it is necessary to lighten the work load. First of all, the work load must be measured in quantity. Therefore, based on research in the field of ergonomics, TVAL (Toyota Verification of Assembly Line), a method of measuring physical load of muscles, was developed in order to clearly compute the priority order for improvements and the effects produced by the improvements.

Application of the Physical Load Degree

When a worker performs a specific task, at a certain point in time his sense of load reaches the limit of endurance, drawing a logarithmic curve to the duration time as shown in figure 3.2.3. And it is confirmed that a sense of load at that time can be expressed in *Physical load degree L* computed by formula 3.2.1. This concept is applied to TVAL.

$$L = d1 \log(t) + d2 \log(W) + d3 \qquad (3.2.1.)$$

L: Physical load degree; *t*: Work time duration, *W*: Work load, *d1,d2,d3*: Coefficient

Concept on TVAL Values and how to Find Them

It is well known that one of the methods to actually measure the physical load degree is to check the muscle burden using a bicycle ergo meter. According to this experiment, the physical load degree can be found using formula 3.2.2.:

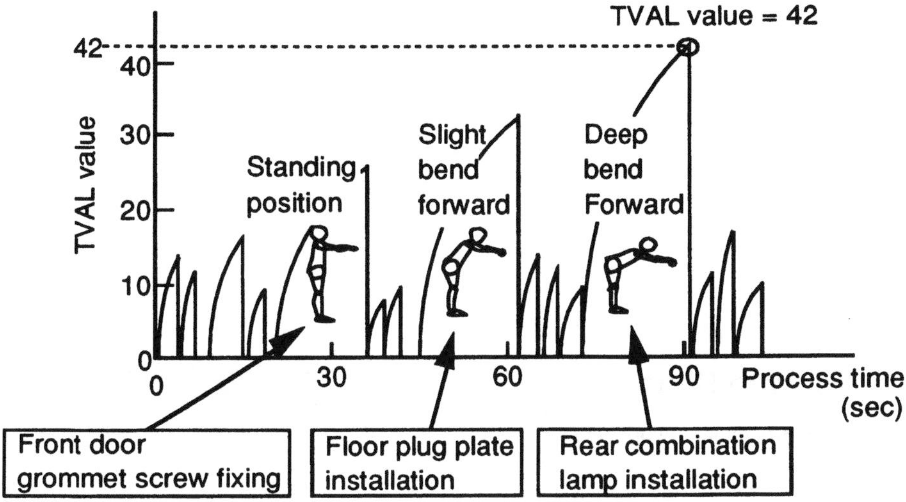

Fig. 3.2.4. Measured TVAL values

$$L = 25.51 \log(t) + 117.6 \log(WB) - 162.0 \qquad (3.2.2.)$$

L: Physical load degree, *t*: Work time duration, *WB*: Load on pedal

Assembly line work, however, involves various burdens and is performed in various postures. It would take a great deal of time to carry out this experiment for each individual operation and to obtain a formula for each case. Muscle burden - determined by a change in the work posture and handled weight - is the major factor for the work load, and can be used as an index for the muscle burden in the maximum muscle load ratios. Therefore, if the maximum muscle load ratios - as measured in the experiment with the bicycle ergo meter and in assembly work - are equal, both of these work loads can then be regarded as equal.

Consequently, the maximum muscle load ratio of assembly work was measured in advance. A conversion table was set up to establish the ratio between this load and the load as measured by the pedal in the bicycle ergo meter, and this value was substituted for WB in formula 3.2.2. Then the result was defined as a TVAL value.

Measurement of TVAL Values

Figure 3.2.4. shows a typical example of the TVAL values found by measuring an assembly operator's work. In this figure, the maximum value is 42. To test the validity of this method, a survey of 500 workers was carried out. As a result, about 80 % of them answered that they felt the work with the highest TVAL value was the heaviest, and it was confirmed that the TVAL values matched their subjective opinion.

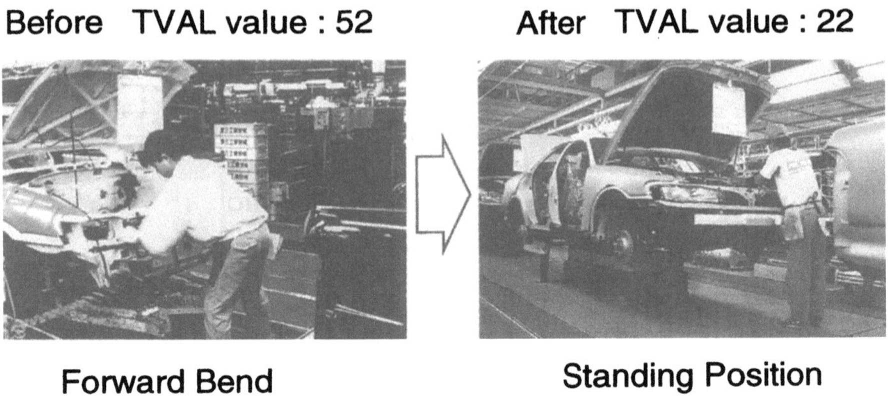

Fig. 3.2.5. Modification of work height

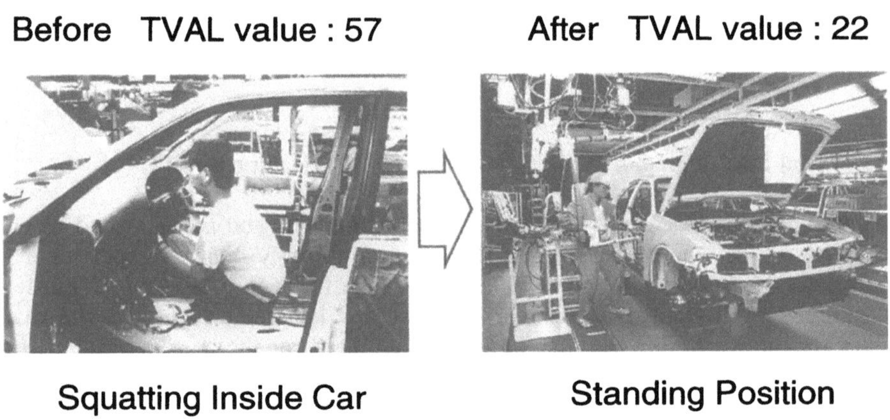

Fig. 3.2.6. Steering Column installed with power-assisted equipment

Example of Work Load Improvements Based on TVAL Values

After TVAL values on more than 2400 kinds of assembly work were measured, tasks with a TVAL value of more than 50, i.e. which were felt the heaviest, were given priority and then, tasks with a TVAL value of more than 35 were targeted to be improved. The model cases are shown in figures 3.2.5. and 3.2.6. One example of the comparison of TVAL values between the result after the above-mentioned kaizen activity in one line and the previous in the same line is shown in figure 3.2.7. This case shows that all of the old processes with a TVAL value of more than 35 have been abolished.

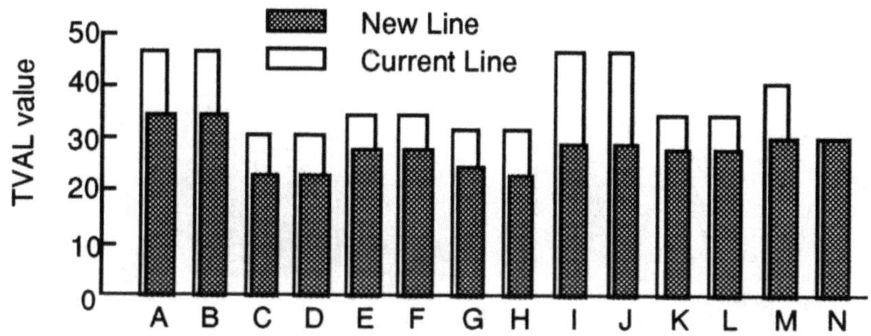

Fig. 3.2.7. Improvement result by TVAL value

3.2.3.3
Employ Automation that People Want to Work With

Automated equipment should be installed near the workers responsible for the assembly work, and its structure should be easy enough to operate and understand. It can be expected that by mastering such equipment by themselves, the workers will gain self-confidence and this will give them a more positive and active approach to their work. Consequently, their morale will be raised. Therefore, the development of *in-line mechanical automated equipment* based on this idea was undertaken. The following describes the basic ideas of this automated equipment.

In conventional automated equipment, a body is, generally speaking, stopped at off-line, and in the case where accurate positioning is required, an industrial television camera (ITV) is used to identify the position of the body and parts. The data is fed back to the servo-mechanism to install the parts to the body.

For assembly workers, however, this system often means huge equipment with a mass of black boxes, and the will to improve this may be lost. Moreover, this system may increase the number of maintenance workers with the necessary expertise. Therefore, the development at this time was aimed at installing automated equipment in the assembly line and performing automatic assembly by synchronizing operations with the body being conveyed at a constant speed. We also aimed at pursuing the realization of automation, using an extremely simple mechanical system. Figure 3.2.8. shows this idea which was introduced for the first time at the Kyushu plant.

A body is positioned on the carrier and conveyed at a constant speed. The parts to be installed to this body are set on the installation equipment pallet which floats with the equipment. First, the equipment mechanically connects with the carrier and advances in synchronization with the forward movement of the carrier. After this, the pallet rises. The parts to be installed are mechanically guided to the body as required by the equipment, finally positioned by the pin and the body hole, and then installed to the body by the automatic tightening device.

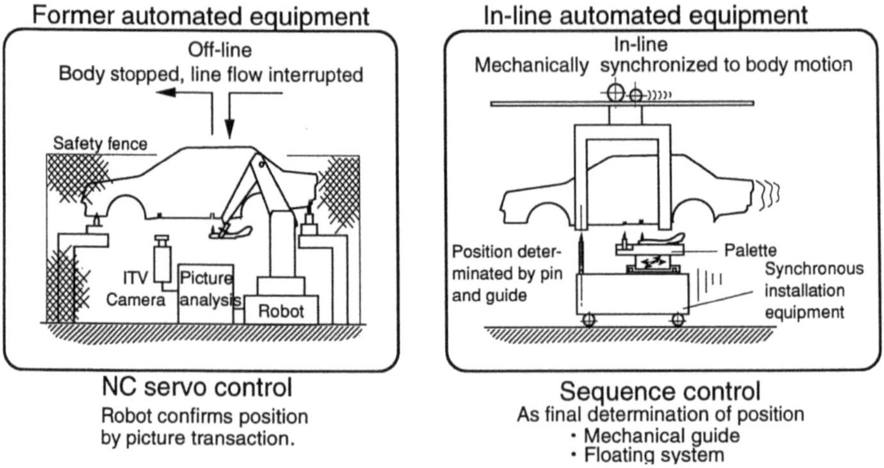

Fig. 3.2.8. Concept of In-line mechanical automated equipment

As an example of this concept of automation, figure 3.2.9. (a) shows the loading and tightening equipment for engine and chassis. Figure 3.2.9. (b) explains an actual example of the simple mechanical system in this equipment. The part called a shock absorber shown in this figure helps the body absorb the unevenness of a road surface. This part is fixed to the body because its shaft has a function to move freely. At this time, we were able to attain our initial purpose by mounting a removable cap on the shock absorber and by implementing a simple structure, this cap is then used to position the shock absorber to the body.

As such easy-to-understand automation know-how was incorporated throughout the plant, many kaizen suggestions were enthusiastically made by the workers responsible for this work. As a result, the equipment has rapidly improved, and even now, it goes on operating with extremely high availability as shown in figure 3.2.10.

3.2.3.4
Make a Comfortable Work Environment

The newly constructed Toyota Motor Kyushu Miyata Plant is located in a hilly green landscape. We made every effort to create a work area which is suitable for a plant blessed with beautiful nature and where people can work comfortably. Our basic concept for creating a working environment is based on the following three points.

- Create a comfortable work environment, e.g. with reduced noise in the plant, a bright and open work area with natural sunlight as basic lighting and appropriately controlled temperature with air-conditioning.

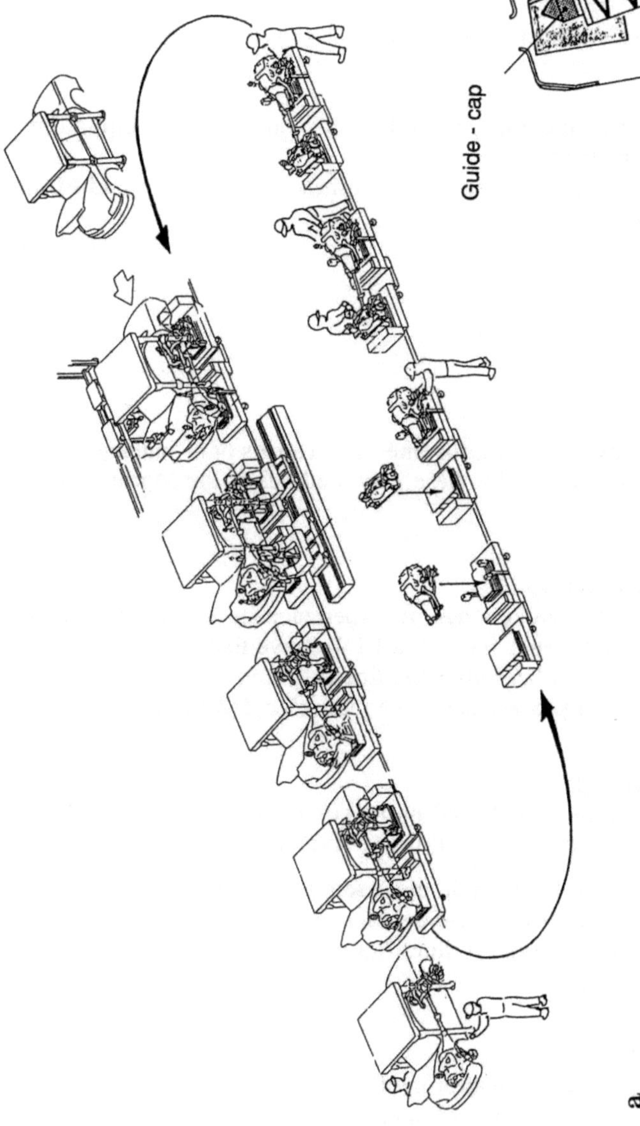

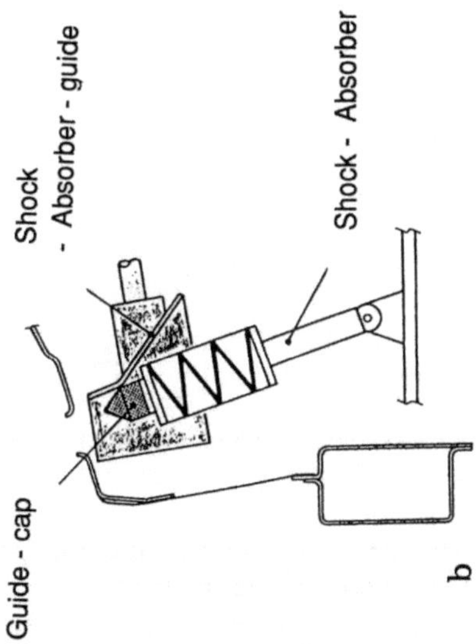

Fig. 3.2.9. Developed loading and tightening equipment for engine and chassis - a) Equipment outlook; b) Guide mechanism

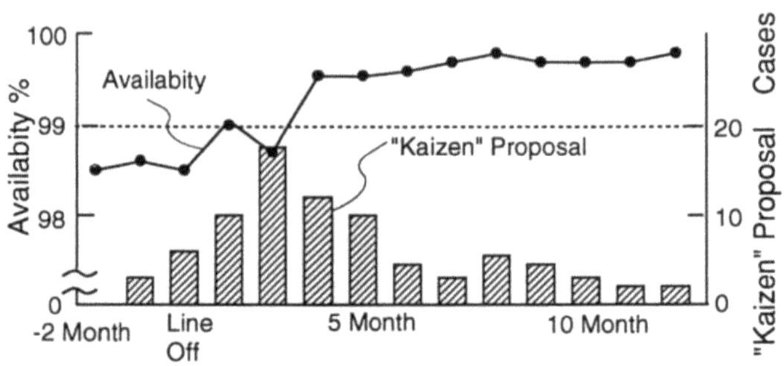

Fig. 3.2.10. Trend of availability and improvement proposals

- Produce a clean and harmonious atmosphere by using color-coordination, both inside and outside the plant.
- Offer perfect amenity facilities and arrange these conveniently and efficiently throughout the plant.

3.2.4 Results

Figure 3.2.11. shows some of the improvements - in terms of efficiency and quality as well as ergonomics - achieved by these new methods at the Motor Kyushu Miyata Plant.

Increase in Productivity and Quality
Although 70 % of the total workers had no experience in the production of automobiles, the production efficiency increased 10 % above that of other conventional plants, (comparison 6 months later after the line-off), and 80 % of the quality defects were reduced in the in-process defective ratio. (fig. 3.2.11. a, b).

Increase in Workers' Morale
The questionnaire on morale, conducted in the company, clearly showed that morale had indeed been boosted and the following opinions were constantly repeated.

- I feel more satisfied with the work in the assembly line than before.
- I came to recognize that I myself had developed and increased my skills.
- After site-practice in the assembly line, I (a new worker) hoped to be assigned to the assembly process.

Decrease in the work load
We developed a quantitative evaluation method for the work load and halved the strenuous work with a TVAL value of more than 35 measured by the method.

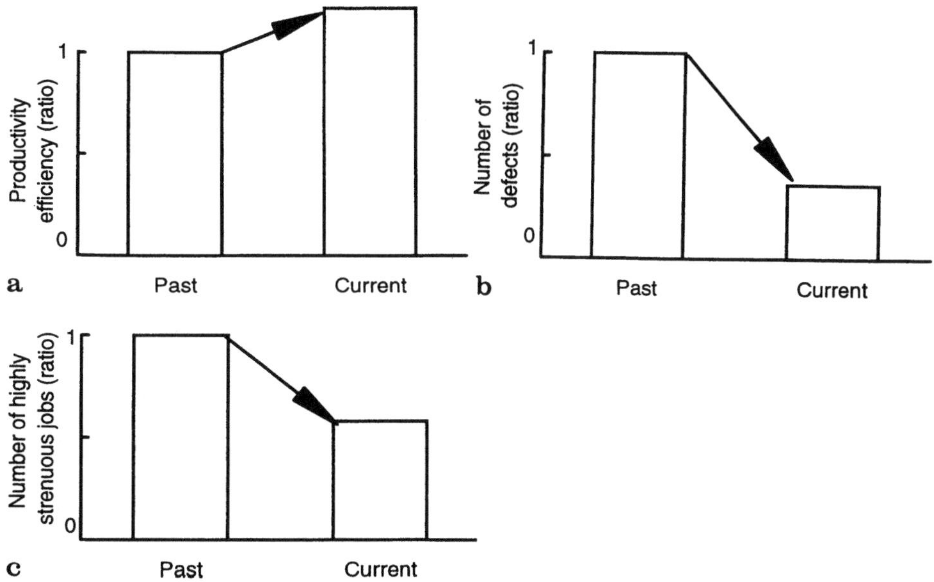

Fig. 3.2.11. Results of the new methods - a) Increase in productivity; b) Increase in quality; c) Decrease in work load

Moreover, we abolished all heavy-burden work with a TVAL value of more than 50 (fig. 3.2.11. c).

At present, in the Miyata Plant, female workers also work on the assembly line. From this, it can be said that we have taken a step forward towards creating an assembly process in which anybody who has a will to work is accepted.

Decrease in cost

Automation raising workers' morale is simple and compact, and requires no special maintenance worker. As a result, we succeeded in building up a low-cost assembly line.

3.2.5 Closing Comment

We described the details and results of the newly developed techniques employed in Toyota Motor Kyushu Miyata Plant in the above. Now, we are applying the concept of a new automobile assembly line, based on this development, to the existing lines.

The technical know-how accumulated in this development can be applied to a large number of fields in the manufacturing industry, as well as to the automobile assembly process. For the coming future, we are striving to create a *pleasure to work and produce*, one of the social themes which the manufacturing industry should tackle.

3.3
Modular Assembly in Mixed-Model Production at Mazda

H. Kinutani

One of the most complicated areas to manage at our manufacturing facilities has been the Trim and Final assembly shop where several different models are produced on what we call a *mixed-model assembly line*. Due to the general shift in the automobile industry to diversified, upgraded and higher quality products, assembly jobs that had been handled manually became increasingly complex. In addition, changes in the labor environment (such as the decreasing labor force and reduced working hours in Japan), the quest for high quality and the need to quickly respond to the customers' needs are all issues which have been increasingly gaining in importance.

Considering all this, we knew that we had to change existing concepts in the final assembly process. We envisioned an ideal trim and final assembly shop based on the question: what should it be like in the future? We came to the conclusion that we needed a simplified assembly shop with a people-friendly environment in which all of the assembly elements (such as vehicle structure, materials handling, process, assembly operation and systems) would be simplified. At Mazda we call this *Simple Base Line* and we implemented this concept in the assembly line at our new Hofu plant.

3.3.1
Conventional Assembly Line

Figure 3.3.1. shows our current trim and final assembly line. A key element of our production system is our mixed-model capability. We usually produce four or five different models using a single, long assembly conveyor. The main feature of this system is high flexibility in production. We can cope with changes in the market or customers' needs very quickly. However, when it comes to further productivity improvements, they will be difficult to be achieved since materials handling, processes and operations are all related to each other in a very complex manner. This is one of the main reasons why it is difficult to automate assembly jobs and it is also the reason why these jobs have been handled mainly by manual operations.

Taking a look at the automation ratio for the entire final assembly line, Japanese auto makers have only three to six percent automation, whereas some Western auto makers have some 10 percent automation. On the other hand, the current shift to further product diversification and upscale products has increased assembly complexity. In addition, quick adaptation to the customers' needs, the

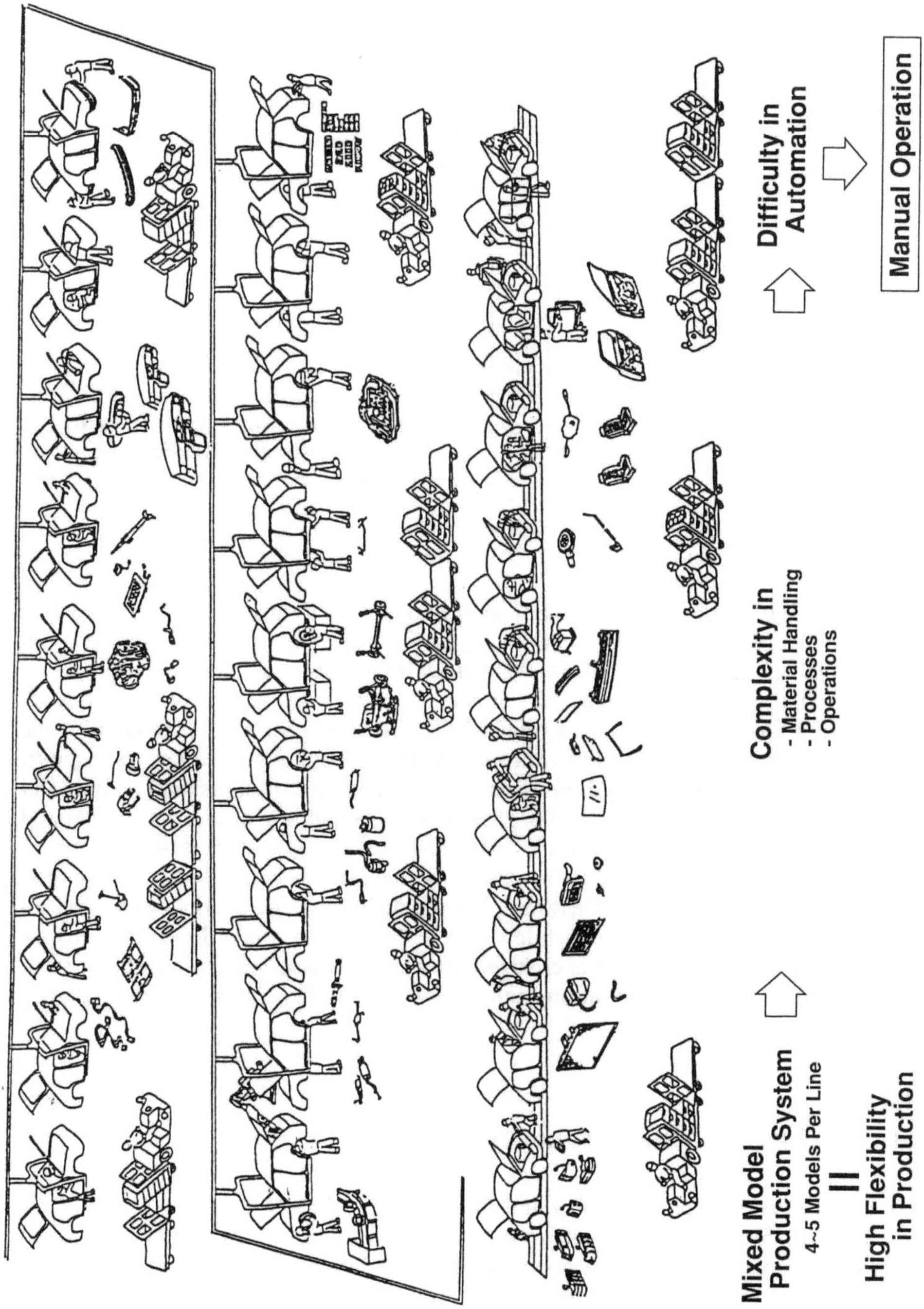

Fig. 3.3.1. Trim and final assembly line

Assembly elements / Objectives (Trim & Final)	STRUCTURE	MATERIAL HANDLING	PROCESS	OPERATION	SYSTEM
Equal work time	○		○	○	
Reduced work time (process)	○	○	○	○	
Simple operation	○	○	○	○	
Easy to manage		○	○		○
Shorter lead time	○	○	○	○	○

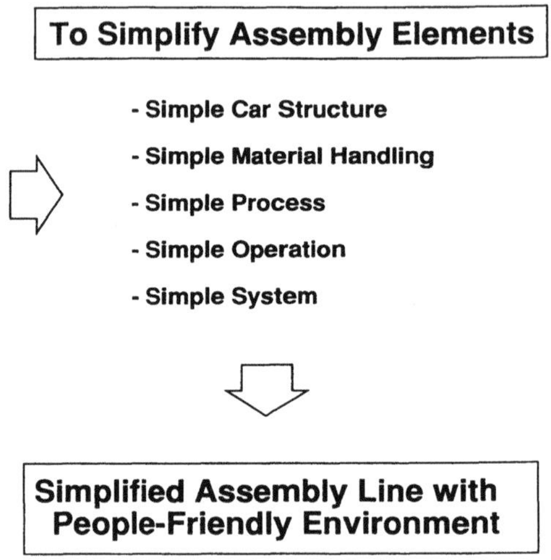

Fig. 3.3.2. Ideal trim and final assembly line

quest for high quality and the labor environment, including the shrinking labor force and decreasing working hours in Japan, are emerging as new challenges. We as auto makers must adapt to these changes with flexibility and speed.

With this in mind, we have also conducted various cost reduction activities on the existing line. However, we recognized that we would come to a standstill if we were to hold on to old assembly concepts. After thorough discussions, we came to the conclusion that we should envision an ideal trim and final assembly line and make it a reality as soon as possible. However, this ideal line could not be realized solely with the effort of the manufacturing engineering people. Because the structure of a car is closely related to the assembly line system, it must also be changed in order to meet this vision. Therefore, we decided to ask the product engineers to develop new vehicle structures while we were developing a new line. The reasoning was, if a vehicle structure is changed after the new line has already been completed, the effect of the innovation cannot be seen until the next model change.

3.3.2 Ideal Trim and Final Assembly Line

We asked ourselves how the trim and final assembly line (fig. 3.3.2.) should look in the future. We concluded that we should have a simplified assembly line which is environmentally-friendly to people and where all the assembly elements, car structure, materials handling, process, operations and systems are simplified.

We named it the *Simple Base Line* (fig. 3.3.3.). The goal of the Simple Base Line is to achieve high quality, cost performance and quick delivery in a mixed-

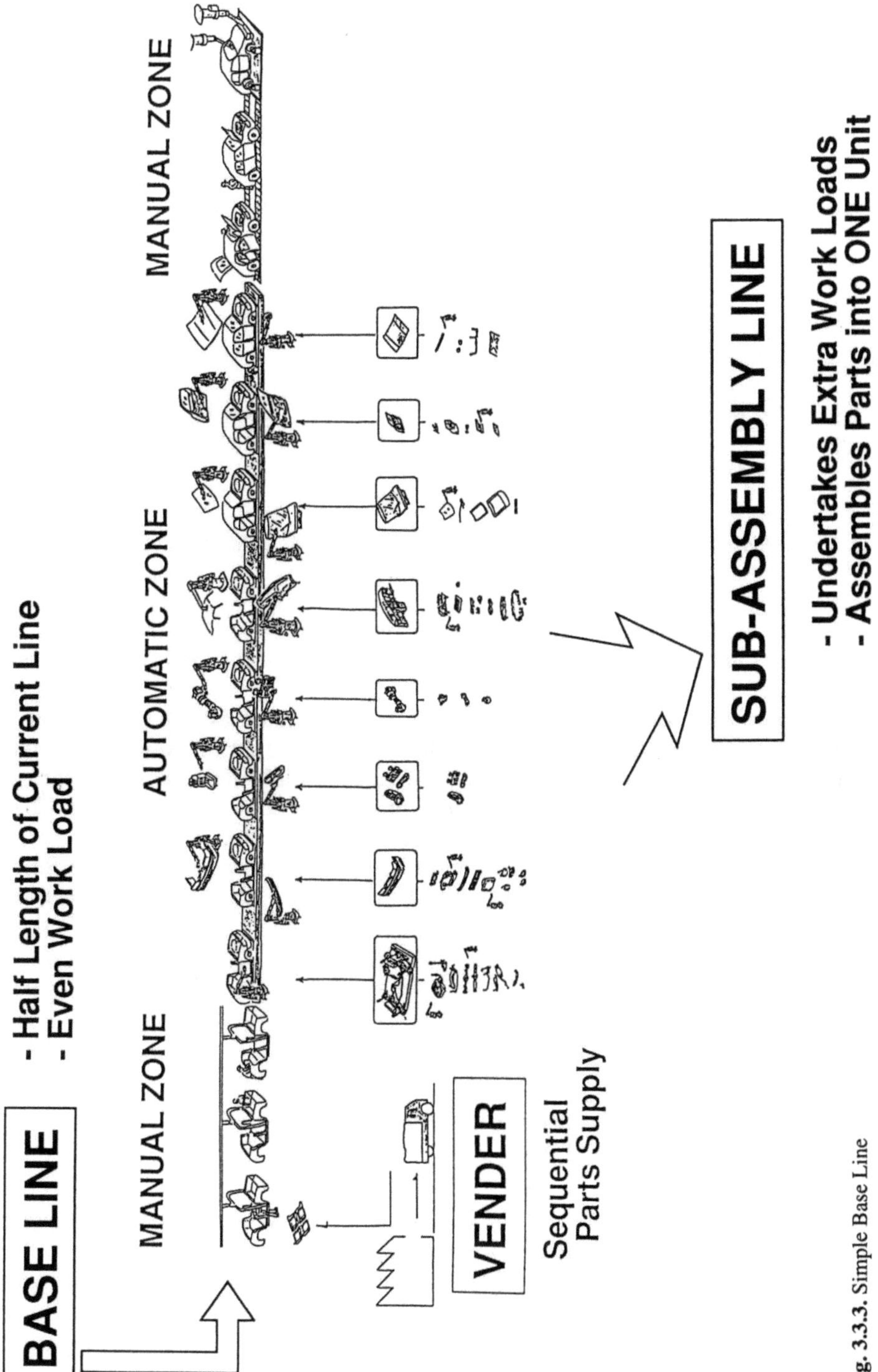

Fig. 3.3.3. Simple Base Line

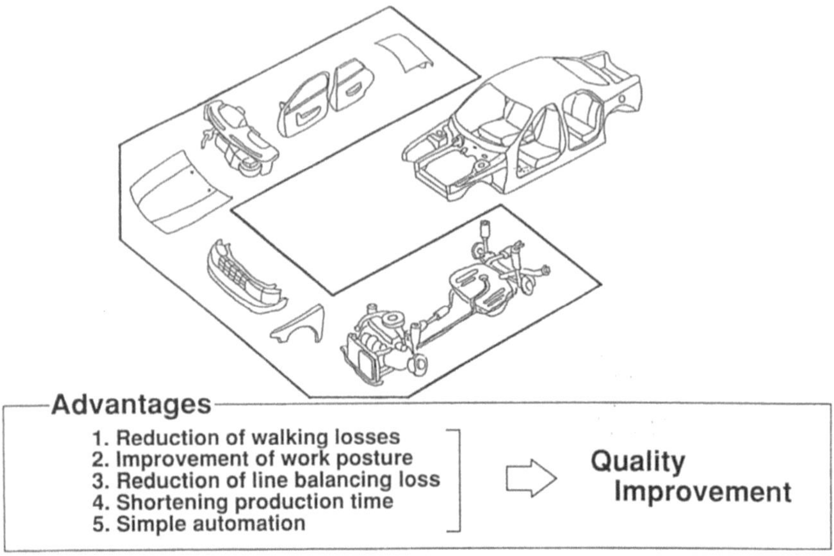

Fig. 3.3.4. Conceptual image of modules (Module: group of parts in functional classification)

model production system despite changes in product needs and the labor environment. Our concept of the Simple Base Line consists of a base line and subassembly lines.

The base line would be half the length of the current assembly line and the work load for all models assembled on the base line would be even. As the work load is different from model to model, the sub assembly lines would absorb the extra work load required for some models through effective line balancing. There, parts are assembled into a unit and transferred to the base line.

We knew that various measures should be taken in order to realize the Simple Base Line, i.e. measures to simplify all the assembly elements. These measures involved:

- reducing the number of parts
- commonizing parts
- modularizing vehicle structures
- using sequential parts supply
- automating heavy duty jobs
- introducing systematic assembly information
- introducing a foolproof system
- others.

We chose modular assembly and automation as two of the most important features. Modules are generally defined as a group of parts, classified according to their functions (fig. 3.3.4.). We believe that there are five major advantages which result from Modular Assembly.

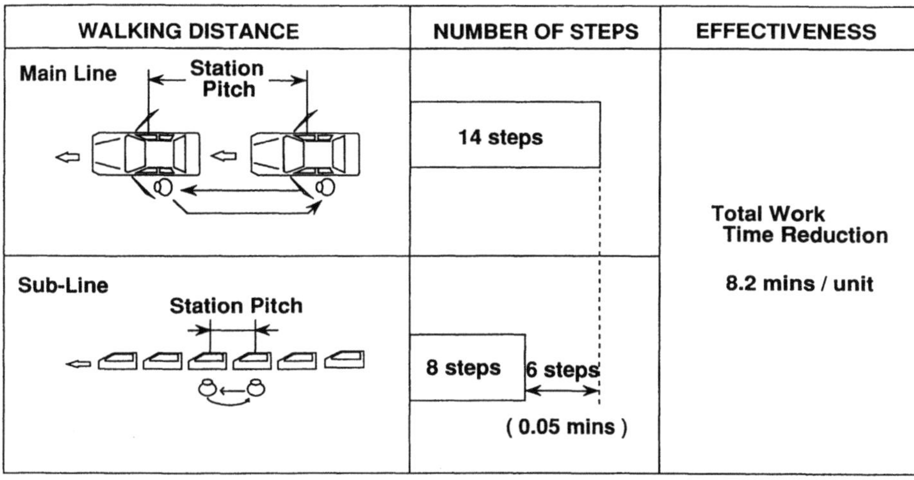

Fig. 3.3.5. Reduction of walking loss - Shorter station pitch at sub-line

3.3.3 Advantages of Modularization

First, in reducing walking distances by introducing a modular structure, we can shift main line assembly jobs to the sub-assembly lines (fig. 3.3.5.). In the case of the door module, as the station pitch length of the sub-lines is shorter than those of the main line, workers at the sub-lines walk fewer steps. As a result, total work time is reduced.

Table 3.3.1. Improved work posture - Work in large spaces and with natural postures

	Ergonomic problem	Work time	Merit of modularization
Hood	Neck bending 48 of 164 workers = 30 %	17 min./unit	- Shortened work time 2.52 min./unit - Lessened fatigue - Simplified work - Improved quality
Lift gate	Looking-up posture Stretching posture 24 of 164 workers = 15 %	6.46 min./unit	

As table 3.3.1. shows, employees' working postures can be greatly improved by implementing modularization. About 45 % of all line workers are in charge of

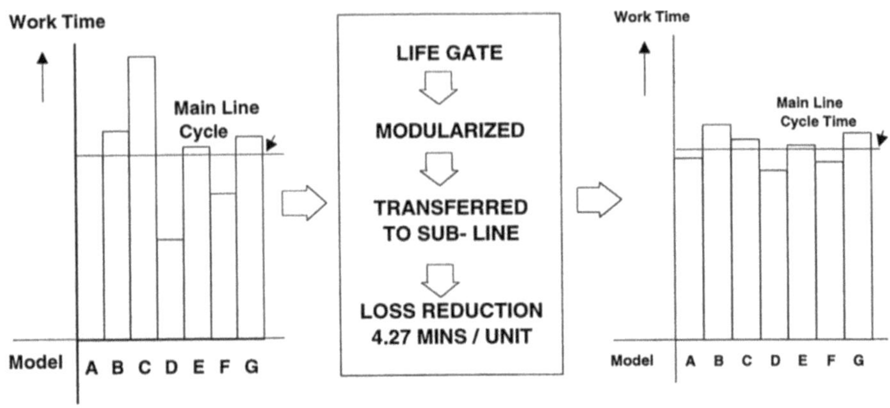

Fig. 3.3.6. Reduction of line balancing loss

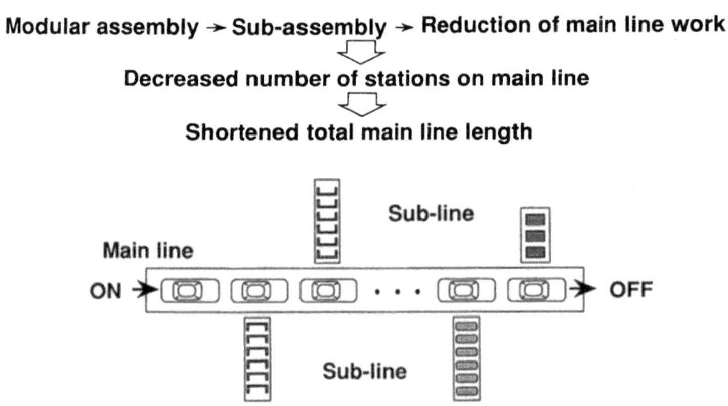

Fig. 3.3.7. Shortened production time

the engine compartment and lift gate operations. Although the working time that is saved is relatively small, by taking off the hood or the lift gate, however, a remarkable improvement in the working posture can be achieved. From an ergonomic viewpoint, this is a vast improvement and contributes greatly to quality, since assembly tasks which had to be carried out "blind" (where the worker cannot see his hands while performing the task) were eliminated.

Figure 3.3.6. shows an example of the different work-time per worker required by the different car models - before and after the introduction of modularization. Work time differences for each model are significant, disturbing the workers' own pace and causing a mental burden. Consequently, the quality of the car will be affected. Ideally, the work time for each of the models should be equal. Here is one example of the advantages of modularization: if the lift gate is transferred from the main line to the sub-assembly line, it reduces the work time fluctuation and thus the line balance of the main line is improved.

The more the work load on sub-assembly lines increases, the less the work load on the main line. This leads to a reduction in the number of stations and the length of the main line is consequently reduced. Thereby, production lead time will be shorter (fig. 3.3.7.).

Modular structure provides a great advantage for automation. For example, take the parts that make up the front end (table 3.3.2.): seven stations in the main line and 12 robots are required to assemble them separately, but if the parts are modularized, only one station and five robots are required. I believe modularization is an indispensable element for investment reduction and simple automation.

Table 3.3.2. Effects of modularizing the assembly of the front end

	Install part by part	Install module
Number of stations	7	1
Number of robots	12	5
Robot type	large, unique	small, common

3.3.4 Activities for Modularization

Although we as manufacturing engineers understood the enormous advantage of modular assembly, it was rather difficult for us to implement it smoothly, because there were many product concerns, such as cost and weight increases, design revision, test standards and so on. As a matter of fact, the product engineers were rather reluctant to implement these changes, even though they conceptually understood the necessity of modular assembly. It is quite understandable that they were also very busy and had to concentrate on the development of current models. This deadlock induced us to explain the necessity of modular assembly to the company's top management.

We presented the vision of the final assembly shop and the concept of the simple base line to them. They fully recognized the benefits of modular assembly and asked us to proceed with its implementation immediately. In order to accelerate modularization, we set up a project team between the Product Engineering Division and the Manufacturing Engineering Division. Figure 3.3.8. describes the organization of the project team, consisting of 21 engineers from the Product, Manufacturing, and Patent Divisions.

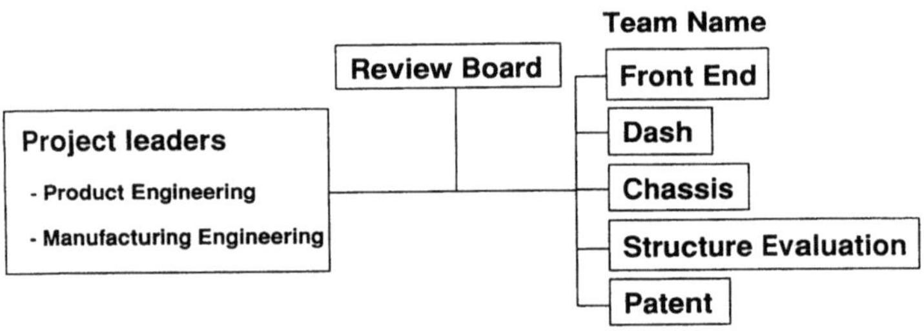

Fig. 3.3.8. Activities for modularization. *Concerns:* 1. cost and weight increases, 2. revision of design and test standards, 3. others; *Presentatioin to company's top management:* ideal trim and final assembly line = Simple Base Line; *Start of modularization:* 21 engineers

Figure 3.3.9. shows the ultimate module structure we envisioned. As we did not have enough resources to develop all of the modules, we started our project activities by concentrating on three modules, i.e. chassis, front end and dash board.

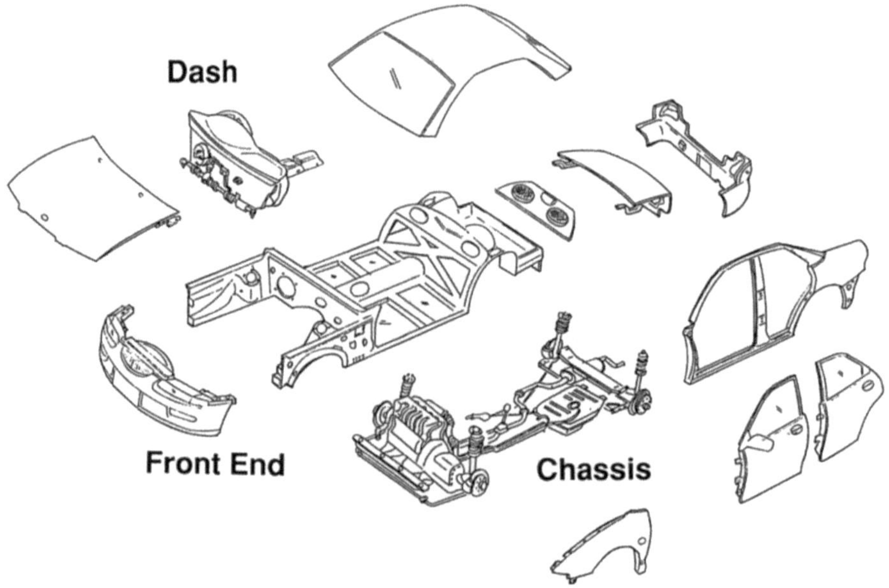

Fig.3.3.9. Ultimate modular structure

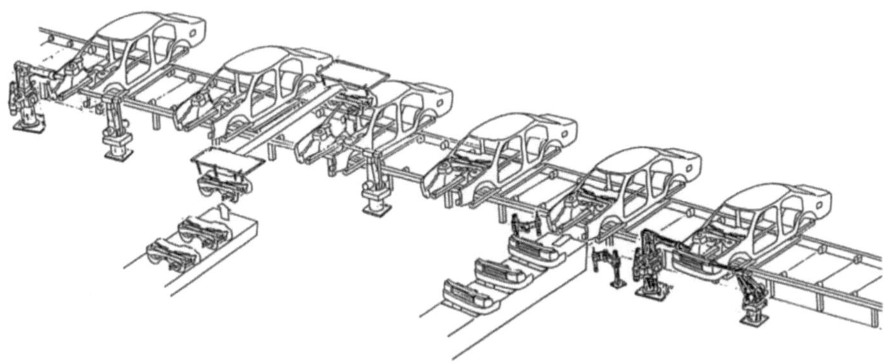

Fig. 3.3.10. Prototype of front end module - Manufacturing engineering: study assemblability to the body; study conveyor and assembly equipment

Project activities may often look ideal, but do not always solve every problem. Our manufacturing engineers offered many proposals to the design engineers, but after they had reviewed engineering standards, only less than half of the proposals were accepted. We fully realized the difficulty in changing the vehicle structure for modularization once it had been established. In the course of this process, manufacturing engineering proposed a total of 119 items to product engineering. Product engineering finally accepted 54 of these suggestions.

Before designing the modular structure for a new model, the product en-gineers designed and made a module prototype using a currently produced vehicle. By using this prototype, they carried out the body strength and vibration tests in order to identify the parts to be redesigned. The manufacturing engineers, on the other hand, studied each module's capability of being assembled to the body, use of the conveyor and need for assembly equipment. During this study, we came to the conclusion that in some areas the body should be conveyed crosswise as figure 3.3.10. shows.

The front-end module structure of a new model, which Mazda produces at the expanded plant in Hofu, consists of about 20 parts including the bumper, radiator, head lamp and so on. Another example of modularization is the Engine and Chassis Module Decking System. In the past, these modules were decked individually to the body by manual operators.

Later, manual decking systems were automated as figure 3.3.11. illustrates. When a body arrived, locator holes were measured by sensors and then the engine and rear suspension were loaded on the body with two different lifts and secured automatically. This method, however, posed a number of problems as far as the conveyor's up-time ratio is concerned. For example, should problems occur during the bolting process, the lift cannot come down until recovery is completed. This forces us to stop the conveyor.

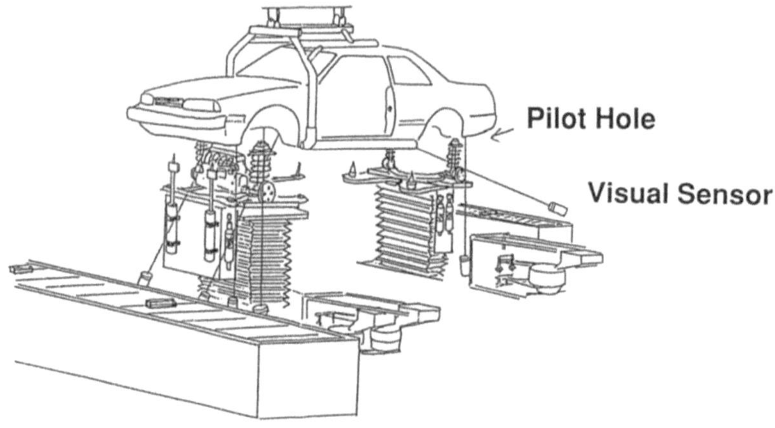

Fig. 3.3.11. Engine/Suspension automatic decking - Concern: Conveyor's up time ratio

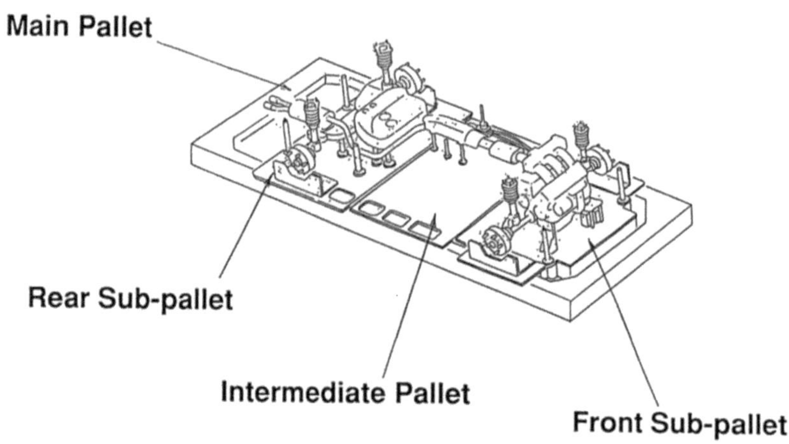

Fig. 3.3.12. Pallet system for modular chassis assembly

After experiencing several problems, we changed our decking concept and introduced what we call the pallet system. As shown in figure 3.3.12., engine, suspension and other underbody parts are positioned on the pallet. Then the body moves down onto the pallet (fig. 3.3.13.). This method is completely opposite to the previous system. This illustration shows a cross-section of the pallet system just after the body has been loaded on the pallet.

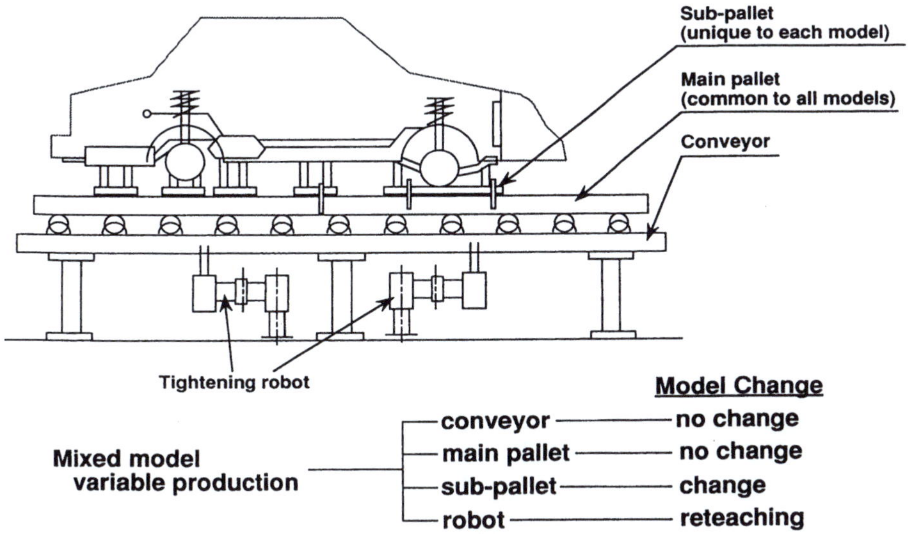

Fig. 3.3.13. Pallet system for modular chassis assembly on the conveyor line

We designed the main pallet to be used with all models. On the other hand, sub-pallets are unique to each model. In this way, we only have to change the sub-pallets for the model mix. Although there are model changes or additions of new models, we do not need to modify the conveyor and main pallets. All we have to do is to change or add the sub-pallets and reprogram the robots. This enables us to reduce capital investment and to shorten production preparation lead time. This pallet system has another advantage as far as the up time ratio is concerned; even if there are some problems in bolting, we do not need to stop the line for recovery. We just send the pallet to the following stations for normal operations. The correction can be done later at the backup stations.

In figure 3.3.14., three different body transfer methods are compared. These are the linear motor, electric trolley and shuttle conveyor. Every method has its merits and shortcomings and we made the best use of them. For the chassis module line, for example, we introduced the linear motor so that we can transfer the pallets more quickly and save space in the pit for robot installation.

Based on the ideas I have just explained, we started developing a new model with a new plant to produce the model as well. Our concept for this new plant is *A People-Friendly Production Facility for High Quality, Upscale Vehicles*. Before developing the product and production line, the Product Engineering Division and the Manufacturing Engineering Division met to decide the hard points, such as locators on the pallet for positioning the body. They kept these "hard points" in mind while developing the vehicle and the production facility. These two divisions also worked together for vehicle structure standardization. For example, we standardized the locator hole pitch and tightening point pitch on those modules for all models to be produced on the new line. In this way, we standardized pallets and tightening tools.

	Linear Motor	Electric Trolley	Shuttle
Transfer Speed	120 m / min	60~90 m / min	60 m / min
Accuracy of Stop Position	◎	◎	◎
Durability	○	△ Potential effect by oil & water	○
Reliability	○	△	○
Interference with Robot	○	△	△
Results	×	◎	◎
Cost (E.Trolley=100)	120	100	80
Station Pitch Changeability	○	◎	×

◎ : Excellent ○ : Good △ : Poor × : Bad

Fig. 3.3.14. Comparison of body transfer methods

3.3.5 The new Hofu Plant

The assembly line of the new plant shown in figure 3.3.15. is divided into two zones. The right side shows the manual assembly zone and the left side shows the automatic zone. This division makes it easy for us to separately control operations. Based on the concept of the Simple Base Line, the number of work stations was reduced to about 60 % of the number required by the existing assembly line at Hofu. Modules, such as front end, dashboard and chassis are installed in the automatic assembly zone.

To provide a people-friendly work environment, we stopped using the noisy chain conveyors and introduced an electric trolley on all lines. Interior colors were coordinated for a comfortable work environment, CRT panels were introduced for the easier pickup of parts and module structures including engine hood were of course added for improved work posture.

3.3.6 Summary

Through the activities I have described, we have completed the mixed-model and variable volume assembly line based on module structures and automation. We

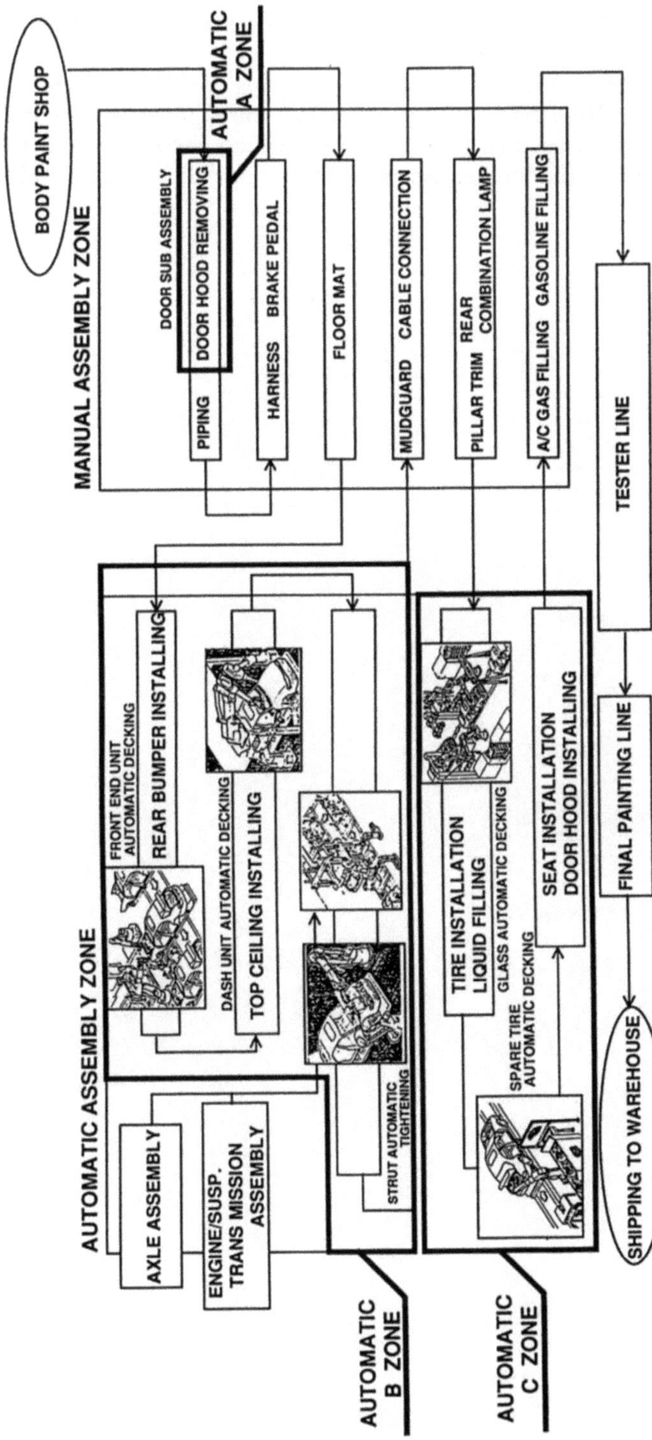

Fig. 3.3.15. The final assembly line of the expanded Hofu plant

have not yet realized the full potential of the Simple Base Line, but I believe we have established the foundation for this concept.

It is generally understood that *simultaneous engineering* is very important, but difficult to carry out. Looking back on my experience throughout this modular assembly project, I have learned that we - as manufacturing engineers - must always envision the future of production systems and become the strongest advocates of our own vision.

3.4 Production of the NSX at Honda: An Alternative Direction for Assembly Organization

K. Tanase · T. Matsuo · K. Shimokawa

3.4.1 Introduction

In the late 1980s, the main purpose of Japanese auto companies was to cope with the problems of labor shortage. The dominant theme in new factories was how to be *human-friendly*. The assembly line at Honda's Suzuka plant No. 3, which was set up in 1989, is one example of this. It relies to a vast extent on automation. Automation, however, is only one way of being human-friendly. The Takanezawa plant is a more recent experiment, it undertakes a different approach towards human-friendliness, using different types of automation technology and software. In this chapter, we will present the approach implemented at the Takanezawa plant.

3.4.2 NSX as an Experiment

The NSX is not only the first real sports car Honda has ever built, but it is also an experiment in the fields of product technology, sales, production and supplier relations.

Product
The NSX cost over 8 million Yen. Its development led to some new product technologies, including a midship engine and a rear-wheel drive system, a full aluminum body as well as proven technologies which had been implemented in other Honda cars, such as the VTEC engine and the anti-lock breaking system. The NSX is Honda's flagship model, implementing Honda's notion of what customers expect.

Sales and After-Sales Service
The new experiment concentrated not only on just product and production technology. It included the development of new methods in the area of sales and services. Although the NSX was regarded as Honda's representative car, the company had never before targeted higher-income customers who were now supposed to buy the NSX. Therefore, a special team dedicated to the NSX was established in the Verno sales channel which is responsible for the car sales. (The Verno channel was set up in 1978 for distributing sports and speciality cars.) As part of the service during the warrantee period, the customer relations team offered *home-doctor-*

services; these included periodical visits or phone calls to NSX owners in order to learn about the condition of the car and to offer advice to owners. This is an experiment on the relationship enrichment between dealers and owners [1].

Traditionally, factories could only gather information about the user or the condition of the car indirectly - via the sales channel. In the case of NSX users, dealers, sales staff and plant people can communicate directly through drivers' training, lectures and other events.

Production

Although the NSX was planned for a very small production scale (25 per day), Honda decided to build a new plant for this car. This permitted the introduction of many innovations oriented towards increasing the attractiveness of work. Because of the small-lot production here, not all innovations can be disseminated to other plants. The challenge was, however, to show the future direction of a plant, which fundamentally attaches importance to combining the capacity of human beings and the characteristics of machines. A more detailed discussion follows in the next section.

Suppliers

For the suppliers of materials and stamping parts, developing an aluminum car body was a new experience. Up to then, they had only small, specified parts. In the case of the NSX, they participated in the development process from the beginning, and supplied the body panels [1]. It should be stressed that the direct, special relationship between plant and suppliers was even maintained beyond the start-up phase. Workers at the Takanezawa plant can hence communicate much more easily with suppliers than their colleagues at traditional plants [2].

3.4.3 Challenges to be met by Production Organization

Although the Takanezawa plant has some unique features, thanks to its small-lot production, it has, in many ways, tried to respond to changes in Japanese society. In this section, we will deal more closely with certain new features of production organization in this regard.

Environment

When the plant started up in May 1990, it was the time of the bubble economy. Domestic demand for automobiles had reached its peak in 1990. The labor shortage problem became ever more serious. Especially the young male workforce, on whom Japanese car makers had depended until then, disliked work under these so-called 3-D conditions (dirty, demanding and dangerous). Furthermore, the average age of workers in the automobile industry had increased in line with general demographic changes in Japan. The shortage of labor, which resulted from the changes in the labor market, induced car makers to emphasize employee satisfaction in their production plants (especially for assembly work).

The Takanezawa Plant

The Takanezawa plant is located in the Tochigi prefecture, north-east of Tokyo. It started production in May 1990. Although it had been planned to produce the NSX only, it is now (1996) producing other cars, too, because of the recent drop in demand for the NSX. Construction costs totaled around 10 billion yen. The site area is 200,000 square meters with 20,000 square meters covered by the two-storey building. The workforce totals 230. Obviously, it is a small plant. The production capacity totals 25 cars per day with a cycle time of 17 minutes. It has a very short line with 141 stations from welding to final assembly. The use of robots is very limited. Only 5 to 6 robots are used in welding and paint operations.

Employees

One of the characteristics of the plant is the age of the employees. The average age of employees is 37 years - much higher than in other plants. The reason for this is that workers with long experience were sought, as it was believed that these workers would be capable of producing a totally new car. 130 of the total of 230 employees came from the Sayama plant, Takanezawa's "mother plant", whilst the remaining 100 workers came from various other Honda plants via in-house relocation programs. The aim of the plant was to maintain a balance between people and machines, i.e. to merge skilled workers (called *craft men* at Honda) and modern technology.

Process Design

Process design followed the principle that each station should be in charge of one function. Therefore, more tasks are assigned to one station than is usual in Japanese plants. The cycle time is hence extended to 17 minutes; this is in contrast to the usual one-minute-cycle in the other plants. Moreover, there is room for a further delay time of one or two minutes, if, for instance, work cannot be completed within the cycle time. Furthermore, the body is transferred by means of pushcarts and not by a conveyor belt, for example.

In the following, we will take a closer look at the welding, paint and final assembly processes.

Welding

In most of the modern car assembly plants, robots perform the welding operations. Especially in the area of spot welding, the automation ratio tends to be over 90% in the newer Japanese plants. In the Takanezawa plant, however, the amount of manual operations has been dramatically increased, and the work performed by robots has been reduced. Robots are used only for operations which workers cannot perform. One example is the spot welding of the underfloor which cannot be done manually because it requires large and heavy welding guns or tooling. Manual welding, however, can lead to problems where welding quality is concerned. Therefore, some operations were mechanized, using robots with a high degree of freedom, capable of handling heavy loads and long cycle time operations.

Fig. 3.4.1. Robot welding (a) and manual welding (b) - Specification of robot drive system: Motor-driven 5-axis; allowable payload: 400 kg [3]

Another answer to the quality problem is the tilt positioner which permits the optimal welding position in the manual welding of the three dimensional body (fig.3.4.1. and 3.4.2.). The function of this tilt positioner, which permits the selection of any welding position, is to facilitate the safe and accurate spot and MIG welding of the inside of the body and the underbody. Thanks to this instrument, the body can be rotated freely through 360 degrees in order to allow welding with a more convenient welding posture. As a consequence, the strain and health hazards felt by workers when bending to an unnatural posture or when working underneath the car body are eliminated. (Having to perform welding work in an unnatural posture increases the risk of getting burnt by sparks). Moving the car instead of the welding gun solved the problem of handling heavy welding equipment. As a result of these measures, welding operations became much easier and welding quality became more reliable.

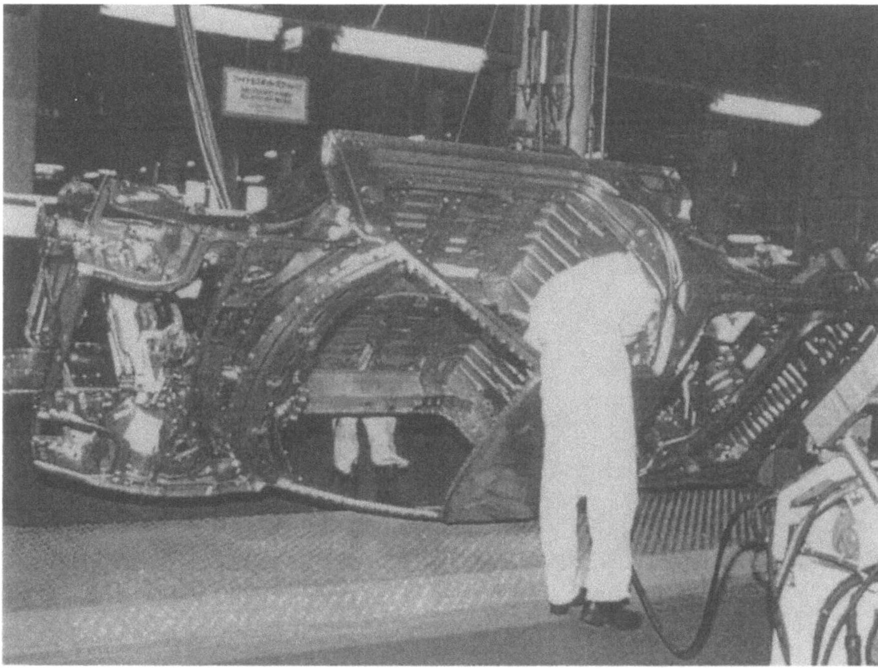

Fig. 3.4.2. Tilt positioner [3]

As already described above, Honda hence intentionally separated the processes of manual welding and robot welding. The aim was not to maximize the degree of automation, but to make the best use of both man and machine. With regard to manual work, various instruments and tools were introduced in order to make work easier.

Painting

In the painting process, emphasis was placed on manual work rather than mechanization. An example is the base coating process (fig. 3.4.3.). Skilled workers perform this operation, which is important for the appearance of the car, in order to achieve a higher quality tint. Furthermore, water-based paint is used to improve the smoothness of the surface, and this also contributes to a better working environment. Robots are used, however, for the clear coating process in order to obtain an even more vivid luster. It is one of the few operations where robots were introduced. The robots move synchronously with the conveyor. In this way, the best of the workers' skills and the precision of robots are combined. As a result, not only the final appearance of the product is improved, but also the workers' pride in their achievements.

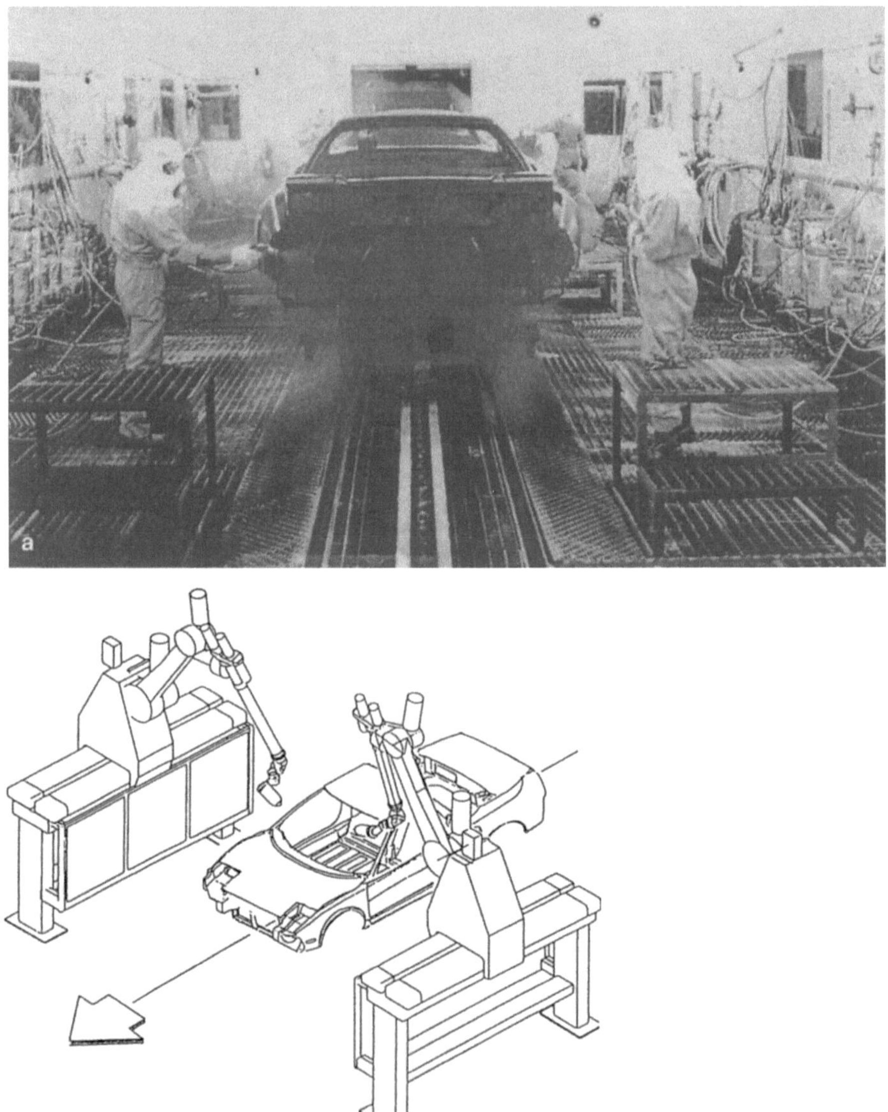

Fig. 3.4.3. Painting: Base-coating process (a) and clear-coating process (b) [3]

Final Assembly

Different types of transfer mechanisms are generally used in assembly plants, depending on the requirements of the work stations (e.g. for operations underneath or inside the car body). The Takanezawa plant, however, has abandoned the conveyor belt completely and uses *pushcarts* to transfer the bodies. A variety of tools and devices were developed in order to accommodate the specific requirements of the different work stations.

Among these stations, the mounting of engine and chassis is of specific importance for the proper functioning of the car. Working postures within a limited space were hence eliminated, so that workers can concentrate on assembly opera-

Fig. 3.4.4. Flexible body hanger [3]

Fig. 3.4.5. Piping and wiring [3]

tions and quality assurance. Honda has developed and installed motor-driven overhead hangers for this specific purpose (fig. 3.4.4.). This hanger can be vertically adjusted in eight steps for the installation of engines and components. Workers can hence perform their work without excessive effort by vertically adjusting the hangers as required. This improves quality and reduces workload.

A second example is the body rollover system which Honda introduced in the piping and wiring process (fig. 3.4.5.). In this process, pipes and wires are connected to the engine in the middle of the car, and the radiator and battery at the

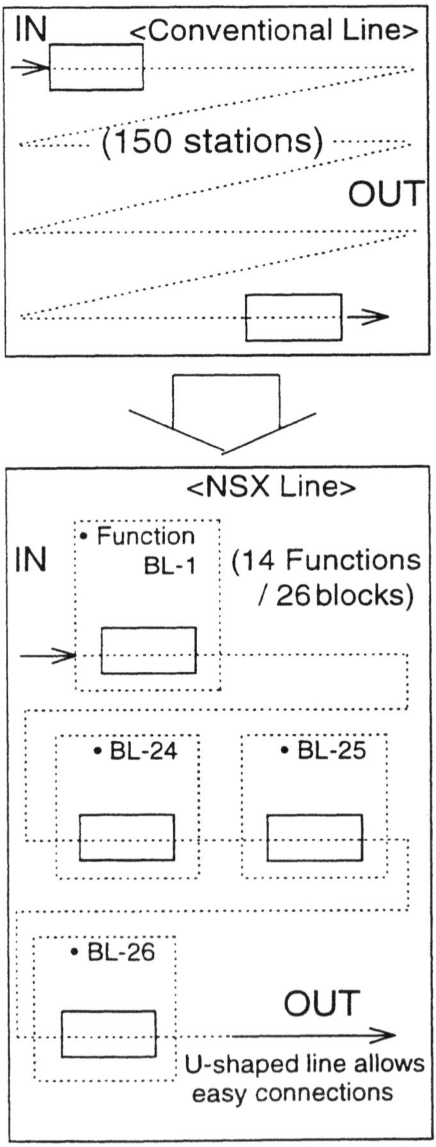

• Conventional line: priority is given to labor efficiency and output volume.

New line: the aim is to balance product quality assurance by functions and work attractiveness

14 functions divided into blocks

1. Piping of underbody parts
2. Brake piping
3. Harness installation
4. Instrument panel /heater
5. Front mask / rear fitting
6. Front suspension
7. Rear suspension
8. Air bleeding
9. Engine compartment
10. Windshield installation
11. Liquid injection / door adjustment
12. Interior fitting
13. Subassembly
14. Adjustments of cars

Fig. 3.4.6. Assembly [3]

front end. These operations must be carried out underneath the body. In the conventional process, the job is performed under an overhead conveyor with workers having to work in a strenuous position with their arms working overhead. In the Takanezawa plant, the body is rotated through 90 degrees, and with the underbody now standing in front of the workers, the piping and wiring operations have become easier.

In conventional mass-production plants, processes are divided to the maximum extent possible in order to increase productivity. Most of the Japanese plants hence have around 150 stations in final assembly (fig. 3.4.6.), and operations are thus becoming more and more monotonous. Taking advantage of the small-lot production at the Takanezawa plant, experiments were carried out with a different process architecture. First of all, the car was divided in to 14 functions which were allocated to 26 sets of work stations (blocks), and the process was set up accordingly. The purpose of each block was related to a specific function of the car, so that each block must ensure the proper functioning of its unit at the end of the process. Moreover, the process features a *U*-shaped layout, so that communication between stations becomes easier. Due to these measures, it was clear that each function unit was responsible for quality assurance, so that workers felt as if they themselves were managing the quality of the car.

Communication with the User

Suitable measures were taken at the Takanezawa plant in order to give plant employees the opportunity to talk directly to NSX users (fig. 3.4.7.). This form of communication serves two purposes. One is for the users to understand the performance of the car. The other, is to reconfirm the quality of the product and to find out points for further improvement. Before this, there had been very few opportunities for direct communication between plant and users. The plant then had to rely on sales channels in order to acknowledge the evaluation or requests by users.

The Takanezawa plant, together with the sales division and staff of the Verno sales channel, now offers drivers' training programs at the Suzuka circuit owned by Honda. Aided by this program, they can find out what appealed to the user most, and they can also learn about specific user requests which had not been possible before. Honda did not consider the possibility of communication only for upmarket cars. It is considered a very important aspect that workers are informed directly about the evaluation of the car they have made.

Communication with the Product Development Division and with Suppliers

In order to implement in-process functional quality assurance, a plant must communicate frequently with the development division. If problems of quality or product design arise, development people come to the shop floor and try to solve the problem in cooperation with workers.

Furthermore, workers at the Takanezawa plant can talk directly with the suppliers producing the parts to be assembled at the plant. If there is a problem regarding availability, quality, delivery or parts design, workers can call the person responsible for the production of the part or, if necessary, those who are in

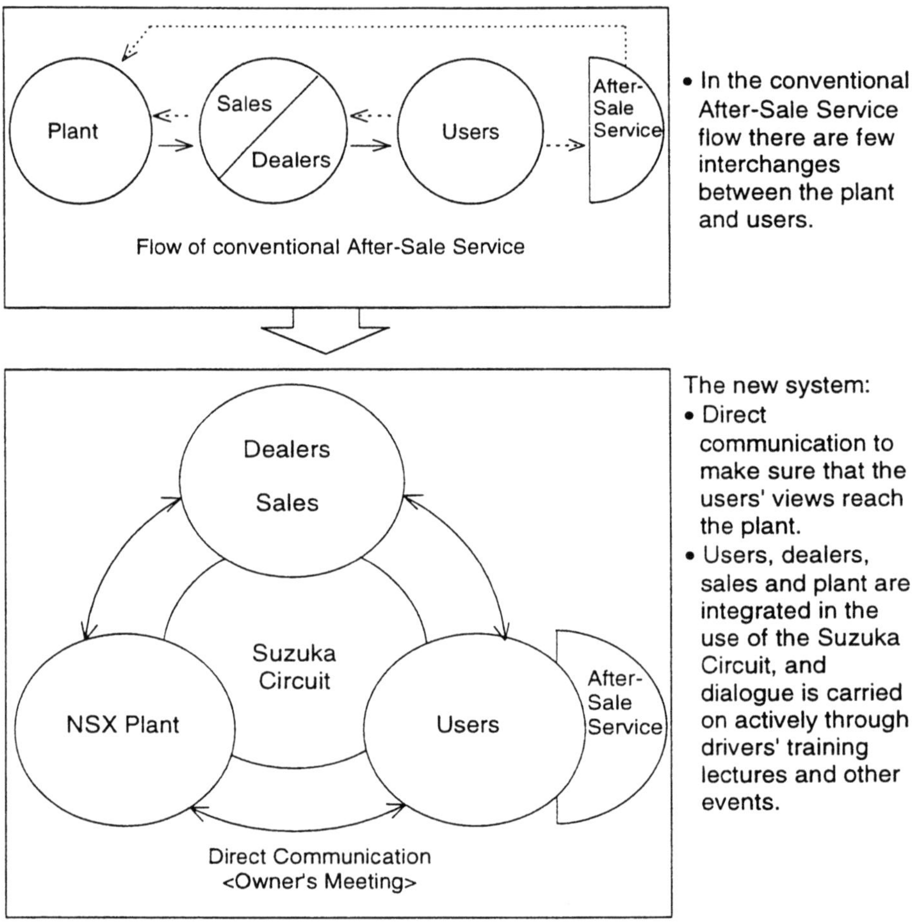

Fig. 3.4.7. Communication with users [3]

charge of the parts design. For this purpose, the name of the supplier, the division and the person in charge at the part supplier, a photo and a telephone number are put on a list which is available at each work station [2].

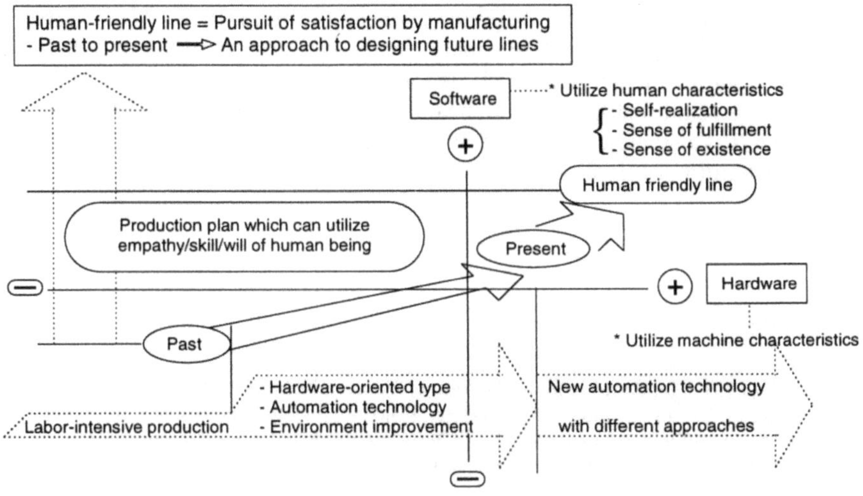

Fig. 3.4.8. Designing human-friendly lines [3]

3.4.4 Transfer

For Honda, a *human friendly* assembly line means a line which gives workers the feeling of *happiness of manufacturing*. In this regard, the conventional approach, where the availability of machine or hardware are given priority over automation, was inadequate in order to fully utilize workers' potential. Takanezawa was the first plant which attempts to use the best of workers' abilities. Although there were specific conditions, i.e. high product price and small production lots, the approach and the knowledge gathered on the basis of this experiment will be transferred to the next generation of production lines for car manufacturing.

3.4.5 Conclusion

Figure 3.4.8. shows a schematic summary of Honda's approach. Until now, the company had given priority to hardware and its utilization. In contrast to this, the experiment described here emphasized that work should be carried out by people, and instruments and tools should be developed in order to make the people's work easier. Body positioners and hangers were examples of this concept. Automation was introduced only for those operations where problems with heavy loads or quality occur when performed manually, such as floor welding and painting. Automobile manufacturing is not only labor-intensive, but also requires complex equipment. The relationship between machinery and workers is still an unresolved problem. The Takanezawa plant is an attempt to adapt different concepts found at other Japanese automobile factories in order to tackle this problem.

3.4.6 References

1 Komiya K (1991) Honda no Kunou (*Distress of Honda*). Kobunsya, Tokyo
2 Shimokawa K (1993) Honda NSX Tochigi Kojo no Insyou (*Impression of Honda NSX Tochigi Plant*). Nikkan Jidousya Shinbun (*Daily Automobile News*), 20 Dec 1993
3 Tanase K (1994) Basis of NSX Line. Unpublished Paper

3.5 The Development of an Intelligent Body Assembly System

T. Naitoh · K. Yamamoto · Y. Kodama · S. Honda

3.5.1 Introduction

In recent years, the environment surrounding the automobile industry has changed drastically. Customer tastes and market needs have diversified. Globalization, such as complementary production arrangements among domestic and overseas plants, as well as parts procurement from vendors around the world, has advanced greatly. Labor shortages due to the trend by some workers to shun primary and secondary industries as well as an aging workforce have become a serious problem. In this harsh and rapidly changing environment, keen competition has existed in the automobile industry, and for this reason, especially in Japan, over 20 new models are introduced each year. Normally, a new model change requires 2 and a half years of production preparation lead time, more than several tens of billions of yen and a cumulative total of several thousands of en-gineering man-hours. A new model change is truly a crucial event for an automobile company [1].

A key point for automobile makers is hence to supply customers with a variety of new products which meet the market needs in time. A new cooperative approach among product development and production engineering divisions, called Concurrent Engineering, which can reduce the lead time required for a new model change, a Flexible Manufacturing System that permits the production of a variety of product types on a single production line and other new approaches are highly desired in order to achieve the goal.

In this situation, Nissan has developed an Intelligent Body Assembly System (IBAS) that makes a new model change or a new model addition possible at a minimum of time and cost.

3.5.2 The Current State of Flexibility

Automobile assembly consists of four main processes: stamping, body assembly, painting and final assembly. The body assembly process has the lowest level of flexibility. The process is illustrated in figure 3.5.1.

In the body assembly process, a vehicle body skeleton is assembled with 300 to 350 stamped parts while passing through sub-assembly lines and then a main assembly line and undergoing approximately 3,500 to 4,000 spot welds. Ba-

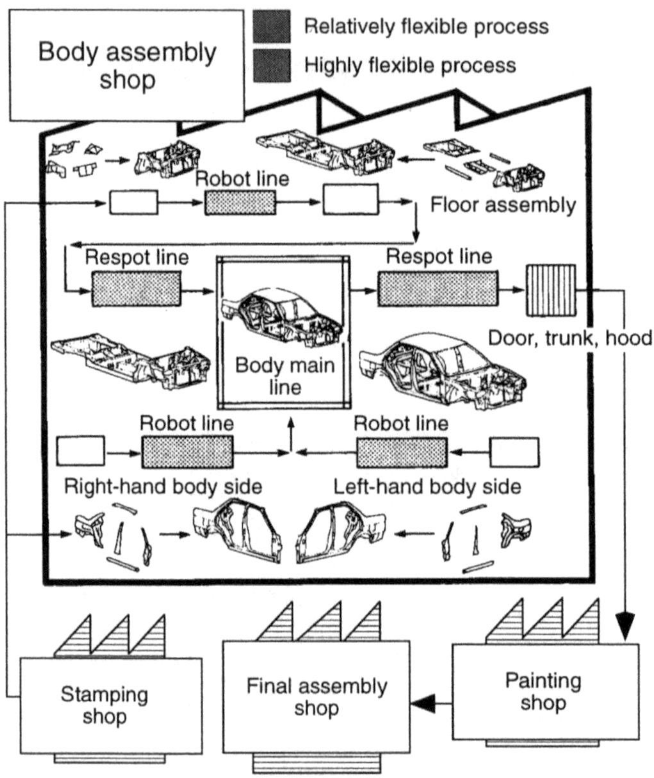

Fig. 3.5.1. Body assembly process

sically, each main or sub-assembly line is made up of 2 parts: the first one is the positioning and preliminary welding process, commonly known as framing, and the second one is an additional welding process, called re-spot welding.

In the process of re-spot welding, the flexibility of the production system has already been achieved by the use of welding robots and tool changing devices [2]. On the other hand, panel parts have to be positioned with high precision during the process of positioning and preliminary welding.

Therefore, positioning jigs, which are mechanically rigid and custom-designed, are normally used for each type of vehicle. The jigs or fixtures used in the main body assembly line are the largest and most complex ones. In the case of a model change, 9 to 12 months of tool design and manufacturing, plus large sums of capital investment are usually required in order to prepare a body main line. Figure 3.5.2. shows an example of a conventional body assembly jig for 2 types of vehicle bodies. It has a revolving device with 2 surfaces, each containing a dedicated fixture for one type of vehicle body. Turning through 180 degrees, this device adapts itself from one vehicle type to another. The limited flexibility of such lines is clear because they are based on a rigid mechanical structure. We identified the body main assembly process as a bottleneck in terms of flexibility,

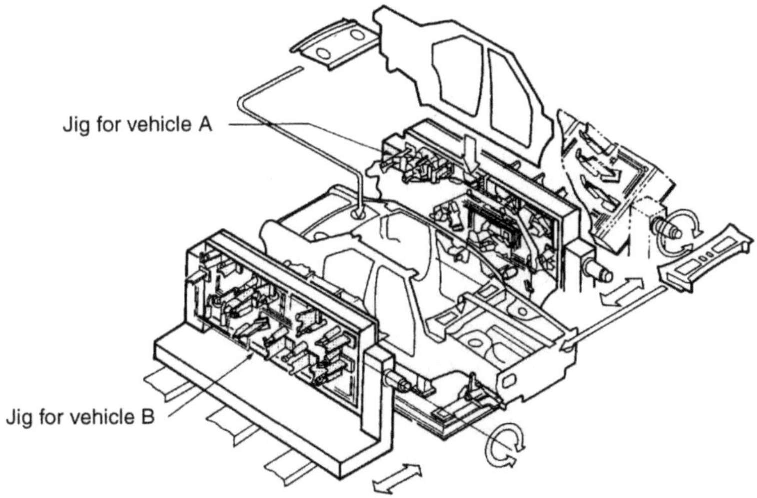

Fig. 3.5.2. Conventional body main assembly jig

and selected it as the first target in the development of a new production system, i.e. IBAS.

3.5.3 IBAS Concepts

3.5.3.1 System Requirements

In the development of IBAS, we pursued the following goals.

A production system that can handle a variety of products.
It should be possible to produce any of Nissan's vehicles by just changing the robot manufacturing programs associated with the product type. The completely multi-mixed product flow, i.e. every single product on the line should be of a different type, should be attained to maintain the constant work load in the succeeding processes, such as painting and final assembly. The basic product output is around 20,000 cars per month, and the cycle time is 45 seconds.

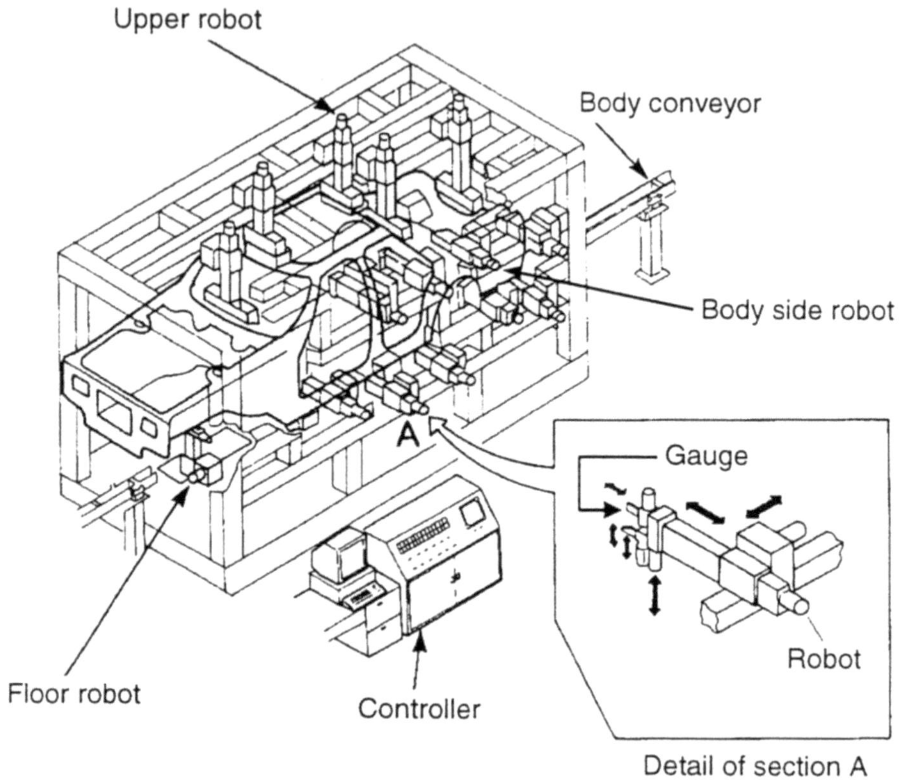

Fig. 3.5.3. Outline of the NC locator

A production system that enables high product quality with high system availability at the same time.
It should have an autonomous product quality function in order to constantly produce high quality and high accuracy vehicle bodies that meet customers' demanding needs. An engineer should also be present in order to support operators, helping them to eliminate the problem within the shortest time possible in the case of a failure, hence keeping system operation ratio and availability high.

A production system that is economically efficient.
The initial investment should be as low as that of a conventional system, and the cost for a new model change or a new model addition should be no more than 20% of that of a conventional one. The system should be as compact as a conventional one, otherwise, it cannot be installed in all of Nissan's plants immediately.

A production system that can reduce the production lead time.
It should fit in the Concurrent Engineering based on CAD data, so that production lead time be reduced significantly.

The system that meets these requirements is truly a "data-based" and "programmable" system with a minimum number of dedicated and inflexible elements.

3.5.3.2
Basic System Design

At the positioning and preliminary welding cell on a main body assembly line, 7 main parts are put together to form a main body skeleton. These parts are first positioned accurately and fixed by gauges and clamping devices at the locations. Normally, approximately 60 sets of gauges and clamping devices are needed in order to position every type of vehicle model, so that it is impossible to mount every fixture for all types of models to be produced on a single fixture base. We came to the conclusion that every gauge and clamping device should be programmable, i.e. robotized, so that it can fit to any vehicle model. We started working on the basic design of this "programmable" fixture and at the same time, listed the basic technical elements necessary to implement the idea. Our further study led to the concept of a programmable fixture (called the NC locator) which consists of 35 positioning robots and 16 welding robots, thus constituting a robotized fixture with a set of 252 servo axes. The NC locator is designed to be able to produce any of Nissan's vehicle models. The basic design of the NC locator is shown in figure 3.5.3.

3.5.4
Configuration of the IBAS Body Main Line

The outline of the IBAS body main line is illustrated in figure 3.5.4. The line has an overall length of approximately 65 meters and a maximum width of 10 meters. The cycle time of the line is 45 seconds and the maximum monthly production capacity is over 20 thousand units.

The flow begins with the loading of parts by the automatic parts setting device at the first cell. A total of 7 major parts, such as floor, body side panels and roof are provisionally located without being welded. These components are then transported by a conveyor to the NC locator cell, where they are accurately positioned and welded at approximately 50 spots. This is followed by an in-line measurement cell where major parts of the vehicle body are measured by robots carrying a laser measurement system. The body is then conveyed to the following cells where additional re-spot welding is performed by welding robots in order to ensure the required strength of a vehicle body. Other components of the line, such as a transfer system, automatic parts setting devices and positioning jigs in

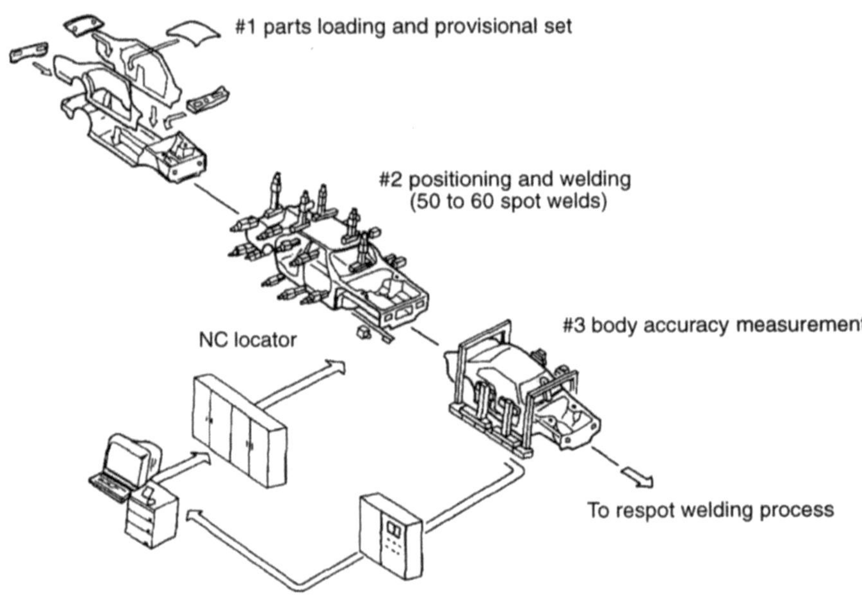

Fig. 3.5.4. Basic configuration of the IBAS body main line

the cells other than NC locator cell are also designed as programmable units with servo axes and other actuators. Figure 3.5.5. shows the front view of the NC locator.

3.5.5 Fundamental Technologies of IBAS

1) *Compact and High-Stiffness Robot*
In the NC locator cell, a total of 51 robots had to be located in a single area as small as that occupied by a conventional body main positioning and welding fixture, and each of them had to move without colliding. Therefore, each robot, as a basic module of the NC locator, had to be compact. On the other hand, we discovered in the course of experiments that a maximum repulsive force of 1.96×10^3 N could be acted at the end-effectors, i.e. gauges and clamp units, mounted on the positioning robots by the panel parts of a vehicle body during positioning. For this reason, each positioning robot was designed according to the Cartesian type, so that it could resist an external force of at least 2.94×10^3 N acting at the end-effector. Sensors, transmissions, gear reduction systems and other components were selected in order to ensure a repeatability factor of 0.03 mm. Figure 3.5.6 shows an example of a positioning robot.

Fig. 3.5.5. Front view of the NC locator

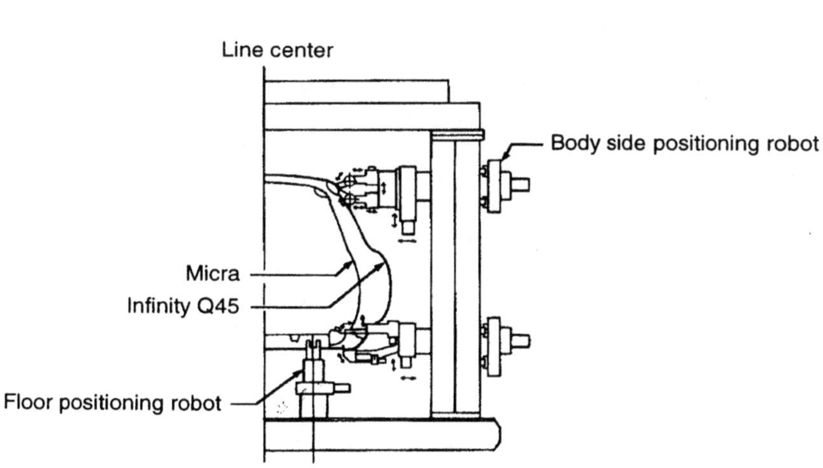

Fig. 3.5.6. Basic module of the positioning robot

2) *NC clamping Device*
We aimed at eliminating gauges and clamping units which were exclusively designed for each type of vehicle body. We found that there were a lot of vehicle body locations where section lines of each vehicle type were similar among different vehicle types. We developed and applied a "NC clamping" device to such positioning locations. This device has a maximum of 7 degrees of freedom, and is capable of adjusting itself to the section lines of a vehicle which have to be fixed.
In the case of other positioning locations where section lines are very different among different types of vehicle bodies, we applied a newly developed clamping device on which 8 sets of thin gauges and clamping units, each of which is specially designed for one vehicle type, are mounted.
3) *High Absolute Accuracy Assurance of Robots*
Normally, positioning jigs for a vehicle body are designed to feature an accuracy of more than ± 0.1 mm. The NC locator had to ensure this accuracy in order to be able to serve as a positioning jig. It was an enormous challenge for us to eliminate robot-specific, internal and external errors and to keep the (relative and specific) positioning error of each robot within the tolerance limits. In the manufacturing process of the NC locator, the accuracy of each component of a robot was measured and modified before and after assembly. Once all the robots were mounted on the main frame structure, they were measured using a 3D camera module, and robot internal errors, such as zero points of internal robot sensors, link length and joint axis directions, as well as external errors, such as robot location, were calibrated using our own calibration system. This approach ensures the desired accuracy of the NC locator.
4) *Control System for a Multiple Robot System*
The main features of the control system of the NC locator are its matrix control configuration and comprehensive modularization. While its structure is designed on hierarchical principles, it performs distributed processing for all of the different functions. The control system of an NC locator has to manipulate 51 robots with a total number of 252 axes and manage over 1000 input and output signals. It is made up of a station sequencer, sub-controllers, unit controllers and servo controllers.
1) The station sequencer sends commands to the sub- controllers at its downstream end, according to a given operating mode of the entire line. An operation unit is connected to it and displays the status of every robot, so that the operator can treat a set of 51 robots as a single robot.
2) The sub-controller is responsible for the management of a group of robots. It manages the unit controllers at its downstream end and sends them start signals, corresponding to the operating mode determined by its higher-level station sequencer. It also processes signals from a teaching unit connected to it. A further function is the management of communication with an electronic work station (EWS). In the case of a system failure, detailed information is made available.
3) The unit controller works like a conventional robot controller without a teaching unit or operation unit. It stores teaching data, gives instructions to

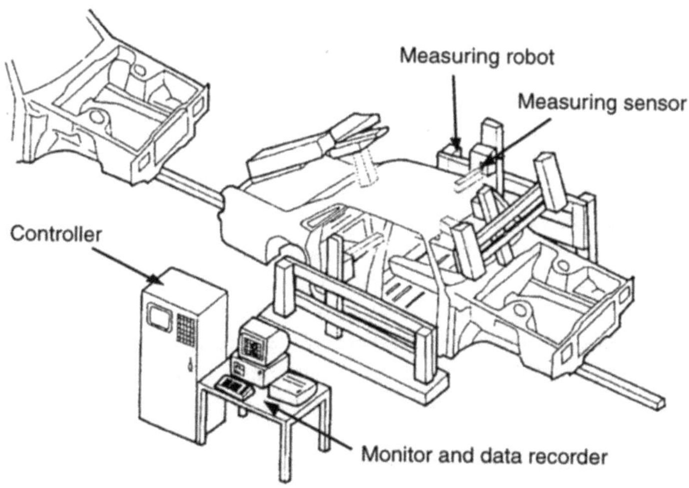

Fig. 3.5.7. Body accuracy measurement system

servo units according to the teach data when a start signal is sent by the sub-controller, and it manipulates input and output signals.

4) The servo controller takes controls of each axis of a robot, according to the instructions given by its higher-level unit controller. Unlike a conventional servo amplifier, it is connected to a set of servo motors by switching devices and drives them one after another, thereby significantly reducing the capital investment and space requirements of the system.

The interface connecting each control unit of the system is based mainly on a serial link in order to reduce cable and wiring requirements. All types of data - from operation commands and robot teaching data to information collected for the operation monitoring and failure diagnosis system which will be mentioned in the following section - are transmitted via the serial links.

5) *In-line Measurement System*

The system consists of measuring sensors, robots, controllers and a data processing unit. The basic configuration of the system is shown in figure 3.5.7.

The points on the body, such as standard holes for body assembly, holes for installing various trim parts and the locations at window, door, hood and trunk openings, are measured in order to check the body accuracy. Generally, the measurement of holes can be carried out by 2D sensors. However, cross-section measurement of a location at openings involves that of curved surfaces and determination of virtual points. For this reason, flying-spot sensors developed by Nissan itself were employed.

The flying-spot sensor is based on the triangular measurement technique. First, a laser beam is directed at the object to be measured, then the reflected beam is received by an optical camera and the position of the object is calculated. A rotatable mirror is also positioned between light source and light receptor. By rotating the mirror in order to move the laser beam like a light-

house beam across the surface of an object, we can obtain its cross-sectional shapes. The measurement accuracy of the sensor totals ± 0.1 mm.
The measuring data is sent to the EWS and processed there. The data is compared to the CAD data of the carbody, and if the difference exceeds the tolerance band (which happens mainly during trial production) the EWS returns an alarm to the NC locator cell in order to halt production. After production has been automatically stopped, appropriate action must be taken in order to keep the accuracy within the tolerance band. The accuracy of the main body, however, can be affected by numerous factors, such as changes in the mater-ials used to manufacture the parts, changes in the condition of stamping dies and other tools due to problems or aging, as well as changes in environmental parameters. It is hence not immediately possible to identify the underlying cause and decide upon the most appropriate action on the basis of the data measured. Further examinations are hence necessary in order to implement a fully intelligent body accuracy feedback system.

6) *Operation Monitoring and Failure Diagnosis System*
The NC locator is a combination of complex mechanisms and control devices, so that we introduced an operation monitoring and fault diagnosis system. The control system of the NC locator constantly monitors the operating status of the cell, such as the current position of each robot servo motor. If a breakdown or error is detected in the NC locator, the control system notifies the connected EWS which serves as its management computer. The EWS displays details of the abnormal situation, identifies the probable cause of the fault in an interactive process and suggests the optimal corrective measure to the operator. If the cause of the failure is believed to be in the control system, an internal self-diagnosis program is activated in the EWS in order to identify and pinpoint the underlying cause of the trouble using the knowledge base. If the cause is suspected to be in the mechanical system, directions for cause investigation procedures and judgment criteria are displayed on the monitor, and the cause of the problem is identified by interaction between system and operator. It also maintains fault history records and gives information in order to schedule periodical maintenance activities.

7) *Off-line Programming System*
An off-line programming system is indispensable in order to schedule and optimize the movement of each robot and to prepare the robot programs to be downloaded to the actual cell. Our own CAD system for vehicle body assembly tooling is fully used to plan the optimum gauge layout, to identify the section line and spot welding locations and to prepare 3D data of a robot, a tool and a vehicle body for graphic simulation. The path of each positioning robot is specific for each location, and it is determined in a straight-forward manner by simply giving a location to be positioned. The optimized path of each welding robot is determined by graphic simulation of the entire cell. Every collision is strictly checked and the cycle time is accurately estimated. We receive the robot programs for the entire cell as a result of the simulation pro-cess. We found that this kind of "data-oriented CAD" approach was truly effective in significantly reducing the lead time for a model change, because it permitted simultaneous product development and production engineering. An example of a graphic simulation of an NC locator is shown in figure 3.5.8.

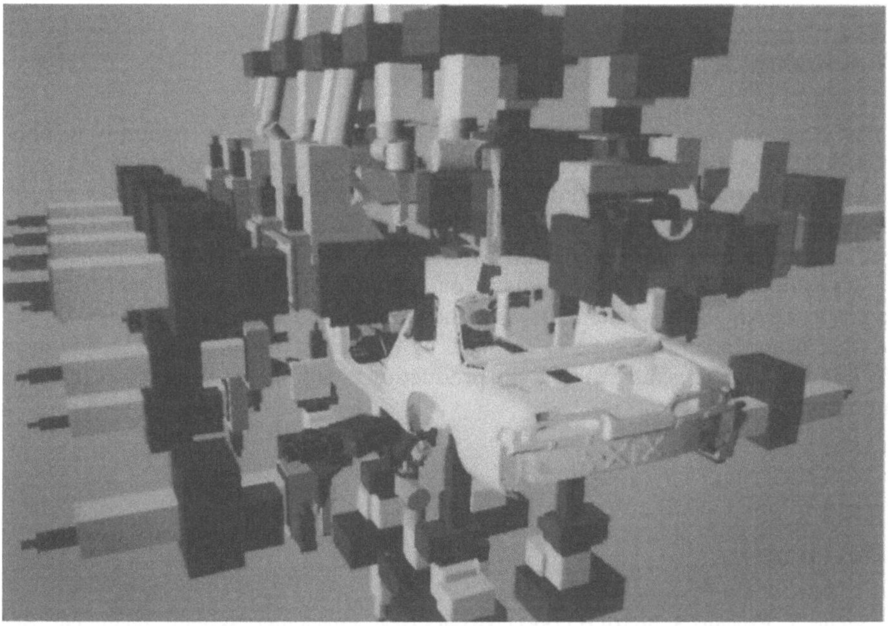

Fig. 3.5.8. Graphic simulation of the NC locator

3.5.6
Benefits of IBAS

Currently, Nissan has 13 IBAS body assembly lines operating in its domestic and overseas plants. We consider the benefits of IBAS to be as follows:

1) It permits up to 8 types of vehicles to be produced on a single line, whereas a conventional line can only produce 2 types of vehicles.
2) It can reduce capital investment for a model change or for the introduction of a new model by 80%, compared to a conventional line.
3) Its Cp ratio (Process Capability Index) is 20 to 30% higher than that of a conventional system. This high quality can make an important contribution towards customer satisfaction.
4) It reduces the lead time for the tooling of a body main line to a minimum of 3 months, whereas a conventional system needs approximately 1 year before launching. This allows us to make a production plan that will accurately meet the needs of the market.

3.5.7 Conclusion

A new concept for a programmable data-based fixture for body assembly has been proposed. This concept has been implemented as an Intelligent Body Assembly System. Some of the technical details of the development of IBAS are described in this paper. First introduced at Nissan's Zama plant in 1989, the use of IBAS has been extended from the body main assembly to other sub-assembly processes at the Kyushu and Murayama Plants. Considerable benefits have been achieved thanks to IBAS. This system fits into the trends of Concurrent Engineering and Flexible Manufacturing Systems, as well as Computer Integrated Manufacturing. We believe that it is also useful in production systems for products other than automobiles.

3.5.8 References

1 Sekine Y, Koyama S, Imazu H (1991) Nissan's New Production System: Intelligent Body Assembly System. Proc of 1991 SAE International Congress and Exposition, Detroit, February 25 - March 1, 1991

2 Yamauchi Y (1988) Application and Evaluation of Robots in Nissan. Proc. of The International Symposium and Exposition on Robots, Sydney, November 6-10: 189-209

3.6
The Opel Production System

P. Enderle

3.6.1
Reasons for a new Production System

Excellent quality and high technical standards are among the most important features of German automobiles. Having these attributes alone, however, is not sufficient for auto producers to remain competitive in international auto markets in the long run. Increasing competitive pressures – particularly in the Single European Market – now call for new concepts which contribute to cost reductions and increases in productivity. Production must remain the center of attention in this process because it is in this sector that the company generates most of its added value.

A critical review of cost structures and key productivity data shows that change is urgently required in order to remain competitive. After all, according to the 1990 summer study, conducted by the Massachusetts Institute of Technology (MIT), Japanese automakers require only 17 hours to assemble a vehicle, compared with an average of 36 hours in European plants.

Within our European Manufacturing Organization, Adam Opel has additional cost effects which highlight the need for change. A comparison of labor costs in Opel's Rüsselsheim plant and the production plants of Vauxhall in Luton and Ellesmere Port illustrates the problem: the current average hourly labor cost in Rüsselsheim is approximately DM 50, whereas in Great Britain, it is 30-40 percent lower.

Unlike most of the other European automobile companies, Opel is not just a national manufacturer but forms part of the total European production network of General Motors Europe. This network consists of 15 automobile and component plants in nine different countries. Opel plants, therefore, not only compete with other manufacturers in Europe, but also with foreign production plants in GM's European network.

3.6.2 Characterization

The key to future economic success is lean production. The introduction of more streamlined procedures is putting an end to the era of extremely labor-intensive production, which began in Europe 70 years ago with the introduction of assembly line production methods.

Lean production: this calls for a combination of the specific features of manual skill and the benefits of mass production. Manual skill ensures a high level of flexibility and high employee capabilities, whereas mass production on the assembly line ensures benefits such as rapid throughput times and low unit costs.

Opel is one of the first European automobile companies to practice a gradual transition from traditional mass production to a completely new production system.

People rather than machines are in the foreground of the second industrial revolution since the introduction of the assembly line. The future success of the company will be determined by the commitment and personal initiative of its employees, and effective collaboration between every corporate division – from purchasing to sales and from designers to assembly workers.

This particular aspect is a foundation of the Opel Production System: our employees are integrated into decision-making processes by means of team work, continuous improvement, and an innovative quality concept. Each individual can make a personal contribution to optimizing the efficiency of production procedures, eliminating nonvalue-added activities and achieving further improvements in the quality of our products. This taps the huge human potential for implementing improvements which would not be possible in any other way. After all, who is more familiar with the details of his work than the person who regularly performs the work?

The Opel Production System is not just limited to production itself. On the contrary, it embraces the whole organization. We need each employee to take the lead in developing ideas and actions for improvements. Everyone shares responsibility – and each improvement suggestion is valuable, as it contributes to identifiable improvements in productivity, quality and cost.

Five major features characterize this new Opel concept:

1. The Team Concept: Team Spirit and Communication;
2. The Continuous Improvement Process: Ideas and Innovations;
3. Standardization and One-Piece Flow;
4. Just-In-Time Production: Inventories are significantly reduced. New systems ensure materials are delivered only as required;
5. Quality Concept: Customer Satisfaction and Zero-Defect Principle.

There is no alternative to the new Opel Production System. The use of lean and more efficient production systems can be a major contribution toward increasing our competitiveness in the Federal Republic of Germany.

3.6.3 Eisenach Plant: A Model for the Opel Production System

The Opel Eisenach plant, which was opened in September 1992, was designed to operate according to these principles. After years of study and various trials, Opel is implementing lean production methods in Thuringia, in all production areas. Opel is following its own course with the implementation of lean production instead of copying Japanese production methods, taking into account experiences gained from Model Shops in Rüsselsheim, Bochum and Kaiserslautern plants and the expertise of its Technical Development Center. At the same time, Opel utilized and optimized the knowledge gained from General Motors joint ventures with Japanese manufacturers. CAMI and NUMMI, two joint venture companies in Canada and the USA, are among the pioneers of lean production methods.

3.6.3.1 Team Concept

Employees are the foundation of the Opel Production System. The company aims at utilizing the experience and technical knowledge of its employees far more than in the past, and at the same time, increasing employee motivation and work satisfaction.

By establishing a working team concept, Opel gives its employees the opportunity to structure and improve their working environment themselves.

Opel was one of the first automobile companies in Germany to introduce a modern and future-oriented form of work organization in order to achieve this: team work. It is a first critical prerequisite for implementing lean production. Team work enables each employee to assist in structuring the production process and to contribute his or her own ideas. More personal responsibility, more improvement opportunities, better working conditions and greater job satisfaction – these are the benefits of the team concept.

The Opel team-work era began as early as 1989 with a pilot project in the Bochum 2 plant. The trial phase was a complete success and production areas increased productivity by more than 18 percent. Therefore, a clear goal was set to introduce team work into all Opel plants by the end of 1994.

Right from the start, team work shapes the daily working routine of employees at Eisenach. Employees in each shop form teams with six to eight members and assume quality assurance, maintenance and material supply functions in addition to pure assembly work. This extended scope of responsibility represents an enrichment of the working conditions for employees. Now they not only have the opportunity to take on more responsible work, but they can also utilize technical skills in a number of new areas.

This helps to reduce the impersonal nature of the work: each employee can easily see that his personal contribution is necessary for the success of the team and the company.

Each team is headed by a team leader who is elected by the team members. He is responsible for representing the interests of the group in other areas, as well as for ensuring that the standardized operations are followed and reviewed constantly in order to identify further improvements. Finally, the team leader is also responsible for holding meetings once a week. This responsibility includes determining the items to be discussed and worked on by the team and also providing all required advice and information to the team.

Summarizing meeting results is important because communication is a prerequisite for successful team work. Team discussions encourage immediate problem solving and thus enhance the satisfaction and self-confidence of the team members.

Management also has a new role in the Opel Production System. Everyone is involved in the new principles of open communication, and leaders assume the function of advisors for the teams. Therefore, managers are frequently approached on any matters concerning the normal work day. There is also an ongoing dialogue between the Plant Management and the Works Council at Eisenach. Employee representatives participate in all the important Management Board Meetings and are, therefore, involved with decisions from the very beginning.

An outward sign of the company-wide team spirit is the Opel Eisenach uniform. By wearing the uniform, Eisenach employees demonstrate that they are part of a large team and that they are all pursuing the same common objectives.

3.6.3.2
The Continuous Improvement Process

The Continuous Improvement Process (CIP) is an integral part of the team concept and has already been in place at Opel for some time. It involves employee suggestions and ideas which can easily be put into practice and which, in sum, contribute to a significant increase in productivity.

Because of their intimate knowledge of their immediate work environment, employees themselves know - better than anyone else - of the problems in the production process. Opel utilizes this practical knowledge and, as part of CIP, it motivates employees to examine work procedures and work flow for activities which are nonvalue-adding. A huge potential has thus been tapped which has yielded real increases in productivity. CIP will never end since many changes provide opportunities for further improvements.

CIP measures relate to all areas of the production process: improvements in materials, a more easily assembled product design, optimization of machinery and equipment, critical observations on the working method, reorganizations of the work place and the standardization of working procedures are the potential results from CIP.

To facilitate the immediate implementation of suggestions made through CIP, Opel Eisenach has established special workshops in every area where employees are able to put their ideas into practice. These workshop areas are called "Kaizen" which means a "change for the better."

Continuous improvement can also be recognized externally. By making a total of 11,194 suggestions in 1993, employees of Opel's east German operation ranked second in the annual awards for employee suggestion plans announced by the

Federal Institute of Economics (DIB). As a matter of fact, Opel Eisenach was ranked number one among similary sized companies throughout Germany.

Opel employees are certainly among the most ingenious and innovative in Germany. With a total of 34,715 submitted suggestions in 1993, Opel was ranked top in the category for "companies with more than 20,000 employees".

3.6.3.3 Standardization and One Piece Flow

Work standardization is among the most important elements of the Opel Production System. To ensure a high level of stability and consistency in the production process, the objective of standardized work is always to carry out the same sequence of operations in the same way. This is a key prerequisite for a constant high level of quality.

Opel employees analyze their work and are continually optimizing the operational sequence – on their own initiative. The standardization process is carried out in three stages:

1. Specification of the cycle or "takt time": the cycle or the "takt time" respectively is the maximum duration of the individual operations for the assembly of a vehicle in relation to the working time and the number of automobiles to be produced.
2. Analysis of the work sequence: Opel Eisenach employees prepare Standard Operation Sheets, which enable the sequence of activities to be specified for each work station, and present the results in a diagram. The aim is to have each group analyze its working methods and its work flow for any wasteful activities. In addition, employees evaluate each movement and hand motion in the search for improvements in each individual operation and the optimum utilization of each team member's time.
3. A Continuous Improvement Process begins after the initial analysis of the operations. Team members optimize movement sequences and combine various activities to obtain an improved and more efficient work sequence. At the same time, they prepare new Work Distribution Sheets (fig. 3.6.1.) to show the working, walking, handling and machine times for every operation, always comparing the „takt time" available. The standardization process also includes regular reviews of safety measures at the work stations and a continual reduction in accident risks.

Each team leader displays all standardized work sheets for his team so that all employees can review any new operations at any time. In this way, the standardized operation sheets become the basis of all team training. Team members discuss further improvements and details in regular team meetings. They estimate material quantities required for optimal line presentation. By calculating how much material they require for each shift, teams can dramatically reduce the quantity of materials.

An additional feature of the Opel Production System is its one-piece flow: automobile components are not completed batch by batch, but piece by piece. This method replaces the traditional mass production of automobile components

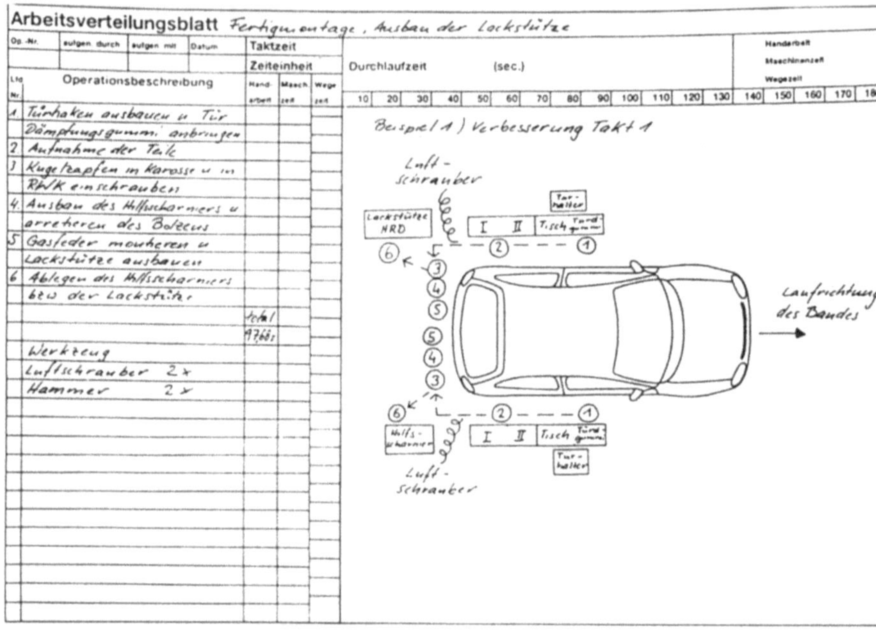

Fig. 3.6.1. Work distribution sheet

in batches which are then stored and transported to production lines. With one-piece flow, experts integrate the pre-assembly of certain vehicle components into the production process and, consequently, only those parts are produced which are required for the next production run.

One-piece flow is also advantageous for a number of other reasons:

- Fewer electrically-driven conveyor belts have to be installed. Working areas can be reorganized as employees always complete just one part in sequence. Opel has established work cells in which the individual assembly stations are not arranged in a line but in a U-shape. Employees' walking times between the work areas are thus reduced to a minimum.
- The activities in the work cells enable employees to carry out various functions. The monotony of traditional conveyor belt work is now a thing of the past. Greater employee satisfaction also has a noticeable effect on the quality of work.
- Production processes are smoother and more flexible thanks to one-piece-flow production and the versatility of employees. Due to the low volume of inventories in circulation, pre-assembly areas are able to react quickly compared to any conventional production processes.

An example of a one-piece-flow concept in the Opel Eisenach plant is the pre-assembly of fuel tanks. Whereas other production plants may produce components for inventory and store them alongside the assembly line for later use, the pre-assembly of the tank is integrated into the production process at Opel. The plant production control system informs employees automatically about the number of fuel tanks required, and these are then completed in line sequence just prior to installation in the vehicle.

3.6.3.4 Just-In-Time-Production

A basic principle of lean production is to produce the right products in the right quantity and at the right time. The Eisenach plant receives delivery of supplier components just in time. This reduces inventory requirements to just a few hours.

One of the basic prerequisites of Just-In-Time-Production is an exact planning of production sequences. At Opel Eisenach, a production plan is developed which specifies model type, colors and options well in advance. Additionally, the plan levels major differences across the schedule period to provide a balanced model mix which meets actual customer demand for the plant.

Thanks to this flexibility, production flow remains constant, even if the volume of individual model types changes at short notice. Production specialists refer to this coordinated type of production planning as "level production" – and it forms the basis for all lean production methods.

Level production not only reduces the risk of certain models being overproduced, but also allows the production process and flow of materials to be effectively controlled according to requirements. Instead of producing and storing specific components in large quantities on a batch basis, suppliers are able to synchronize their production with the Opel Eisenach plant and to produce only the required quantity of components.

Because of the level scheduled production sequence in each area of the Opel Eisenach plant, the use of highly efficient and cost effective logistics systems is possible. With the help of the Eisenach Material Information System, suppliers receive hour by hour information on materials required. Freight companies working for Opel Eisenach have devised special route plans and call at various suppliers in sequence on each route. The routes are planned so precisely that the components reach the Opel plant just in time. This reduces in-plant inventory requirements to just six hours.

Seats and bumpers reach the Eisenach assembly line without any intermediate storage. There is a continuous exchange of data between the production information system in the Opel plant and suppliers. Just two hours before a body rolls into the final assembly area, the seats and bumpers required for the specific vehicle are ordered by the SILS computer (supply in line sequence). These components reach the plant on a special trailer, are immediately off-loaded automatically and transferred to the installation point in the Assembly Shop in line sequence.

Opel is one of the first automobile manufacturers to implement the just-in-time principle for rail transportation as well. Up to 35 trains travel daily between the various European assembly and component plants of General Motors, and transport approximately 95 percent of all freight within the production network – automobile components with a total weight of a million tonnes a year. The Eisenach plant is integrated into this logistics network. Trains from Zaragoza (Spain)

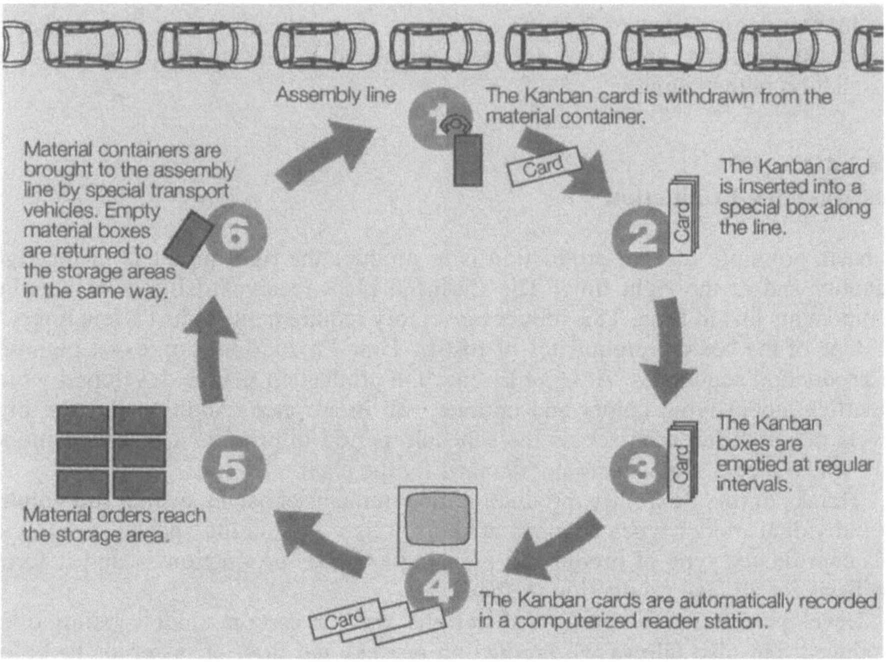

Fig. 3.6.2. The Kanban Principle

reach Thuringia at regular intervals. They also bring the components required for the Corsa production in Thuringia just in time: transport schedules and production sequences are coordinated in such a way that most components are immediately forwarded from the plant rail dock to the production line without any intermediate storage.

Opel has also introduced the cost-saving, just-in-time principle to its material flow systems inside the Eisenach plant. Employees control material supplies themselves with the aid of special cards or light signals – according to specific requirements and right on time.

Production operates on a "pull" basis in every production area. Employees control the material flow themselves and only order those components which they actually require.

This is done on the Kanban principle (fig. 3.6.2.) – Kanban literally means "display card." Special order cards (Kanban cards) are affixed to each material container, showing details such as part number, destination, quantity, storage location and line use location. As soon as an employee withdraws material from the container for the first time, he inserts the Kanban card into a special box near the work station which is emptied regularly. The return of the card to the storage area is a signal for new material to be delivered to the line. Special transport vehicles deliver the requested material to the appropriate work station.

The Kanban principle offers many advantages:

- The material flow is constant, reliable and flexible.
- The quantity of circulating material remains small and manageable.
- Material can be packed in small, manageable containers which can be stacked in shelves alongside the work station in ergonomically favorable positions. The distances employees have to walk are thus also reduced.

For components which have to be transported in special containers because of their size or weight, Opel has introduced an ordering system based on the so-called "Andon principle" (Andon = light signal). Switches are installed in the production areas which enable operators to request additional material at the press of a button. Control lamps are illuminated on a nearby display panel when the operator activates the switch. This is a signal to bring the required parts to the appropriate work station. The Kanban card for this part near the control lamp is brought to the storage, and the material is then delivered to the line via fork-lift, while the card is returned to the display panel for future use.

Space is saved by the call signal system because the only material located line-side is the amount actually required. In addition, containers are placed closer to the work stations and, consequently, the walking times of employees are reduced significantly.

3.6.3.5 Quality

The Opel Eisenach plant has also made considerable progress in quality as a result of team work, standardization and just-in-time production.

The quality concept of Opel is geared to customer satisfaction: each employee is simultaneously the "customer" and "supplier" of another employee, and the objective of top quality can only be achieved if each employee "partner" delivers perfect goods and perfect workmanship. Actual customers – drivers who have purchased Opel – form the final important stage of this customer/supplier relationship. Their satisfaction is the prime objective of the Opel quality concept.

The Basic Maxim is to avoid errors from the very beginning. Employees in every production area practice the zero-defect principle. Opel has introduced this principle in order to minimize the cost of additional quality checks and to reduce unnecessary repairs. It is based on the fact that no deviations will be accepted from the quality standards, and that no defective part or work is passed on to subsequent processes. The same obligation applies to all Opel suppliers.

Typical causes of any defects are eliminated at Opel Eisenach in a number of ways:

- On the one hand, the risk of human error, such as forgetfulness, misunderstanding or lack of knowledge, are reduced by regular education and training courses specifically related to the employee's job. On the other hand, visual aids, such as the display of standardized work sheets, ensure greater understanding in the individual assembly stations.
- Technical defects, which are normally attributable to machine and equipment failures, are mainly avoided by self-controlling systems. Automatic production tools are designed in such a way that any deviations from the quality standard

are immediately identified and production is stopped. Specialists call this "autonomation": machines are equipped with systems which enable them to stop production if there are any deviations from the scheduled procedure. Employees are informed of the problem by means of optical and acoustic signals.

True to the principle of not passing on any unsatisfactory parts to subsequent processes, each employee has the right to stop the line at any time and to ask colleagues for help with solving any quality problems which have arisen. For this purpose, there are two cords in each station in the Body and Final Assembly Shops – so-called "Andon" chains (Andon = light signal). If an employee pulls the yellow chain, an optical and acoustic signal is activated, requesting help from the other members of the team. If the red chain is pulled, the assembly line is stopped until the employees have solved the problem.

Large panels on the production lines – the Andon panels – display information on the production status and make the internal procedures transparent for all employees. In this way, each employee's sense of involvement and responsibility for the process is heightened.

Regular spot checks (audits) of finished vehicles are important elements of the Eisenach quality concept. Specialists from Quality Assurance carry out detailed checks on Astras and Corsas produced on each shift. All errors are identified and communicated to each area in a morning meeting. Three and a half hours later, a representative of the team reports on specific action taken to avoid such errors in the future. Executives and managers also take part in this second quality review and discuss possibilities for further improvements with the employees involved.

3.6.4 First Results

Excellent workmanship and higher productivity are concrete results of the new Production System. In Eisenach, employees require only about 70 percent of the average production time of other European plants. In addition, space, manpower requirements, warehousing and transport capacities are about half of those in traditional plants.

Opel has been practicing lean production in its Rüsselsheim, Bochum and Kaiserslautern plants for some time. Employees in these plants have formed Model Shops which have provided an introduction to the Opel Production System.

Opel employees in the Model Shops in Rüsselsheim, Bochum and Kaiserslautern are learning and testing the new production procedures which are gradually being extended to the whole company. Initial experience has shown a great potential for innovation, when the knowledge and experience of employees are utilized to improve production procedures.

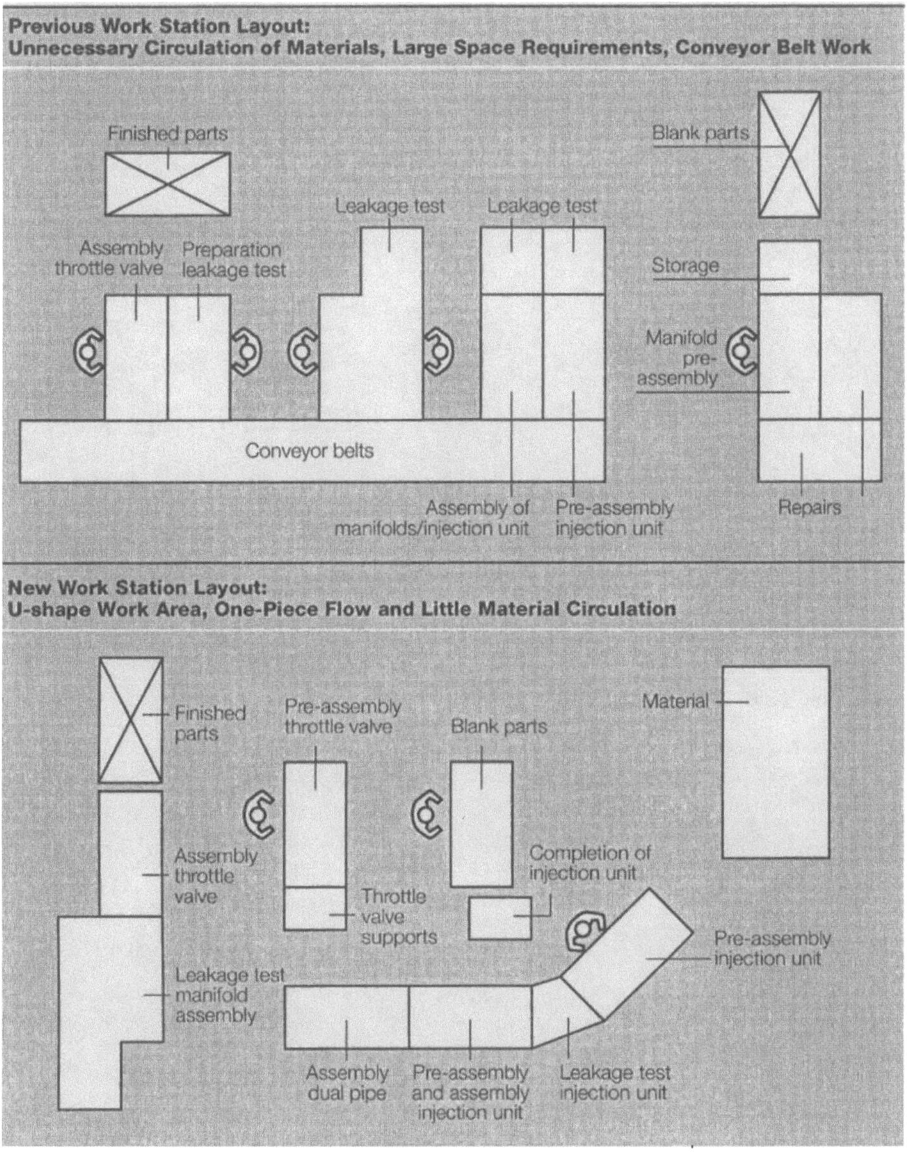

Fig. 3.6.3. Kaiserslautern, pre-assembly of intake mainfolds

In its Rüsselsheim plant, Opel utilizes a material supplies system based on real requirements by means of Kanban cards. Production planners have also introduced one-piece flow in the instrument panel area for Omega models and in the exhaust pre-assembly area on the Vectra/Calibra line. Instead of storing instrument panels and exhaust equipment at lineside, these components are pre-

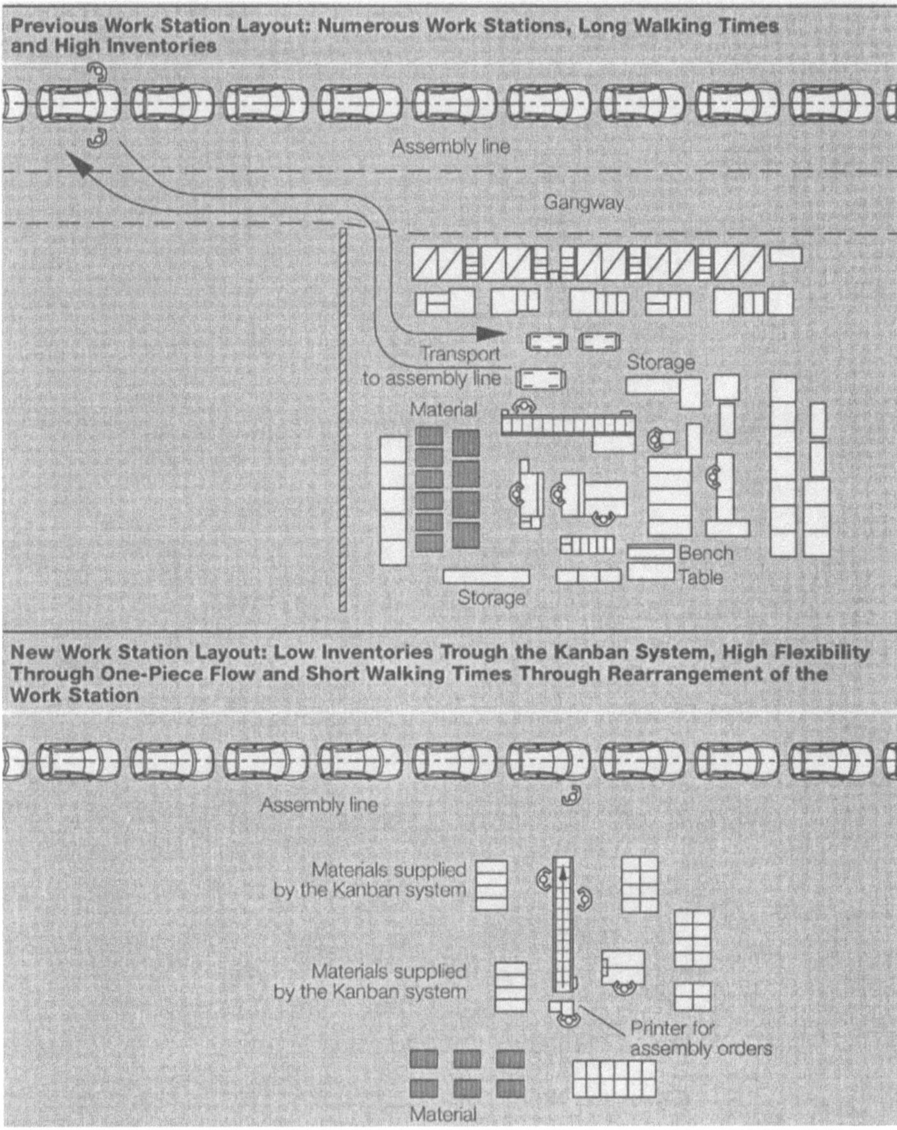

Fig. 3.6.4. Bochum, pre-assembly of radiators

assembled just in time. Space requirements and inventories have been dramatically reduced as a result.

Employees are integrated into this change process from the beginning. In one-week workshops, they make suggestions for improving work procedures, and through the Continuous Improvement Process they continue to submit new proposals for further optimization of procedures. Thanks to CIP, a storage area of more than 200 square meters on assembly line 1 in the Model Shop was found to be unnecessary. Now, production planners can use this space to combine various operations in order to reduce walking times and to achieve a more leveled utilization of the employees.

The success of these lean production methods is already apparent in line one's Model Shop. Thirty-four employees previously assembled chassis and bodies for 310 vehicles in each shift – today a team of the same size produces an additional 15 Vectra models.

With the help of lean production, employees at Kaiserslautern have improved production of the intake manifolds for 16-valve Astra, Vectra and Calibra engines in their first Model Shop (fig. 3.6.3.). On the strength of numerous CIP suggestions, the conveyor belts, which connected the individual work stations for the manifold assembly, have now been dismantled.

Instead, employees set up work cells in which the assembly stations are laid out in a U-shape. Each employee is thus able to sequentially perform his work, to carry the component from the work bench without major walking time and, finally, to send it on for further processing This one-piece-flow production has not only increased flexibility, but the circulation of materials has been reduced by up to 95 percent in the Model Shop. As a result, production costs have fallen by 19 percent and there has been a 22 percent saving in work space.

Lower inventories on the assembly line, shorter walking times for employees and more productive work procedures – these are the positive results which Opel Bochum employees have achieved with the new production system. Fuel tanks, radiators, bumpers and ABS units are pre-assembled in line sequence in the Model Shops of the Astra plant (fig. 3.6.4.). The main objectives of these changes, which were largely based on CIP suggestions by employees, were a reduction in walking times and in inventories. The results achieved to date are very encouraging. By using Kanban principles it has been possible to reduce walking times by 20 percent and lineside material stocks in final assembly by 70 percent.

Employees have now set up three Model Shops in the Bochum 2 plant where engines, transmissions and other components for Opel models are produced. By organizing work areas more efficiently, combining various operations and developing a new materials concept, the 40 employees in the MacPherson strut assembly area have made significant improvements within a short period of time. Space savings of 350 square meters, a 60 percent reduction in materials circulation and an 18 percent reduction in walking time are the result of this. Further improvements are also expected, as additional CIP suggestions are implemented.

3.7 Platform and Modular Concepts at Volkswagen – Their Effects on the Assembly Process

B. Wilhelm

3.7.1 Introduction

Platform and modular concepts as adopted by Volkswagen are future-oriented strategies for enhancing the range from the customer viewpoint, and at the same time they are the basis for the technical and organizational implementation of a marketing philosophy adjusted to customer requirements.

They are essential foundations for increased market orientation and innovation, for the exploitation of new production potentials through new forms of automation and new organization structures, and for further quality improvement. They also help redraw the relationship with suppliers and swiftly put into practical effect the recycling concepts urgently required.

Platform and modular concepts are a part of Volkswagen's core strategy for the maintenance and improvement of its competitive position in the future.

3.7.2 The VW Platform Strategy

The *platform strategy* is Volkswagen's strategic approach and a synonym for a group-wide (i.e. covering the four brands of Volkswagen, Audi, Seat and Skoda) standardization and differentiation strategy for product development, production process and procurement.

Platform strategy is a superordinate structuring principle comprising on the one hand the actual platform with floor group, drive system, running gear and the unseen part of the cockpit. The salient feature of the platform is its application across the brands. (fig. 3.7.1.)

Complementary to this is the visible part of the vehicle - that part which is characteristic of the individual brand, and which is referred to as the "hat". Design of the "hats" is the responsibility of the brands, whereas the platform enjoys group status.

Since a platform defined in this way accounts for some 60% of development and production costs, the concentration on only a few basic platforms results in a large number of synergetic effects, arising primarily from the larger unit numbers per platform type.

On the product side, for example, lower development costs, lower development risks, and a higher rate of innovation are the result. In production too, a

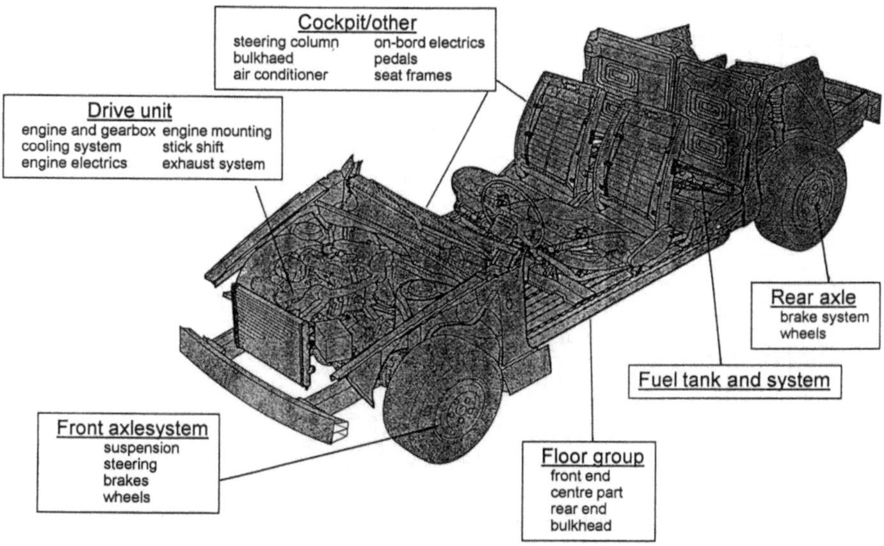

Fig. 3.7.1. Definition of a platform (Main Elements)

large number of improvements are registered. In the vehicle-producing factories, for instance, the introduction of the platform strategy permits a more flexible and cross-brand utilization of facilities. This in turn leads to lower fluctuation on the level of plant utilization, and a more even manner of production - an important factor for high quality and high productivity. Furthermore this kind of flexibility increases the scope for increased inter-factory competition. In the single-part production facilities, this leads to lower unit costs as a result of lower tooling and investment expenditure.

Modularization takes place so to speak on the secondary level, its objective being a stronger differentiation of the product from the customer standpoint.

To define the term: By *module* we mean a complex assembly forming a closed function unit which permits specific differentiation and which, as a consequence of defined interfaces (function, geometry), can be developed, manufactured and assembled independently. Such an assembly must be interchangeable and/or capable of alternative installation, it must represent an efficient unit in terms of production and logistics, and it must require a minimum of modifications.

Modules are broken down by the following structuring principle: First-order modules are assemblies which are installed directly in the body, such as the cockpit. Second-order modules are subassemblies, such as heating systems or air condition systems, which together make up first-order modules. Third-order modules finally are subassemblies which, as a unit, go to make up a module of a higher order. (fig. 3.7.2.)

On the production side therefore the effect is primarily an identifiable improvement in the manufacturing process, and enhanced manageability of the increased product variety which results in terms of logistics, quality and cost.

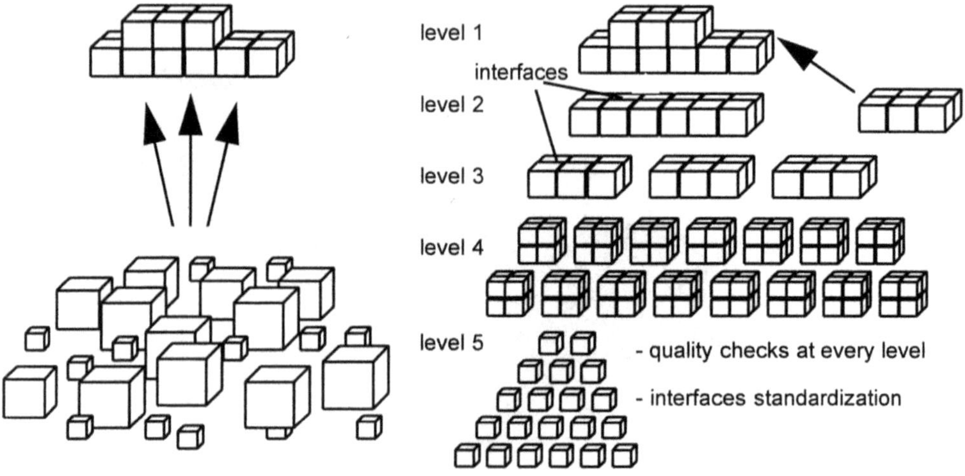

Fig. 3.7.2. Modularization in assembly - structuring principle

The main advantages of a modularization strategy of this kind are:

On the product side
- the possibility of producing variants in line with market requirements, with a limited effect on production structure and assembly process
- only small increases in parts proliferation, and a correspondingly relatively low level of expenditure for product modifications and new variants.

On the production side
- shorter throughput times through the hiving off or parallelism of pre-assembly stages with synchronized delivery, resulting in lower material inventories and warehousing costs
- higher productivity and quality through automation
- modular factory structure.

An additional advantage lies in the possibility of integrating suppliers more closely into the added-value chain: the main point being that the suppliers are also involved as development partners. In this way their role is moving increasingly away from that of a "simple" parts manufacturer and towards that of a high-level assemblies supplier.

Disadvantages are conversely identified as being the additional interfaces created, and the in part distinctly higher harmonization requirement in the definition and development phase, but also during start-up and subsequent series production.

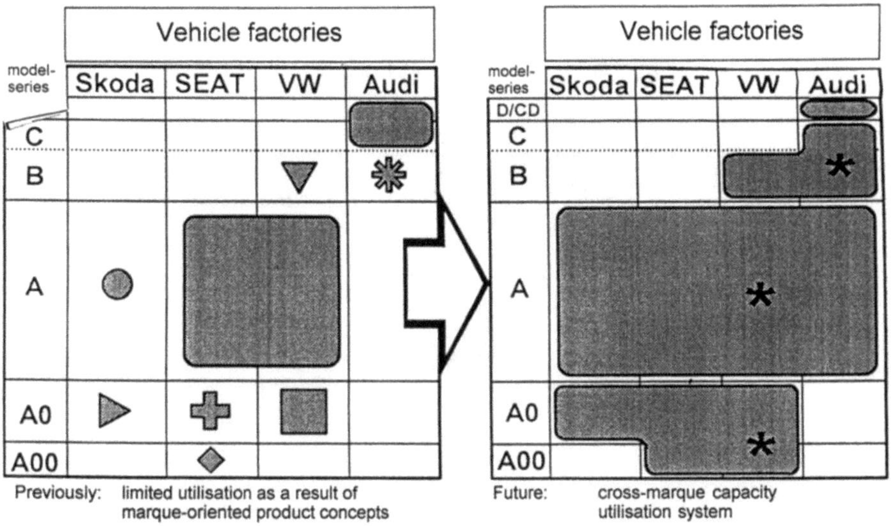

Fig. 3.7.3. Capacity utilization vehicle manufacturing

3.7.3 Practical Implementation

The implementation of this strategy logically has considerable effects on the 20 or so vehicle-producing factories of the VW group world-wide. The problems arising in this context are essentially the allocation of the various vehicle types to the individual assembly plants, that of platform production to the "platform-supplying plants", and the design of the global interlinked manufacturing system, including the integration of the suppliers into this system.

The practical advantages - shown here with the example of the *vehicle - production plant* allocation - speak for themselves: the previous, more brand-oriented product strategy with its 17 or so different platforms permitted only a relatively low level of flexibility in the allocation to plants. The reduction to four platforms by contrast not only permits a marked increase in unit numbers per platform type but also a non-brand-dependent, flexible utilization of the indi-vidual vehicle-producing factories in line with market requirements. (fig. 3.7.3.) Furthermore, considerable cost reductions are made possible by an optimization of material flows within the global, interlinked production system.

Fig. 3.7.4. Modular car structure

3.7.4 Modular Concepts

With the start-up of the Golf II in 1983, Volkswagen achieved a quantum leap in terms of *modularization* and automation. The designation *Hall 54* at the Wolfsburg plant - a synonym of successful automation of vehicle assembly - has become a yardstick for process innovation. A saving of approx. two hours in production time and a marked improvement in quality and working conditions, due to the increase in automation from 5% to 25%, were the outstanding results of this development. An essential element of this was the modularization of the product and its design emphasis on ease of assembly and automation (fig. 3.7.4.), which was also reflected in the modular structure of the factory shop (fig. 3.7.5.). This course was consistently further pursued in the case of the current third-generation Golf, particularly with reference to the front-end and cockpit modules.

These are modules having an especially high degree of integration of functions. Above all the cockpit is a highly-innovative product concept (fig. 3.7.6.). Not only does it combine all the electrical functions, the steering and braking system including ABS, and the air-conditioner, but also - as a final touch - the cockpit module comprizes the entire facility for either RHD or LHD application. In this way Volkswagen succeeds on the one hand in offering customers a large number of functions in line with their equipment requirements. On the other, the effects of these in production can be largely limited to the pre-assembly phase, leaving the assembly lines unaffected.

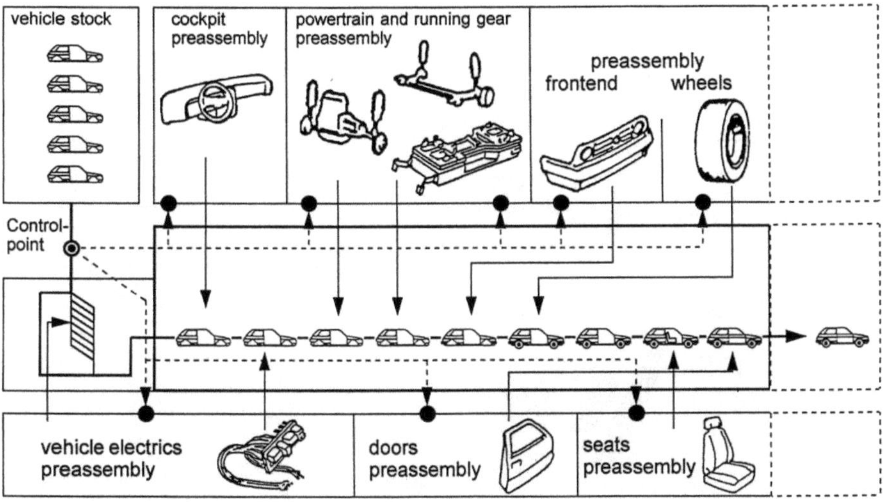

Fig. 3.7.5. Effects of modularization on factory structure (—— parts flow/ ----- information flow)

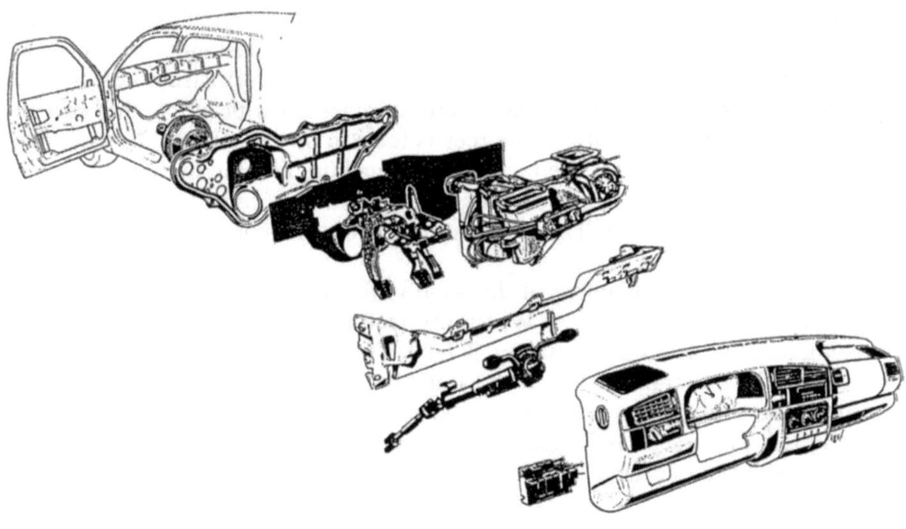

Fig. 3.7.6. Cockpit modul

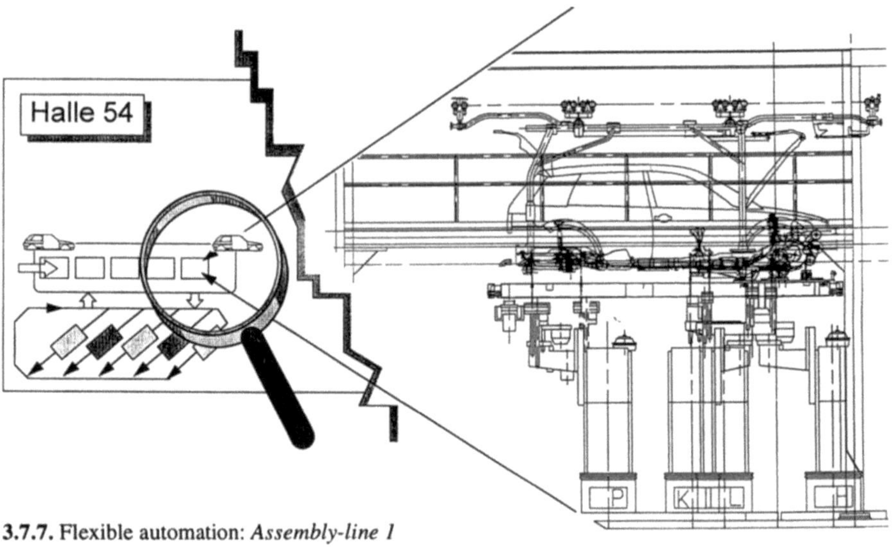

3.7.7. Flexible automation: *Assembly-line 1*

Characteristic for such large modules is automated pre-assembly of individual submodules combined with a residual element of manual assembly. Thus at the same time Volkswagen achieved a further step in the direction of a modular factory.

An interesting aspect emerges in this context: a factory structure of this kind permits, in a relatively simple way, the introduction of, for instance, new production technologies, generally also in conjunction with new work structures. If such innovative steps are implemented on a current large-series product, this has the advantage that the deployment of resources is focused on the introduction of these new technologies. In this way, the increased expenditure and the higher risk inherent in a parallel introduction of new technology and new products in the start-up phase can largely be avoided. The result is a steeper initial curve, and higher quality.

Volkswagen pursued a course of this kind in the introduction of flexible automation on Assembly Line 1 at the Wolfsburg factory, which ideally complements the existing large-scale automation (fig. 3.7.7.). Typical features of this production concept are a flexible transfer line with assembly robots for the installation of fuel lines, fuel tank, brake pipes and the complete running gear and drive unit. Also, the decoupled pre-assembly stage divided up into individual assembly cells. Transport of the running gear and drive unit modules is via a driverless transport system, and the workforce is organized in teams.

The flexibility benefit of this system has proved itself unreservedly, especially for the introduction of the Golf III. In this case, Volkswagen took the opposite course - because of the maturity of this technology, which had been introduced few years ago, it was possible to concentrate fully on the new product. The upshot was a new vehicle start-up, without any losses worth mentioning, at a high quality level. In the meantime, Volkswagen has introduced several of these systems - some with small modifications - in various factories.

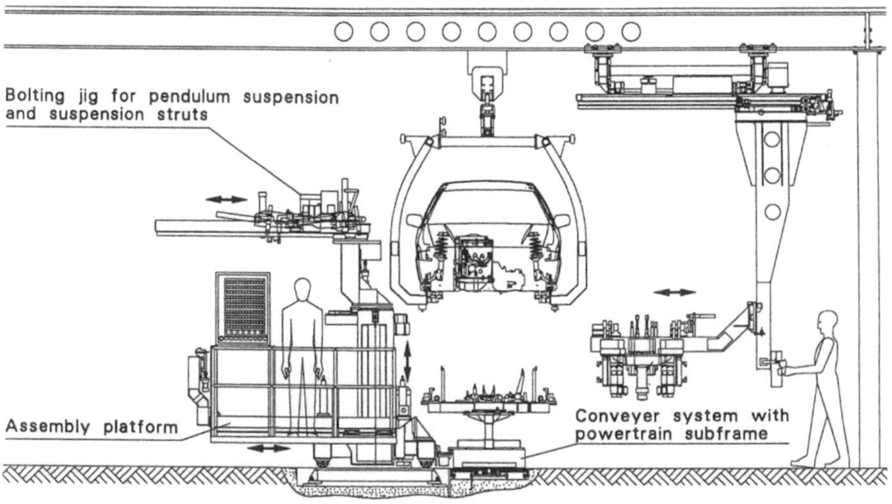

Fig. 3.7.8. Semi automated powertrain assembly

3.7.5 Future Developments

Comparing the development of the automation level in body shell shop, paintshop and assembly, it can be stated that it was not until the start-up of the VW-Golf II that assembly technology achieved a noteworthy increase in assembly and thus a significant improvement in productivity and quality.

If one further compares the productivity and capital investment of German, European and Japanese manufacturers, the following can be stated: European and especially the big German manufacturers have only partially succeeded in catching up with the Japanese, despite massive technological input. The main reasons for this have been the perhaps excessive emphasis on technology, the neglect of opportunities to optimize the organization side, and the framework conditions prevailing at the production sites.

In future, greater attention must be paid to organizational and above all logistical aspects, without however relinquishing the advantages inherent in the deployment of innovative automation systems. This will however require the consistent further development of modularization, and a product design approach which takes account of the demands made by automated production. From the current viewpoint, the following trends can be discerned:

In the automation and mechanization of assembly there will be a further expansion of flexible automation concepts, applying primarily to preassembly of submodules, and also a move in the direction of low-cost concepts. In the latter case the idea is essentially to apply, instead of fully-automated solutions, concepts which have a favourable cost-benefit ratio based on an optimum combination of man and machine. Examples of this are part-automated assembly systems, one such being illustrated in (fig. 3.7.8.). Here the operative controls the complex positioning and joining operations manually, while the technology handles the lifting, for example, some of which is heavy, and the bolting operations. This approach was used for the first time on the new Polo production lines in this form, and is increasingly used in small series production. A further development is the mechanization of ancillary activities such as materials feed, handling and so on, the objective being an investment which as far as possible is independent of the vehicle type.

In terms of manufacturing structures, the trend towards *segmentation* is increasing. Basically, segmentation means the breakdown of production volume into batches by product variants and frequency, and the targeted allocation of suitable technologies and manufacturing structures in accordance with the cri-teria of flexibility and productivity. What this actually means in terms of assembly automation becomes clear from a consideration of the development during the last 15 years (fig. 3.7.9.). The first noteworthy installations of automation technologies in vehicle assembly occurred in the manufacturing of large unit numbers with comparatively few variants.

These took the form of classic, dedicated conveyor systems. The first generation of Hall 54 at the Wolfsburg Volkswagen factory is typical of this. The next development step - the above-described flexible mechanization of Assembly Line 1 - comprizes flexibilization of these assembly systems by means of robots, driverless transport systems and a cellular structure of the preassembly areas for the manufacturing of middle-range series with numerous variants. The latest development stage is represented by the partially-mechanized assembly systems already mentioned (cf fig. 3.7.8.).

The main advantages of this segmentation strategy are above all to be found in the combination of the strengths and reduction of the weaknesses of the individual system solutions. These are primarily the high productivity benefit and the low specific investment costs of dedicated systems with which - this applies essentially to Volkswagen plant in Wolfsburg - some 70 to 80 % of vehicle volumes can still be produced. The disadvantage of the lower flexibility is balanced out by the specifically utilized for production of the multi-variant middle-range seriens, accounting for 20 to 30 % of production volumes, the increased investments remain within acceptable limits. The precondition for this is the consistent implementation of the standardization philosophy described at the outset.

A further main emphasis is the design of the factory structure. The development here is in the direction of networked manual or automated assembly cells, up to so-called "fractals", or partly-autonomous factory sections, i.e., independent organization units incorporating all functions and competencies (fig. 3.7.10.). A further development in supplier relations is so-called "insourcing", which is the incorporation of suppliers into the factory itself. The aspect of assembly organization will undergo a marked upgrading: team concepts, a stronger process organization and cross-specialist training in a balanced linking of people and technology are the salient characteristics of this. Finally a modularized product structure of this kind is also the basis for an effective and practical recycling strategy.

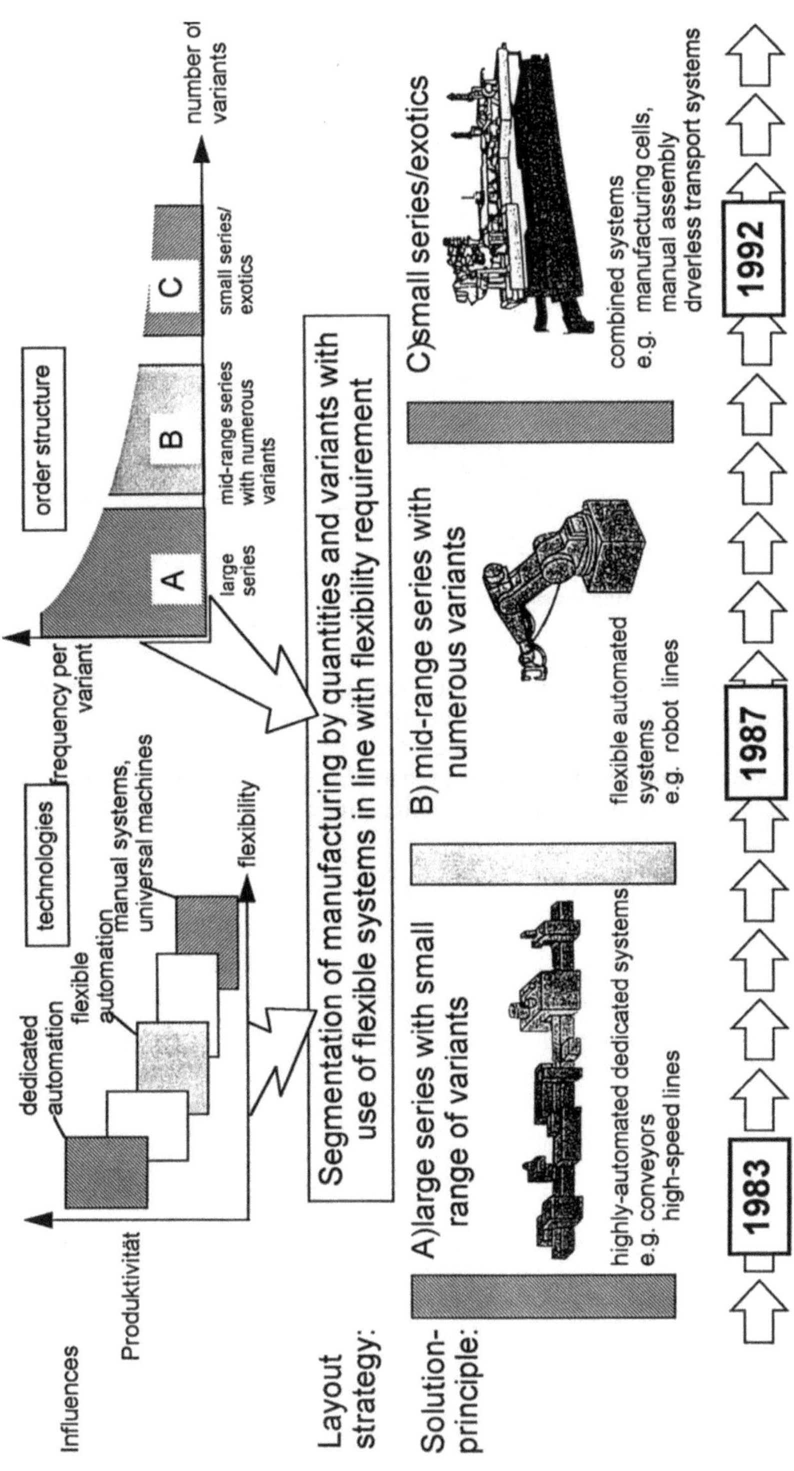

Fig. 3.7.9. Principle of *manufacturing segmentation*

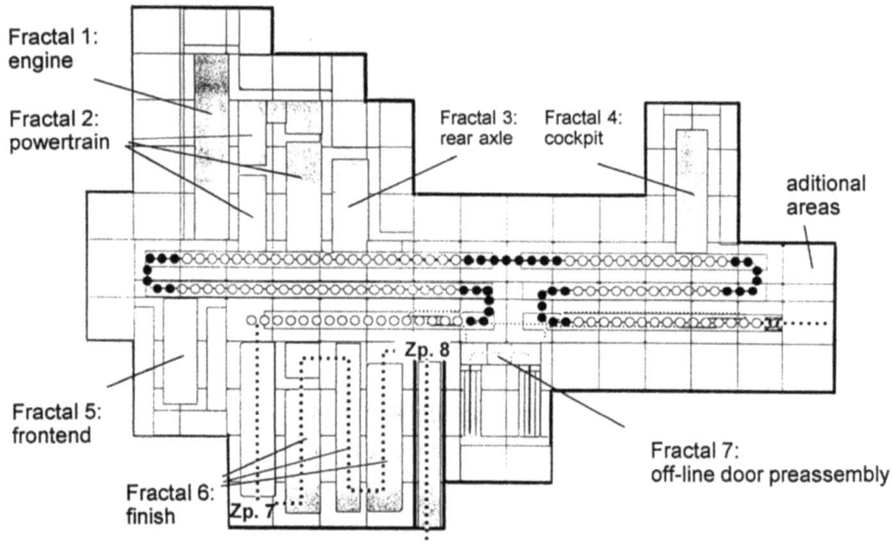

Fig. 3.7.10. Fractal structuring in assembly

Platform and modular concepts are the more effective, the more efficiently the core business processes of product creation, manufacturing, and procurement and marketing function. Particularly with regard to a shortening of the time taken up by product development and production preparation, modularization is an essential precondition for the simultaneous engineering process.

3.8
Automation at Renault: Strategy and Form

F. Decoster · M. Freyssenet

At the end of the 1970s, Renault decided to automate its plants in order to increase productivity and flexibility, to improve quality and to replace strenuous and repetitive tasks with more highly skilled work on automated machines. Automation was intended to help the company face the challenges arising from a diversified, competitive and rapidly changing market, as well as to resolve the work crisis.

Stamping, welding and painting lines were then rapidly automated, reaching high levels of automation by the mid-1980s. However, since the early 1980s, there has been a trend towards reducing the rate of automated operations, particularly in welding shops. Automation of final assembly was introduced at the beginning of the 1980s with the robotized assembly of windscreen, rear window, powertrain and rear axle. This has now been extended to almost all the underbody parts to seats, and to the instrument panel, steering wheel and bumper subassemblies.

Initial optimism about automation's ability to respond to the new constraints of production has given place to increasing realism (at Renault as at other European and American manufacturers, even though Renault may have realized this earlier than most other vehicle producers), and automation is now combined with other measures in order to reach these objectives. An automation strategy was hence gradually developed, which now appears to be durable.

3.8.1
Automation Within a Strategy for Assembly

3.8.1.1
The Level of Specialization of Factories and Final Assembly Lines

Almost all final assembly lines in French factories are single-model with two exceptions: one at the Flins factory which assembles the Clio and the Twingo, the other at Sandouville which assembles the Safrane and the Laguna. In contrast, the lines at Spanish and Belgian factories are more polyvalent and less automated.

Yet, regardless of whether lines are single-model or multiple-model types, they handle all versions of their models, including special vehicles (four-wheel-drive, for example). Some problems are caused by this diversity, especially on the automated work stations. For the time being, it has been decided to try to resolve these difficulties within an assembly line context and at a reasonable cost. The assembly of some versions, such as convertibles and prestige vehicles is sub-contracted.

3.8.1.2
The Assembly Process

Through the modularization of the vehicles, the assembly process is characterized by the preparation of sub-assembly work on secondary lines followed by further assembly on the main line. The entire process features a fishbone structure. Some small sub-assemblies are made at fixed work stations next to the main line. These work stations are often staffed by workers classified as disabled by a company doctor (French law requires every company to offer 6% of its jobs to disabled people).

This fishbone structure permits pushing back diversity onto the secondary lines. The trend is to maximize standardization of assembly operations on the main line, especially by designing identical location points and modes of attachment for different vehicles.

There are buffer stocks between the main and secondary lines, even though attempts are being made to reduce them at junction points. The main line itself is divided into several sections due to the introduction of an automated zone for interior assembly operations.

Manual operations are, to the maximum extent possible, grouped by *function*. The sections thus formed are not separated by buffers. However, they end more and more often in a quality control station and a re-work station. Each section is staffed by a team (equipe de travail).

Workers are often equipped with *trolleys* on which small components are stocked, thus avoiding unnecessary movement and loss of time. These trolleys move along with the assembly line. At the end of their respective work areas, they return to their starting point, either automatically or under operator control.

Some sub-assembled parts and some other parts are in sequence, i.e. their arrival at the station where they are to be assembled coincides with the arrival of the ad-hoc vehicle on the main line. The worker does not have to sort through the stock in search of the appropriate part. The parts delivered in sequence are the large components, complex components as well as sub-assembled parts made on the main line. Some of these parts are delivered on a just-in-time basis from the suppliers (seats, for example) or from Renault's mechanical components factories.

3.8.1.3
The Choice Between Manual and Automated Operations

Final Assembly

It has been decided to automate some sections of the main line whilst retaining a manual approach for the secondary lines on which sub-assemblies are prepared manually, because diversity remains too complex for the company to implement a complete automation of operations.

On the main line, the assembly of the dashboard/steering wheel and underbody parts, some heavy parts - such as seats, windows, and bumper subassemblies - is almost fully automated or semi-automated. The elimination of physically strenuous tasks (overhead work or handling heavy loads) has been a criterion for investment decisions, particularly given the increasing age of the workers. Moreover, mechanical aids have been set to assist the assembly of heavy parts: batteries, wheels, spare wheel.

Stamping and Welding Shops

The stamping lines were automated during the 1980s. In the body shop, the level of automation fell because of the cost of investment, as well as the difficulties experienced with a high degree of automation. The actual level of automation results from the need to automate the welding of a certain number of spots (around 60 - 70%) where automation leads to a higher level of performance, and the need to automate some tasks in order to improve working conditions. Figures 3.8.1. and 3.8.2. clearly show the evolution of automation and investment level in the welding shops opened since 1992.

3.8.1.4 The Way Forward

The principle of the assembly line with continuously advancing vehicles is retained in the manual sections. For two future vehicles, plans exist to use a support which can be vertically adjusted, so that the vehicle can be placed at an appropriate height for the workers. Moreover, the workers on sub-assembly lines are often carried along with the line, so they no longer have to walk while working.

In the automated areas of the final assembly shop, the line now moves intermittently (stopping at each work station). In the body shop, the lines also move step by step and the transfer is automated. Some sub-assembly parts are welded in robotized *islands*, where robots are used to handle parts and perform welding operations.

3.8.2 The Form of Automation

3.8.2.1 Problems Encountered and Solutions Implemented

The first automated assemblies were on a continuously moving line, i.e. not in stationary positions, especially for the set of the underbodies. Problems of adjusting and synchronization turned out to be very complex. For this reason, all the work units today move step by step.

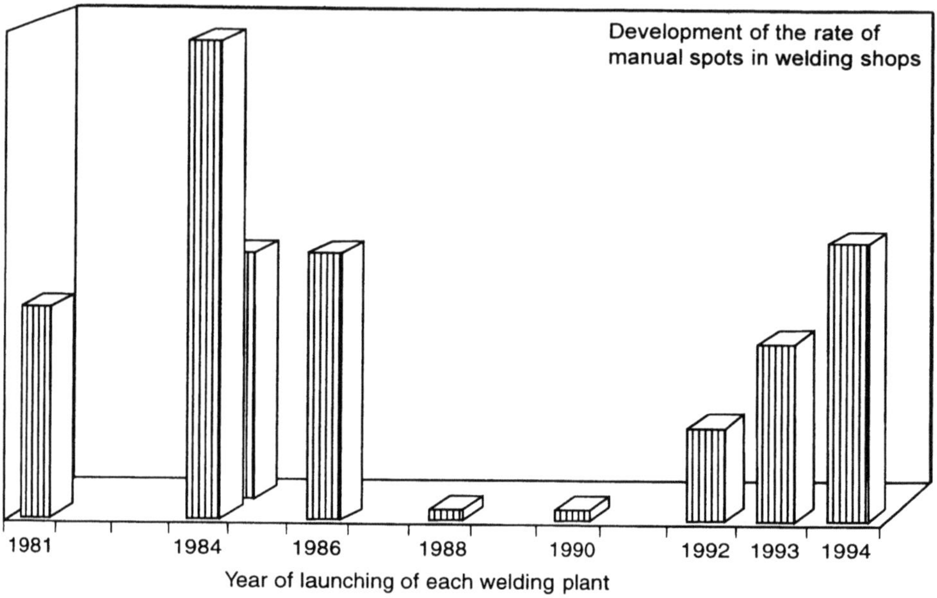

Fig. 3.8.1. Development of the rate of manual spots in welding shops. Source: Renault, Direction des technologies de production

The poor positioning of parts during the berthing process under the body and during automatic fixing has long since been a source of costly line stoppages. The product design and production engineering departments gradually tackled the constraints due to automation of assembly by modifying the design of parts and their positioning. The modularization of vehicles into sub-assemblies is another means of standardizing their placement, no matter how these vehicles differ.

In terms of design, the main development has been to establish *project* structures. In the Production Engineering Department, the position of a project manager for the launching of the product was introduced. This manager leads a team made up of people from production engineering, reassigned from their departments, and people assigned from the factories involved in the project. The first projects to benefit from this structure were related to the development of the PK gear box, the G-motor, the Twingo and the Laguna cars. An intermediate form was implemented for the development of the Safrane and Clio cars.

The involvement of workers in the process design teams was reinforced by the assignment of *socio-technical engineers* who, since 1991, have been members of the Production Engineering Department. They are assigned to the process design manager and are responsible for assuring that ergonomics, work organization and training (in liaison with the Personnel Department) are integrated into technical

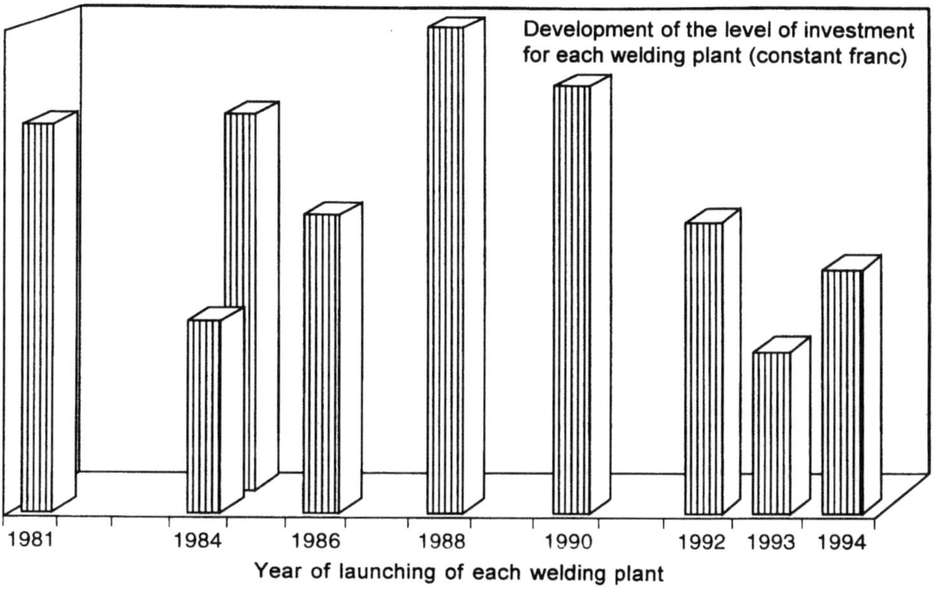

Fig. 3.8.2. Development of the level of investment in welding plants (constant francs). Source: Renault, Direction des technologies de production

decisions. Generally speaking, working conditions are increasingly being taken into account. For example, the growing concern about the reliability of technological systems is reflected in various forms of design aimed at ease of maintenance.

3.8.2.2
Make or Buy Automated Equipment?

At the start of the 1980s, Renault still designed its process equipment, only subcontracting feasibility studies for alternative technical choices and construction from time to time. Since the mid-1980s, the Production Engineering Department has provided a list of specifications describing the functions to be fulfilled. Design is performed by the mechanical engineering companies which have been consulted. Renault then selects the company which proposes the best costs and required times and which guarantees the highest degree of reliability.

3.8.2.3 Separate or Mix Manual and Automated Operations ?

Both solutions can currently be observed. The principle is to put manual subassembly operations and automated assembly as close together as possible in order to reduce in-process stocks. This, however, is not possible in all factories. Some factories, for example, have two levels. Preparation takes place on the ground level, and the main line is arranged on the upper level, along with the automated assembly zone. All the factories were built before 1974 and are not fully appropriate for the new process.

3.8.2.4 Type of Automation

The cycle, its start and its supervision are automated. Theoretically, any unacceptable fault causes the production line to stop automatically. The operator is informed of the area where the problem has occurred by a video screen. He must then identify the exact location in order to decide what to do. If he is unable to find it within a given time, he calls in maintenance, which is located nearby and attributes priority to the repair of automated segments. In general, stoppages are caused, as noted above, by the inadequate positioning of parts, which is particularly frequent with less common versions (four-wheel-drive cars, for example). Better design of parts and of their positioning, along with improved machine design, is necessary to eliminate such stoppages.

3.8.2.5 Design of Automated Work Stations

At different stages of the design and installation, new process equipment is subject to various procedures for analyzing reliability, accessibility, safety, ease of maintenance and critical paths. This is based on various techniques, such as failure mode and effect analysis (FMEA). Nevertheless, due to a lack of transparency in the overall process, operators may face problems of visibility and „readability“ when trying to identify the problem. Diagnosis and action hence take more time. Therefore, consideration is given to a simpler and "better-to-read" design of the machinery.

The new final assembly process is planned without repair-bays, where, traditionally, the defective vehicle could be removed, repaired and then placed back on the line. The defective vehicle is removed at the end of the final assembly line in order to prevent stoppages. The defect is then corrected.

3.8.3
Work Organization

3.8.3.1
The Creation of a Category of Operators for Automated Equipment

Automated sections are controlled by operators who have received about 4 months of training, particularly in order to be able to intervene and deal with simple stoppages. If they are unable to repair the equipment within several minutes, they must call a maintenance worker. Due to the design of the equipment, they are generally unable to check the correct setting of parts before they enter the automated section. In theory, the sub-assembly line, or the upstream work station, should supply an impeccable product.

Operators document the length and apparent causes of stoppages on a video display unit. The data is subsequently analyzed, either by the unit leader or by the maintenance team or by some specialists, so that reliability can be improved.

Each operator belongs to a *Basic Work Unit* (Unité Elémentaire de Travail: *UET*), consisting of some 20 people, even though this number varies from factory to factory. If the automated areas are physically separated from the manual work areas, the operator belongs to an UET consisting only of workers assigned to the area: i.e. operators just like him or workers who are in charge of the remaining, non-automated operations (some placement of parts, rework etc.) of this area. In return, in cases where the preparation work is near the automated area, the operator is a member of a larger UET which includes both machine operators and preparation workers belonging to the manual area.

3.8.3.2
The Basic Work Unit (UET : Unité Elémentaires de Travail)

Since 1994, all of Renault's European factories and services have been organized in Basic Work Units (UETs). The unit leader belongs to the hierarchical line management. He is on the first level of a hierarchy which has been reduced from seven levels to five (plant manager, production manager, department manager, shop-floor manager). He is responsible for a team of some twenty workers. Each unit is organized around a homogeneous activity, undertaken within the same space and time. Relationships with other teams have a customer-supplier form. Simple management indicators help evaluate performance and serve as a basis for improvement activities.

The responsibilities of the UETs increase with growing experience. At the beginning, they include part of the peripheral activities: in particular, minor maintenance tasks and systematic quality control. The unit leader initiates "development plans" and has individual interviews in order to reach an agreement with all workers with regard to all objectives.

Members of shop-floor UETs are trained to be polyvalent (i.e. capable of working at various work stations) and self-supervising (i.e. capable of checking the quality of their own work). The development of their skills towards primary maintenance, problem-solving, passing on of know-how and reduction of costs and lead times depend on the capability of the UET to develop this kind of dynamism.

3.8.3.3
Production / Maintenance Relations

In the Welding Shop

Automation of welding shops during the first half of the 1980s was an opportun-ity to organize work in order to modify production/maintenance relations. In some factories, maintenance workers were selected to become operators. Their resistance to this assignment, and the under-utilization of their skills led to the abandoning of this form of organization. Elsewhere, mixed teams were created, made up of operators (formerly unskilled workers trained for around 4 months in the fields of control and primary maintenance), maintenance workers and a shop-floor technician, with the objective of improving operations in the welding section for which they were responsible. Since the generalization of the UETs has been decided, the future of these innovative forms cannot be fixed. The formal definition of the UETs does not plan to assign production workers and highly skilled maintenance workers to the same UET.

In Final Assembly

In final assembly, this division is sanctioned by the creation of maintenance UETs. However, the dynamism of progress and work enrichment which the UETs may engender, may lead operators gradually taking responsibility for secondary maintenance tasks for which maintenance workers are currently responsible. But it seems hard to overcome these limits.

Maintenance workers currently depend on the shop-floor manager and the Production Department manager for the daily organization of their work. If called to an automated section, they must respond immediately. The problem is that they are often asked to respond to stoppages due, not to broken down machines, but to inadequate positioning of parts which the operator is unable to detect because of the complexity and "darkness" of the section. This situation is seen to be a temporary problem. Inadequate positioning of parts will disappear through improved quality of work by operators and better design of parts and tools.

3.8.3.4
Working Conditions

Figure 3.8.3. concerns final assembly operations. One can see the progress made in terms of physical working conditions: postures assumed and effort made. Progress is essentially due to the combination of three inter-related developments: ergonomics of work stations, automation of operations and modularization of the product.

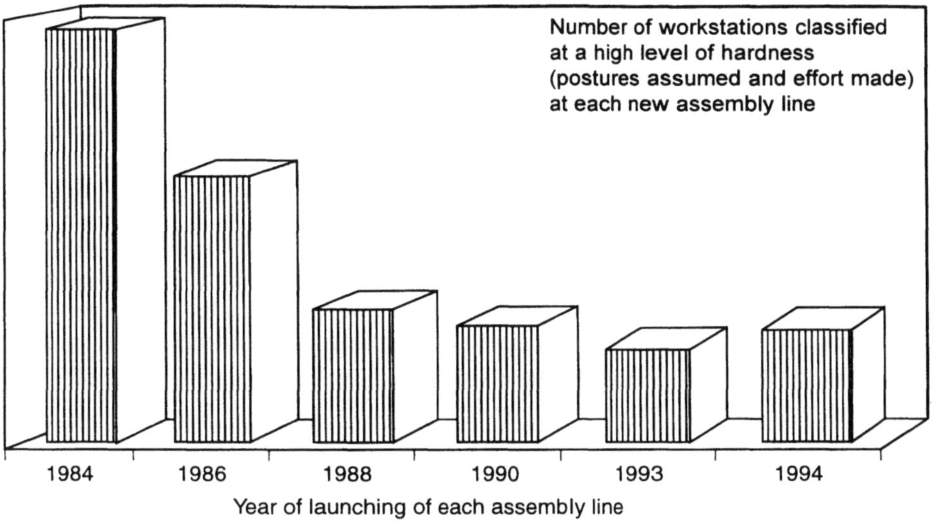

Fig. 3.8.3. Number of work stations classified as a highly strenuous (postures assumed and effort made) on assembly lines. Source: Renault, Direction des technologies de production

If the growth in productivity and quality is the essential objective for Renault, the improvement of working conditions remains a prime concern. Today, Renault does not face a loss of interest in industrial work - like in the early 1970s and like the Japanese at the end of the 1980s - but faces a significant aging of its work force. For this reason, Renault favors the easing of physical burdens and is less concerned about reconsidering the work content for its manual workers, as seems to be the case in current Japanese factories.

3.8.4 Discussion

The final assembly system at Renault is characterized by strong modularization of the car into large sub-assembly parts, a just-in-time process with 60 cars an hour as a target, a rate of automation compatible with investment opportunities and corresponding to an acceptable level of equipment availability (around 60 - 70% in the welding shops), and an attempt to eliminate "non-value added operations", mainly by creating Basic Work Units (UETs) for workers.

The strategy and the form of automation implemented in the mid-1980s may still be adjusted, and even modified, for several reasons. The rising average age of the work force - this rate could be reduced, but not stopped - poses a problem, not only with regard to physical working conditions, but in relation to cognitive capacity. The need to make industrial jobs more attractive for young workers with a higher level of education than former generations should lead to changes in the design of automated equipment in order to make it more intelligible. The development of team work (UETs) requires each complete and coherent sub-assembly of vehicle parts to correspond to one particular team, so that the team can then be responsible for it. In some cases, this should mean that manual and automated operations are combined within the same team. Finally, three factors should reduce the dysfunctions that still hinder the launching of new vehicles: the development of a project-based organization, the simultaneous design of product and process - made possible by the July 1994 merger of the Design Department (the product designers) with the Production Engineering Department (the process designers) - and finally, the participation of the factories at a very early stage of development projects.

3.9 Building Capabilities in Assembly Automation: Fiat's Experiences from Robogate to the Melfi Plant

A. Camuffo · G. Volpato

3.9.1 An Evolutionary Approach

The need for competitiveness in automobile assembly is driving firms toward more cautious and selected investment in flexible automation technology. During the 1980s, a number of car makers were enticed by the myth of computer integrated manufacturing, but recent surveys conducted in Japan (Fujimoto, in this book) and by the International Motor Vehicle Program of MIT (MacDuffie and Pil, in this book) show that the approach to automation has changed. The automation of manufacturing remains a long term trend, but in the short term economic and social factors shapes it in peculiar ways.

This chapter illustrates the automation strategy of Fiat Auto describing its evolution from the experiments of the 1970s driven by industrial relations conflicts, to the "pan-technologist" philosophy underlying the *Fabbrica ad Alta Automazione* of the 1980s, to the more realistic concepts inspiring the *Fabbrica Integrata* organizational model of the 1990s.

3.9.2 Evolutionary Phases of Fiat's Automation Strategy

The field research[1] on which this paper is based shows that the history of Fiat's automation strategy is curiously non-linear and consists of three evolutionary stages: "pioneering" rigid automation; "super" flexible automation; and "realistic" integrated automation.

The chapter analyzes each phase focusing on the relationship between new and existing technologies, the impact of these on efficiency and flexibility, and the related implications on organizational variables, human resource management and labor relations.

[1] It consisted of two visits for each of the following plants: Rivalta, Mirafiori, Cassino, Termoli and Melfi. Plant managers (at different levels) and union representatives have been interviewed (total of 100 hours). The data presented in the chapter have been elaborated on information kindly provided by Fiat Auto Co.. Time series have been constructed on the basis of the data provided by the production engineering corporate staff. The chapter does not deepen the analysis on stamping and painting.

The three-stage process can be characterized as follows. *Within* each stage innovation is prevalently incremental and builds on consolidated know-how, i.e. there is a certain degree of continuity, both in terms of technological homogeneity (a certain technological agenda has been worked out) and goals or objectives pursued. *Between* each stage innovation is to some extent radical, i.e. there is some kind of discontinuity, also here both in terms of technological heterogeneity (technologies different from those previously used are put in place in order to perform a given segment of the automobile manufacturing process) and organizational or market variables taken into account [28,32].

Since 1972, the adoption of flexible automation technology has always played an important role in Fiat's competitive strategy as well as for all the other major car makers (Figure 3.9.1. shows the increasing number of robots installed at Fiat Auto plants in the last two decades). However, this role changed substantially over time, and Fiat's automation strategy has developed and unfolded with different meanings and prevalent objectives.

In the first evolutionary stage (1961-1974), automation (it was still prevalently rigid and hard automation) was applied only to selected parts of the manufacturing process; in the second stage (1975-1988), state-of-the-art (at that time), flexible automation technology was progressively implemented (although not always successfully), in selected plants, in most of the segments of the automobile manufacturing process; in the third stage (1989-1994), the adoption of flexible automation technology has become more realistic and cautious (especially in assembly), subjected to thorough economic evaluation, and strongly interdependent with the design of the organizational variables.

3.9.3 Measures and Methodological Issues

Throughout the chapter, a longitudinal analysis of the evolution of Fiat manufacturing automation is carried out using different measures of automation for the different segments of the automobile manufacturing process analyzed [19]. In body welding, the degree of automation is measured by an automation ratio computed as the percentage of spots automatically welded out of the total number of spots applied to the body. This ratio is quite simple and frequently used by car manufacturers, and allows, to some extent, meaningful inter-firm comparison. Of course the number of body panels and components to be assembled differs across different car models, and the same is true for the number of spots, depending on the class, size and design of the model. Moreover, this measure does not take into consideration seam welding, as the IMVP plant survey methodology does, and does not control for the amount of welding outsourced, that can be relevant especially for some body parts and panels. However, the differences are usually not relevant enough to affect the meaning of this ratio. In engine manufacturing, the degree of automation is measured by an automation ratio computed as the percentage of non manual manufacturing steps out of the total number of operations. In this automation ratio, the key-word is "manufacturing step". Its

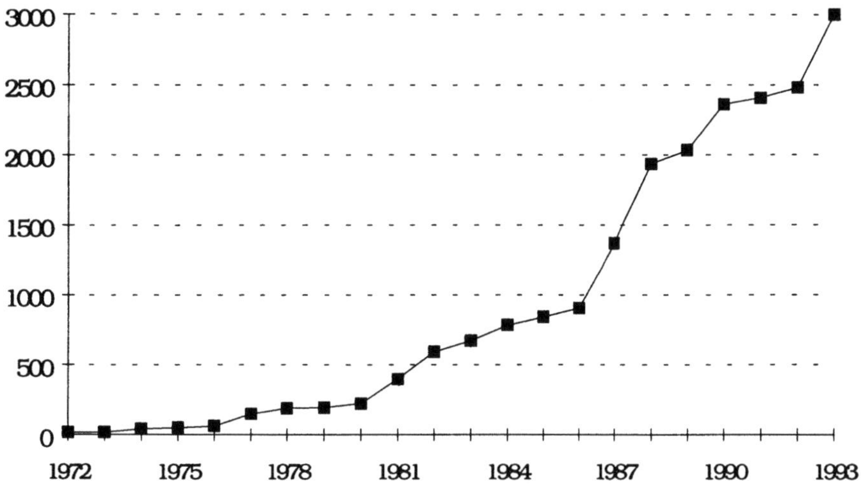

Fig. 3.9.1. Total number of robots installed at Fiat Auto Co. (1972-1993)

ambiguity in terms of contents, length, etc., within the same plant, across plants and across firms suggests a cautious interpretation of the data. However, since the analysis carried out in this paper is longitudinal, i.e. it refers to the same company and plants over time, these shortcomings should not be too relevant. In final assembly, the degree of automation is measured by an automation ratio computed as the percentage of the total assembly time that is not performed by people. The total assembly lead time is calculated as the total time required to perform all the production steps (down times such as those spent by the body or the components in buffers are not included). Also in this case, the ambiguity of the automation ratio is rather high, because of differences in layout, quantity of sub-assembly carried out off-line, car type and model (small. medium, large, luxury, etc.).

3.9.4 The First Phase: "Pioneering" Rigid Automation

Manufacturing automation was introduced in Fiat plants in the early 1970s. By that time, in fact, on the one hand automation technology had become fully available for automobile manufacturing (CNC machine tools, robots, first computer programmability, etc.), on the other hand the social context characterizing manufacturing at Fiat plants was seriously challenging the traditional ("Fordist") organizational model.

To tell the truth there were earlier examples of rigid automation in some sections of the manufacturing process. For example, in 1961, at the Mirafiori plant, Fiat began the automatic welding of the underbody of the *1300* and *1500* model (it was a rigid chain-system). Similarly, in 1966 also other parts of the body were automatically welded (body sides and engine bay), while final body framing was still performed manually, through the so-called *Mascherone* (a three-dimensional fixture on which the different parts of the body were held together for manual welding). Previously, arc welding was used, with people working in extremely uncomfortable conditions (noise and fumes).

In the late 1960s and early 1970s, labor conflicts, low quality, financial imbalances and marketing weaknesses jeopardized Fiat's survival.

More generally, union pressure for union rights, for changes in job classifications and for a revision of work organization increased in the early 1970s. In order to avoid such compelling pressures, in the following years Fiat implemented an automation strategy aimed at:

- reducing conflicts eliminating the most dangerous and tiring manual operations (uneasy working conditions were a major cause of labor disputes);
- reducing union influence on the workers and bypassing union control over work organization (extremely high in the traditional fordist assembly line where social rules determined working standards and speed).

Fiat began investing in robots and other labor saving equipment ("hard automation"). The first standalone robots were introduced at Mirafiori in 1972. The body framing of the *132* model was carried out in part automatically by a multiwelder (the so-called *Mascherone automatico*). A similar solution was adopted also at the Cassino plant for the framing of the *126* model body. In both cases the system was extremely rigid. The equipment was dedicated and non convertible or re-usable for new models.

A further step toward automatic body welding was made in 1974, when the *132* model body framing operation was modified, i.e. articulated into two sections for the *131* model. This robotized welding process was based on a two-phase body framing process, allowed a certain degree of product mix flexibility and was partially re-usable for new line setups.

Despite these refinements, this welding process based on huge multiwelders was neither flexible nor efficient. Fiat therefore changed body welding technology by adopting the Robogate system which was developed at that time by Comau, a subsidiary of Fiat. The Cassino plant can well summarize this discontinuity in Fiat welding technologies. Initially (1972), Cassino produced large volumes of the *126* model, and in 1976 the *131* model went into production there as well. As Mirafiori, Cassino was equipped with huge multiwelders for most of the body welding, including the body framing station. The extremely high number of welds set per station (300) resulted in so high down times that the big multiwelder did not repay the investment as planned. Thus, when the production of the *126* was transferred to another plant, Fiat got rid of the multiwelders. This lesson was a significant factor in highlighting the importance of automation flexibility and in the adoption of the Robogate system [17].

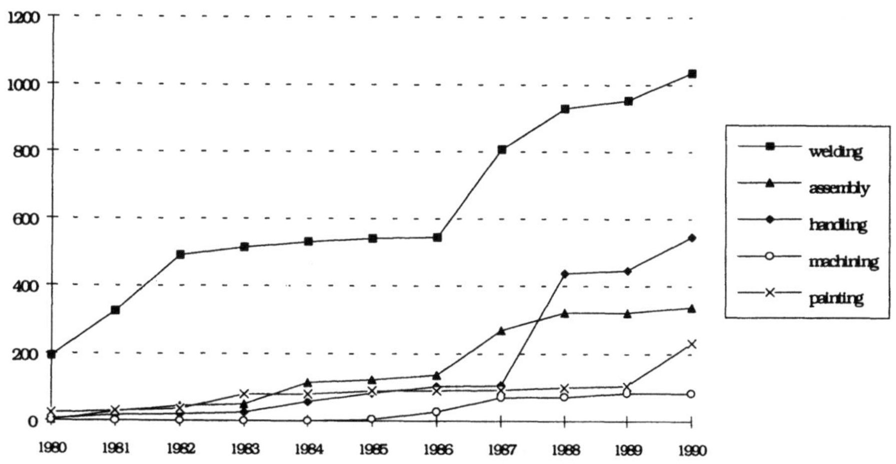

Fig. 3.9.2. Total number of robots installed at Fiat Auto Co. (1980-1990): Breakdown by manufacturing operation

3.9.5 The Second Phase: "Super" Flexible Automation

3.9.5.1 The Premises (and Promises) of "Pan-technologism"

In the early 1970s automation technologies made a conspicuous leap (Carlsson, 1984) thanks to the application of computer control to robots, numeric control machines, transfer machines, PLCs, and Machining Centers. In order to fulfill the market new requirements in terms of product quality and differentiation, the key issue became flexibility. Fiat progressively adopted the new flexible automation technology made available by equipment manufacturers, namely Comau, a subsidiary of the Fiat Group.

Figure 3.9.2. shows the patterns of Robot adoption in Fiat plants since 1980, broken down according to different operations and of the automobile manufacturing process.

But this adoption process was initially focused only on given segments of the automobile production process (first of all body welding, and then stamping, engine components assembly, etc.); only in the mid 1980s some existing plants (Termoli, Cassino) were comprehensively redesigned according to Computer Integrated Manufacturing principles. Furthermore, Fiat engineers prevalently concentrated on technical aspects, increasingly pushing the adoption of state-of-the-art equipment toward the technological frontier, but without fully evaluating the implications of this strategy in terms of organizational structures and management information system complexity.

The theoretical flexibility of automation technologies implemented by Fiat was very high. However, similarly to what Jaikumar (1986) pointed out for American firms while comparing them with Japanese ones, the results in terms of "real" flexibility, market response, quality and efficiency were not so satisfactory.

3.9.5.2
"Islands" of Flexible Automation

The opportunities provided by the greater technology flexibility were first exploited by Fiat with the implementation of the Digitron system at the Mirafiori plant in 1974 for the *131* model assembly line. Digitron was a computer-controlled system of docking (body-chassis *marriage*).

Through computer-control, the mechanical units were automatically loaded on a robocarrier, and conveyed to automated screwing stations where the body, brought around by an air conveyor, was assembled. The Digitron system consisted in the elimination of manual, "hands-up" body-chassis screwing, and fasteners or bolts tightening. Therefore, docking was performed automatically and asynchronously (5 parallel stations), with still-standing bodies.

As already mentioned, the major purpose of these innovations was clearly to solve ergonomic problems by eliminating the more tiring and wearing jobs (i.e. those associated with episodes of industrial conflict). However, Digitron represented for Fiat the first example of flexible application of automation technology (the robocarriers and the bolt/fastener tightening stations could work on different platforms and body types, there was an automatic device for body-engine coupling) to complex assembly operations.

Later, this system also represented a fundamental background of competencies and know-how for the realization of the highly automated factory at the Cassino plant (1988). In 1974 Fiat introduced at the Mirafiori plant also some automatic Tandem-type stamping lines, and in 1977 the first robotized painting stations.

Another relevant discontinuity with the rigid automation implemented in the first phase, took place in body welding, where the multiwelder (*Mascherone Automatico*) was replaced by Robogate. The Robogate system (first implemented at the Rivalta and Cassino plants for manufacturing the *Ritmo* model in 1978 and still used, with updated versions, in all the body welding sections of Fiat Italian plants) theoretically allowed complete flexibility in terms of market response (that version of Robogate could manufacture a relatively wide scope of models), in terms of product lines renewal and update (that version of Robogate could be relatively easily converted and adapted to the introduction of new models), and in terms of process (that version of non-in-line Robogate process could reduce process-breakdowns related costs). Put in practice, the system was designed for five different bodies (but Fiat applied it only for two), 80% of the investment was reusable and adaptable for new product lines, and the non rigid sequence of working-manufacturing operations (computer regulated asynchronous movements) allowed systematic prevention of complete breakdowns.

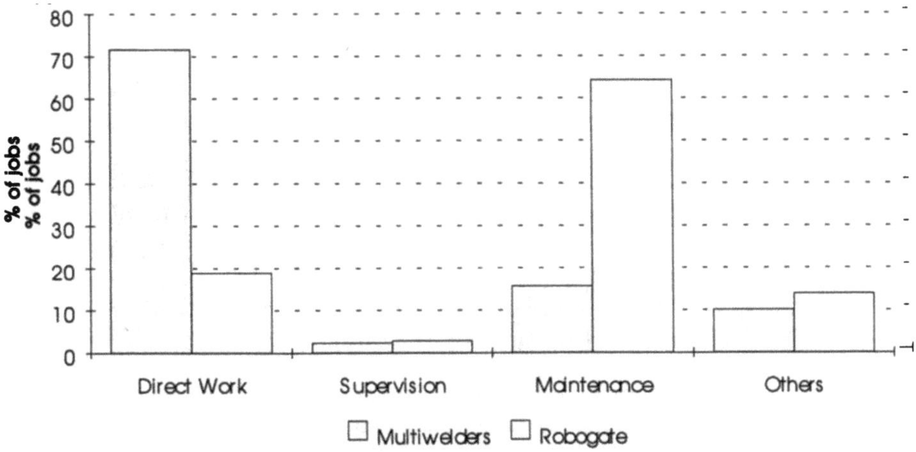

Fig. 3.9.3. Changes of job distribution at the Rivalta plant body shop: Robogate versus Multiwelders. Total number of employees: Multiwelder based system:=247; Robogate based system= 137

The body framing process proceeded as follows: self-standing bodies were loaded on pallets (one for each model). Every pallet was mounted on a computer-controlled AGV: welding was fully automated (100% of spots are automatically welded) and took place as the set robocarrier-pallet-body passes through gates. Robogates were located in parallel, and the body usually entered one for the first stage of framing and another for the second stage. There were two pairs of gates at each Robogate, and these could slide longitudinally into position when the appropriate body type came into the Robogate. Bodies could thus be welded up in random order [17].

The degree of automation provided by Robogate was so high that it required only inspection and service by maintenance workers (indirect workers quadrupled, as direct workers dropped to a quarter) as shown in figure 3.9.3. which illustrates the new job distribution at the Rivalta plant and compares it with that of the previous system based on multiwelders.

An indication of the flexibility of Robogate is that the existing units at Cassino have built 2.7 mm bodies since 1978. Initially they were used for the *Ritmo* (*Strada*) model, and in 1986 the *Regata* was added. The same Robogates were then used to produce the *Tipo* hatchback and the *Tempra* saloon and estate car [17]. This initial version of Robogate was highly flexible from two standpoints. The productive mix could be easily changed and adapted thanks to computer programming and robocarriers (through which bodies can be moved around and welded up in different sequences), and to body automatic recognition in the first framing station where gates and *bilancelle* were automatically positioned in order to block the body and operate on it. Model renewals or restyling could also be easily managed. A new body implies only new computer programs for robots, the set up of gates and a new pallet [24].

Even though these sophisticated technological features were never completely utilized, as the system entailed an excess of costly manufacturing flexibility (especially the 25 robocarriers the initial version of Robogate used were money and space-consuming), Robogate represented worldwide an advanced solution (later on it was installed by Comau in other European and U.S. plants like the GM ones in Zaragoza, Spain and Lansing, US).

A different version of Robogate was implemented at the Mirafiori plant in 1982. In order to improve Robogate cost-efficiency, its computer regulated asynchronous movements, originally intended to prevent breakdowns of the whole system, were transformed into a step-by-step line sequence (it came up that line breakdowns were reasonably manageable). In this version of Robogate the set pallet-body was moved around on floor fixed paths instead of through AGVs. A decreased flexibility in moving was traded off with an increased cost-effectiveness.

This automation strategy went on with the implementation of LAM (*Lavorazione Asincrona Motori*) at the Mirafiori plant in 1979. The LAM system (Asynchronous Engine Assembly) which is still in place today at Mirafiori, is not a high-automation solution (only 25% of engine final assembly operations are performed automatically and approximately 90% of components are assembled manually) and was prevalently targeted at improving working conditions. LAM consists in the partial automation of the automobile engine assembly process: assembly is divided into 10 assembly areas; within each area, a number of assembly operations are performed by 4 to 5 workers located on a working station; strict line interdependence among the assembly areas is therefore eliminated; each worker, in a given working station works autonomously; he or she calls (pressing a button) a computer-guided minitrailer or AGV (42) on which an engine-pallet set is loaded and transferred automatically from in-process buffers; after finishing the relevant set of assembly operations the engine-pallet set is automatically transported back to the inter-operational buffers (each workers trigger it always pressing a button). The change in terms of job distribution associated to the introduction of LAM is graphically represented in figure 3.9.4.

The restructuring of the engine assembly process (in terms of both machining and assembly) implied:

- safer and better working conditions, job enlargement (extended working sequences and larger *takt* - 4 to 8 minutes), and a certain degree of job enrichment;
- increased flexibility; a few engines of the same "family" (1100-1500 cc) were assembled (four models, with 110 versions).

LAM impact on productivity was nevertheless lower than expected partially because of intrinsic features and partially because it was applied to the manufacturing of pre-existing engines (which had been designed at the end of the 1960s and not designed to be assembled with LAM). Afterwards, and with the new engines, the situation improved.

Fig. 3.9.4. Changes of job distribution at the mirafiori plant engine assembly shop: traditional versus L.A.M. Total number of employees: Traditional system=400; L.A.M. system =300

Although LAM does not represent a highly automated solution, it played a crucial role in the evolution of Fiat engine manufacturing process, becoming part of the repertoire of capabilities and technological know how of the firm. At the new Pratola Serra engine manufacturing plant, where a wide and new set of engines is manufactured since 1996, assembly through LAM has been preferred, as inspiring concept, to the automatic line assembly system implemented at Termoli in 1985. An example of high automation is instead the engine head-piston displacement assembly line at the Meccanica 2 section of the Mirafiori plant. This line was set out in 1982 based on Marposs equipment. 95% of operations were performed automatically. Despite its relatively high rigidity, it represented a landmark for later Fiat engine manufacturing plants Termoli and Pratola Serra.

3.9.5.3
The Highly Automated Factory ("the Technological *libido*")

The idea of a fully automated factory was conceived and implemented at the Termoli plant (where the Fire engine for the *Uno* model was produced). For the first time Fiat simultaneously designed and developed the product, the plant and the manufacturing system. The Fire manufacturing system at the Termoli plant

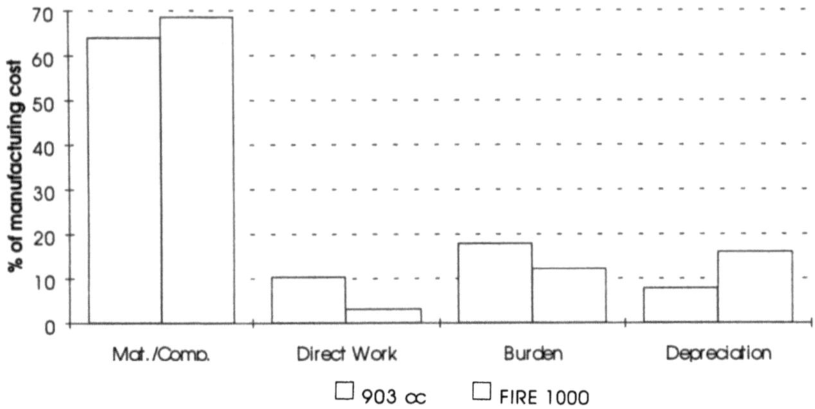

Fig. 3.9.5. Comparative cost structures (as % of manufacturing cost) for fire 1000cc and traditional 903cc engines at Termoli plant. Note: the unit cost of a Fire engine is 10% lower than a traditional 903 engine and its quality and performance higher

produces a wide array of engines, and represented the first, worldwide recognized, example of Highly Automated Factory (HAF - *Fabbrica ad alta automazione*). The automation ratio in the machining section is over 90%, while it is slightly lower in assembly. Computer integration is managed by the S.I.M., an information system that manages simultaneously production planning, logistics, etc.

Although Fiat management considered it as the "natural" evolution of the Mirafiori LAM system, the technological and organizational concept underlying Termoli is rather different.

The degree of automation is similar to other Fiat's plants for the manufacturing and assembly of some mechanic components. For instance, the engine head-piston displacement assembly section is analogous to that at the Meccanica 2 section of the Mirafiori plant. But in terms of engine assembly, Termoli is very different from LAM. Here the degree of automation is much higher (approximately 80% of operations are realized automatically), there are 78 automated working stations instead of 30, and 2 manual working stations instead of 10. Termoli marks a return to sequential and synchronic operations.

The Fire system, focused on the manufacturing of the engine for the *Uno* model, Fiat best selling car of the 1980s, was at the beginning less flexible in terms of product mix and equipment convertibility. But its relatively dedicated lines allowed higher productivity and efficiency (figure 3.9.5. shows comparative manufacturing cost structures for Fire and similar previous engines). This is the reason why it represented an important shift from a somewhat overabundant concept of flexibility to a more moderate one, determined by market requirements in terms of product variety and variability [22: 70].

The key elements of the new plant were:

- the internalization of all the strategic and core manufacturing operations;
- the optimization of material handling and layout;
- the high speed processing (with cycle times lower than a minute) and utilization rate (equipment up-time and capacity utilization were always very high);
- the extensive use of robots (to reduce the direct workforce).

The only manual operations maintained in the cycle consisted mainly in the final assembly of carburetors and filters.

Up to the present, Termoli remains a benchmark (both internal to Fiat and compared with competitors) in terms of automation for engine manufacturing, especially since it has been retooled and enlarged in order to produce the engines for the *Punto* and the other new models. One important recent development (higher degree of automation) probably is the Meccanica 3 section at the Mirafiori plant, where, in 1989, a new robotized shop for gear-box assembly has been set out. This shop has an automation ratio as high as 99% (number of operations carried out automatically on total number of operations), is based on Comau technology (Smart robots), and is very flexible (it assembles gear-boxes for 5 models). However, on the whole, the Termoli plant undoubtedly conferred to Fiat an outstanding technological and manufacturing status, recognized by most of its competitors.

The new Fire engine enabled Fiat to reduce by 10% the list price of its best selling *Uno* model, previously equipped with the traditional 903 cc. engine. Fire was much better also in terms of both quality and performance. This better performance came from such features of the Fire manufacturing system as: reduction of the number of components by 30%, employment cutback by nearly 40%, a halved manufacturing lead time (107.5 min. versus 231.5 of a 903 cc. engine), and a production schedule of 1000 engines per shift on three daily shifts.

Cattero [6] suggests that the organization of work, the job contents and the skill profiles of blue collars changed as manufacturing began, production ramped up and organizational processes consolidated. Figure 3.9.6. shows the changes associated to the introduction of Fire in the distribution of jobs at section 3 of the Termoli plant.

The organization of work at the Termoli 3 plant evolved according to a learning process. The original aims were those of massive reduction of direct workers and strong presence of indirect (mainly service and maintenance jobs) workers, referred to, in the firm jargon, as *meccatronico* [6]. But as this design was being implemented and as production ramped up, learning took place and a new job and skill profile emerged: the system controller or conductor (*conduttore*) [20]. At full capacity Termoli employed nearly 340 system controllers out of 1167 total jobs (almost 30%).

The system controller directly interfaced with a given part of the manufacturing cycle through a computer; he or she performed some residual manual direct operations and some "service" or "indirect" activities (maintenance and setup, manufacturing process testing) [7].

While procedures and parameters were defined by specialists, the system controller had a major role in interpreting and diagnosing the "weak signals" which could jeopardize the smoothness of the manufacturing process. The system controller job, emerged from an organizational learning process at the Termoli

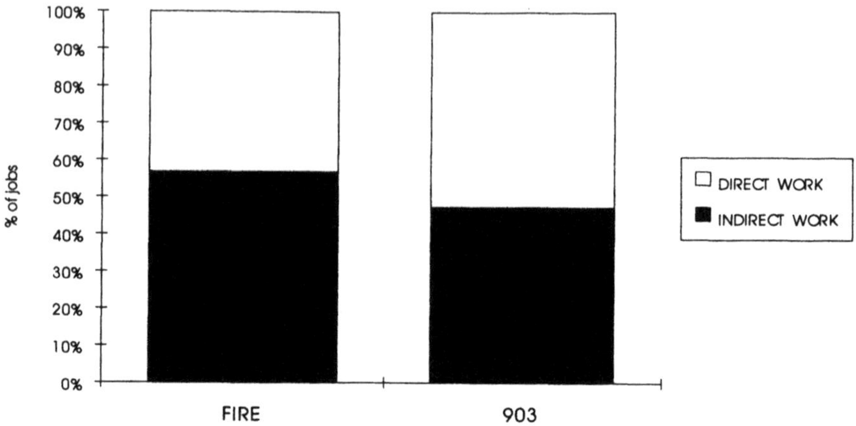

Fig. 3.9.6. Changes of job distribution at the termoli plant: Fire 1000cc versus traditional 903 cc engines. Total number of employees: Traditional 903cc engine system=1582; Fire system=930

plant, was institutionalized with the April 18, 1986 plant-level management union accord. This job and skill profile was later included in the metal workers national contract job classification schemes with the CCNL (National Collective Labor Contract) signed on April 18, 1987. Besides, the idea of having a skilled worker able to conduct and control a given segment of the manufacturing process was later applied by analogy to other sectors like painting, body welding and final assembly. In fact, within the *Fabbrica Integrata* (the new organizational model Fiat implemented in the early 1990s), there are the so-called *Conduttori di processi integrati* and *Operatori di Processi Integrati* (integrated process conductor, IPC and integrated process operator, IPO), two new jobs and skill profiles inspired by the system controller figure.

3.9.5.4
The Analogical Application to Assembly

The Fire engine and the Termoli plant were a great success for Fiat. The management summarized the competencies and skills (in terms of automation technology, organizational structure and human resource management policies) so successfully developed and implemented at the Termoli plant in engine manufacturing. in the HAF - *Fabbrica ad Alta Automazione* - concept. This concept was then applied (by analogical replication) to car assembly operations. As a consequence, the Cassino assembly plant was restructured.

In other words, the set of core competencies associated to the HAF philosophy were adapted to the Cassino plant, where the *Tipo* (and later the *Tempra*) models were originally manufactured and assembled. The *Tipo* was being produced there also before, since its introduction in September 1987, but the managerial decision to implement the HAF in Cassino dates Jan. 1988.

At the Cassino plant[2] the press shop is highly automated and equipped with automatic transfer lines. There are also two plastic moulding shops. Body welding is carried out in two buildings, with all the subassemblies being produced in the first shop and the Robogate framing stations being housed in the second shop. Subassembly is carried out automatically by Comau Smart articulated-arm robots (engine bay, front floor and rear floor) or, in case of small subassemblies, by pedestal welders under robot control. A certain number of laser (61) is used for line welding. Components loading is also partially robotized (Expert Scara-type robots). A tabbing station, where the roof and sides are added, allows the body to self-stand. A conveyor transfers the body to the next shop, where it is dropped on to an AGV and transferred to the Robogate [17].

There are six Robogates in parallel, and the body enters them for two stages of framing. Painting is partially robotized. Robots apply weld sealer, and the application of primer is automated throughout. There are four lines operating in parallel, but only one color paint spraying booth is automated.

There are four parallel lines in the trim and final assembly shop, each working to a cycle of 2 minutes and 30 seconds. The doors are removed as the body enters the shop and placed on pairs on small AGVs to be taken to a separate door fitting line, which is partially automated.

Between the main assembly lines is the fascia/instrument panel sub-assembly area, in which only testing is carried out automatically. Then, the assembly travels on its AGV to the line. At this stage, the four lines converge into two lines for insertion of the complete fascia by robot. The body is held stationary, first at the insertion station (where one robot lifts the fascia from the AGV and places it on a fixture), and then at the tightening station (where other robots take the fascia, slide it into position in the body and tighten bolts and fasteners).

Subsequently, the body is transferred into an overhead buffer area before being routed to one of the four assembly lines again, where the supply of components is matched to the bodies possibly according to customer specifications. When trimming is completed, bodies are routed to the final automated assembly area. Here there are also three lines for sub-assemblies: rear suspension; front suspension; power train. With all sub-assembly work complete, the units are mounted on tubular pallets and transferred to the engine dressing area where some work is done manually.

The mechanic units are then transferred by an AGV to the two docking lines. Docking is completely automated and builds on the concepts pioneered in the 1970s with the Digitron system. The units are installed at three stations. Pallets are first unloaded from the AGVs and moved into position sideways. Then the body is lowered down on to the mechanical units from the overhead monorail. Robots tighten fasteners at the first two docking station; nutrunners operate at the third [17]. The next assembly stations are also mostly automated. Robots install the

[2] In 1995 also the Cassino plant was partially restructured with the introduction of the new "Bravo" and "Brava" models. Some results of this research have been published elsewhere [34].

windscreen, rear glass, headlamp module, bumpers, wheels and tires. Then, a few extra operations and testing is done in the finishing line.

The adaptation of the HAF concept originally developed at Termoli to the Cassino assembly plant entailed implementation problems. It was soon clear that logistics was going to play a crucial role in the whole system, and that the degree of complexity to be governed was too high. The manufacturing system adopted by Fiat was theoretically designed to improve flexibility and market responsiveness. However, on the one hand sub-assembly lines remain inflexible. On the other hand, the total efficiency was reduced by short but frequent breakdowns, usually due to automated assembly equipment.

For these reasons the degree of complexity to be managed, compared with Termoli, was much higher. This complexity in governing the manufacturing process also stemmed from:

1. The high level of product variety; in 1989 Cassino featured a number of versions of the *Tipo* and later also of the *Tempra*, often manufactured on customer specifications; in some segments of the manufacturing process (e.g. welding) flexible automation technology proved to match the required flexibility, in other segments (e.g. components or mechanic sub-assemblies) it did not.
2. The strong presence, notwithstanding automation, of manual operations; in the more manual areas variance and coordination needs are higher and still critical for quality and productivity.
3. The impossibility (or lack of capabilities) to properly manage logistics and information flows, fully taking advantage of information systems to control the manufacturing process (e.g.: some problems such as doors and body coupling after internal trimming had to be solved manually for a long time; the assembly of *Tipo* rear door also had to be realized manually for some time);

Considering that there were many more manual operations than in Termoli (according to Fiat figures, Cassino automation ratio in assembly activities was around 22% -1000 automated operations out of 4500 or, measured differently in terms of percentage of total assembly time carried out automatically, 25%), and, notwithstanding the 146 robots installed in the assembly shop, it is understandable why, in the initial stage, Cassino productivity and quality levels were lower than expectations, even when market conditions were favorable.

The balancing of the assembly line, which originally had a constant cycle time (but then was modified in a step-by-step line), required continuous adjustments as cars with different characteristics and options (called *specialità*, i.e. "specialties") had to be manufactured implying more operations and longer cycle times [29]. Because of this, along the assembly line some operations overlapped with those of the subsequent workers downstream, and this called for considerable flexibility in job assignments, administered by the first line supervisors (the so called *caposquadra*). Examples of "specialty" characteristics were: air conditioning, automatic transmission, sunroof, etc. In order to keep the line smooth, job redistribution required the intervention of off-line "jolly workers" together with "specialty operators". This caused troubles and inefficiencies with line breakdowns. Moreover, the complexity of the manufacturing system entailed problems and variances which had to be systematically absorbed by the workforce (the information system designed to handle the problems often broke down or proved to be inadequate). For example, problems in doors or door-body coupling caused about 10% of cars to advance through final assembly without doors, requiring additional end-of-line operations.

Other problems can be inferred by the emergence of informal work rules and "local" redefinition of procedures. For example while the *caposquadra* organized work and assigned jobs aiming at minimizing balance loss across team member, the workers often revised these assignments considering task interdependence (for instance workers prefer to change "informally" job assignments and do more convenient operations - from a sequential standpoint - even if this implies a higher degree of saturation for themselves).

Plant managers or first line supervisors shop stewards in some cases might redefine (more stringently) time spans for some operations. Another example of informal work rule redefinition can be found in the use of electric screw drivers. They were forbidden for safety reasons (and replaced by air-powered units), but workers kept using them because being lighter and handier they required less strain. The organizational problems encountered, combined with the sales of the *Tipo*, which did not meet Fiat high expectations, prevented Cassino and the *Tipo* from reaching the goals which Fiat expected. Fiat's management nonetheless recognized all these initial shortcomings and systematically improved the plant's performance.

At the end of the decade it had become clear enough that a new organizational model and a different, more integrative approach to flexible automation technology adoption was needed.

3.9.6 The Third Phase: "Realistic", Integrated Automation

3.9.6.1 Re-thinking the Role of Automation

In the early 1990s Fiat had to face a new competitive climate. The success and excellent performance of the 1980s were replaced by recession, an outdated product line (and a certain slowness in model renovation), a too heavy reliance on the domestic car market and on market segments difficult to defend.

All these circumstances moved Fiat to launch, in October 1989, a reorganization and investment process which involved all its operations, the relationships with suppliers and the dealers' network. It was a five-year plan, articulated in a number of projects [34].

Integrative part of this strategy was a new approach to manufacturing automation. Overall, the degree of automation in most of Fiat plants was, at the end of the 1980s, similar to those of competitors, but some of them (Cassino, Termoli) were equipped with state-of-the-art process technologies. In these plants the degree of automation was therefore higher than average [17]. The new approach (the internal jargon referred to it as the *lean machine* approach) consisted in a more cautious and "sober" adoption of flexible technology automation.

3.9.6.2 The Fabbrica Integrata (IF)

As a consequence, Fiat's new approach to competitive manufacturing focused on organizational and human resource related issues by means of the *Fabbrica Integrata* (I.F.) project. The IF originates as a necessary consequence of (and as an attempt to overcome) some of the mismatches which emerged in the Highly Automated Factory (HAF) at Cassino.

As automation ceased to be the only driver of productivity and quality, organizational, industrial relations and human resource management choices (delayering, empowerment, teamwork etc.) are rediscovered as key constituents of the firm's competitive performance.

The IF model was initially designed in 1990. Fiat chose to implement it almost simultaneously in its plants rather than test it in a pilot plant. During 1991 the only partial applications took place at the Termoli and Cassino plants, but Fiat extended the model to all its plants during 1992 (first at the Rivalta and then at the Mirafiori assembly plants).[3]

On the whole, the IF can be seen as a crucial component of the manufacturing paradigm-shift taking place at Fiat, where the relationship between flexible automation technologies and organizational competencies becomes more balanced and complementary. The implementation of the IF is nevertheless still on the way and to some extent troublesome [8].

3.9.6.3 The New Plants

While the IF was being implemented in Fiat plants, the new automation strategy of the Turin-based company emerged. Signs of this new approach can be seen in:

- designing the greenfield, EU-supported new assembly plant at Melfi (South Italy);
- redesigning the Pratola Serra engine manufacturing plant;
- setting out a new assembly line for the "Punto" at the Mirafiori plant.

The full implementation of the Integrated Factory model at Fiat is being reached with two new plants located in Melfi (province of Potenza) and Pratola Serra (province of Avellino) in the South Italy.

The Melfi plant began operations at the beginning of 1994, with the assembly of the new *Punto* Fiat model. At full capacity this plant will produce 450,000 units with 7,000 employees. The plant has a Press shop, a Body shop, a Paint shop and an Assembly shop.

---

[3] A comprehensive discussion of the IF is not an objective of this paper and has already been carried on elsewhere [3,5,34].

In the press shop, Fiat installed multi-station transfer presses (Schuler and Komatsu) to produce all pressings. These have automated handling and quick die-change mechanisms [13].

The press shop connects directly to the body shop. In body welding, small press welders are used at the beginning of several lines to reduce handling, but robots (Comau Smart) do almost all the other welding. Fiat has made efforts to raise the efficiency of its robots, and this is one reason why there are about one third less than at Cassino. An updated "high volume" version of Robogate is built with a shuttle transfer system instead of AGVs, which increases uptime and reduces the space required [13]. In most sub-assembly lines, and for most of the main body line, Daifuku monorails and cradles (self drive type) are used as transport.

There are two body lines running in parallel, which continue to run through the paint and final assembly shops. The sub-assembly welding lines run at right angles to the main body lines, and are arranged as cells. In some of them (rear floor, main floor and underbody), the first station is a press welder.

The body shop welds a lower number of body types than Cassino, but the introduction of new models (such as the Lancia Y) is faster and more cost-effective thanks to the so-called „5X2" concepts. The body shop is scheduled to work for approximately 10 years. It worked for only the *Punto* body type till 1995. Fiat recently introduced a new model (Lancia Y11), with a body type similar to the *Punto*. Thanks to an innovative layout, this new body type is framed on the same line and with the same equipment used for the *Punto*. Only the initial stage of welding is carried out on a different line, parallel to the *Punto* one. As a consequence, the new model was introduced basically without stopping or intervening on the existing body shop. The new line was set up with very little stoppages and down times (weekends in a few months). Painting is completely automated and waterborne.

Driven by economic reasons, the welding shop for the *Punto* line at Mirafiori is slightly different and represents a step back in welding automation. Compared with Melfi and Cassino, the *Punto* line at Mirafiori presents many more manual operations in the upstream stages of welding. For example, the welding of small components and some of the sub-assembly are carried out manually, although in safe and ergonomically optimal working conditions.

This step back in the adoption of flexible automation technology is mainly rooted in economic factors. In fact, the relatively small volumes produced on the line do not allow to break-even the investment required for robotizing some sub-assembly activities in welding.

In final assembly, the lines at both Melfi and Mirafiori represent a reduction in the degree of automation if compared with Cassino. The automation ratio is in fact 6%, and is very high only in docking and final testing. A number of operations that had been automated at the Cassino plant, are carried out manually at Melfi (e.g.: tire and disk wheel assembly, mechanical components assembly onto a pallet the length of a car). Automation is not aimed at substituting manual work, but rather at assisting it.

Assembly lines are spacious and much emphasis is put on ergonomic devices for manual activities (e.g.: 70 degrees car-body tilting for under-side operations; automatic hiking of the body to a comfortable level for operators).

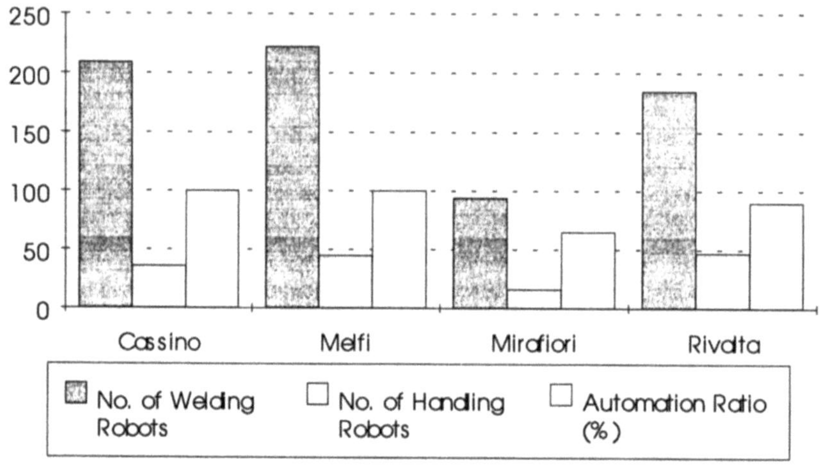

Fig. 3.9.7. Total number of robots and automation ratio for selected Fiat auto plant body shops

This approach to assembly automation is complementary with the new organizational concept (the *Fabbrica Integrata*) and with innovative, cooperative industrial relations [5].

The Melfi plant allows significant productivity gains: manning levels are very low by European standards; compared with Cassino, the total assembly lead time for a body (to pass through) was reduced by about 30%, the work in process material decreased by about 35% and the number of cars per direct employee grew by about 20%.

The success of the *Punto* model and the outstanding results in terms of costs and quality achieved at the Melfi plant are, together with the favorable trend of non-domestic, European markets, some of the variables explaining the excellent financial results recently achieved by Fiat during 1995.

The Pratola Serra plant will produce a family of engines: gasoline and diesel engines with four or five cylinders and with two or four valves, and a range of displacement from 1400 cc to 2400 cc. The ramp up of production is scheduled for 1995. At full capacity the plant will manufacture 800,000 engines a year with 1,300 employees. The scheme of the Pratola Serra plant builds on the experience already matured at Mirafiori and Termoli, but the new plant enjoys many relevant improvements: a more advanced information system for process control, faster machining centers, higher degrees of flexibility.

Overall, the degree of automation is similar to that of Termoli, but in engine assembly the Pratola Serra plant is going to mark a return to the L.A.M. concept of parallel assembly. According to Fiat managers, this will allow higher flexibility, matching the wide array of engines that the plant will produce.

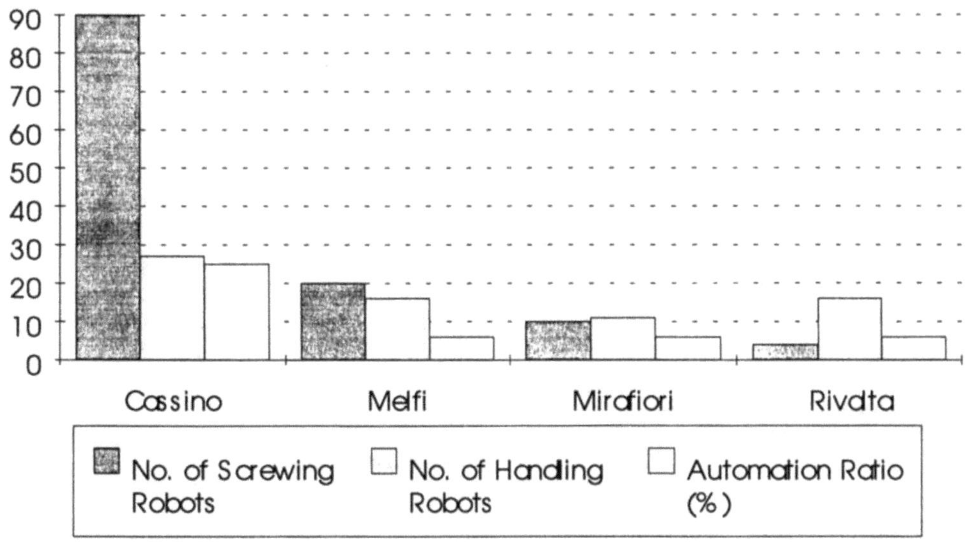

Fig. 3.9.8. Total number of robots and automation ratio for selected Fiat auto plant assembly shops

Summarizing, this third evolutionary phase of Fiat automation strategy marks a discontinuity in the approach pursued by Fiat to manufacturing automation. The lesson learned from the success and problems of the highly automated factory suggested a more cautious approach to automation, especially in assembly. Hence, a different automation strategy has emerged, where organizational and economic variables are more strictly integrated with the technological aspects, and where people become the driving force of productivity and quality improvements, while technology supports them.

Figures 3.9.7. and 3.9.8. summarize the present situation of manufacturing automation in some of the Fiat plants. The number of robots and automation ratios in the body and assembly shop of the most recent plant (Melfi-1993) is significantly lower than those of the immediately previous one (Cassino-1988), which, are nonetheless higher than those of the oldest plants (Rivalta, 1978).

3.9.7 Conclusion

Figure 3.9.9. summarizes three findings of this chapter and sketches the three evolutionary phases of Fiat's automation strategy, pointing out the type of linkage between existing and newly implemented technologies (continuity, incremental

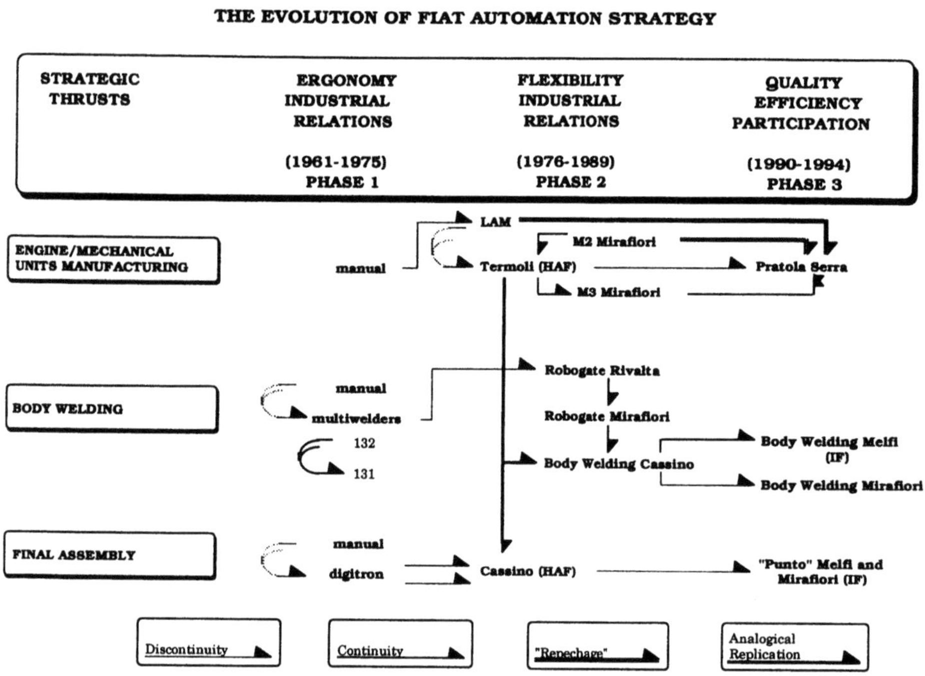

Fig 3.9.9. An evolutionary map of Fiat's automation technologies: phases, linkages and strategic thrusts

innovation, adjustment or replication within each phase; discontinuity, radical innovation between each phase) and Fiat's main strategic thrusts.

Wrapping-up, the longitudinal study of Fiat Auto plants presented in the chapter suggests that, in the case of Fiat, the implementation of automation and the development of the related know-how have a cumulative and path-dependent nature [12,27,33]. In fact, the manufacturing systems progressively put in place in a firm's plants usually represent the basis on which new concepts and solutions are developed. These can result either from internal research and accumulation or from the import of "extra-mural knowledge" [10].

Furthermore, as suggested by the competence-based [26] or dynamic capabilities based [25] views of the firm, the manufacturing equipment used in a firm's plant is the result of a learning process, based on the internal development, external acquisition, imitation, analogical replication, combination, and selection of capabilities [21,30].

Investments in manufacturing technologies are, in part, non-reversible choices, represent long term commitments [16] and hence, sunk costs. Moreover, the knowledge incorporated into technologies becomes integrative part of a firm's repertoire of capabilities. Parts of it can be retrieved over time becoming modules of original technological solutions.

The pressure posed by new, competing equipment (developed by suppliers or by competitors) or by other factors (e.g. the possibility to set out greenfield plants, revolutionary changes in product design, innovative suppliers' strategies or suppliers-OEMs relationships) can induce non-linearity in the automation strategy, i.e. major changes and inconsistencies between the present and the newly implemented manufacturing technologies. Similarly, the necessity to imitate competitors that successfully implemented organizational paradigms based on lean manufacturing [35] in order to respond to the "regime of variety" [11] can cause a mis-match between the existing and the "desired" technological trajectory of a firm.

Finally, technological variables, economic reasons and social constraints imply wide differentiation of automation implementation among the different segments of the automobile manufacturing process, and hence multiple, although intertwined, evolutionary patterns of manufacturing technologies at the firm level.

Different strategies underlie the three evolutionary stages identified. While in the first phase Fiat's automation strategy was defensive and strongly related to industrial relations problems, in the second phase Fiat pursued the technological utopia of the unmanned factory wagering on the potential of computer integrated manufacturing. In the third phase, the automation strategy is complementary and more integrated with the new organizational and human resource strategy.

3.9.8 References

1 Adler P S (1992) Technology and the Future of Work. Oxford University Press, Oxford
2 Baum J, Singh J (eds) (1992) Evolutionary Dynamics of Organization. Oxford University Press, London
3 Bonazzi G (1993) Il tubo di cristallo. Il Mulino, Bologna
4 Camuffo A, Volpato G (1996) Dynamic Capabilities and Assembly Automation: Organizational Learning at Fiat Auto. Industrial and Corporate Change, (forthcoming)
5 Camuffo A, Volpato G (1995) The Labor Relations Heritage and Lean Manufacturing at Fiat. The International Journal of Human Resource Management Vol 6 No 5
6 Cattero B (1992) Inseguendo l'integrazione. Il percorso verso la Fabrrica Integrata alla Fiat di Termoli. Politiche del Lavoro Vol 3 No 2
7 Cerato L et al (1987) La fabbrica: camminando con l'innovazione, SIPI, Rome
8 Cerruti G, Rieser V (1992) Fiat: aggiornamenti sulla Fabbrica Integrata. Quaderni di ricerca IRES No 1
9 Ciborra C, Lanzara P (1984) Progettazione delle nuove tecnologie e qualità del lavoro. Franco Angeli, Milan
10 Cohen W M, Levinthal D A (1994) Fortune favors the prepared firm. Management Science Vol 40 No 2
11 Coriat B (1995) Variety, Routines and Networks: the Metamorphosis of the Fordist Firm. Industrial and Corporate Change Vol 4 No 1
12 Dosi G, Freeman C (eds) (1988) Technical Change and Economic Theory. Columbia University Press, New York
13 Economist Intelligence Unit (1995) Inside Fiat's Melfi Plant: Rivalling Japanese Productivity. European Motor Business 2nd Quarter

14 Fujimoto T (1996) Strategies for Assembly Automation in the Automobile Industry. Chapter 4.1 in this book
15 Fujimoto T (1996) What do you mean by Automation Ratio? Definitions by the Japanese Auto Makers. Chapter 2.4 in this book
16 Ghemawhat P (1991) Commitment. The Dynamics of Strategy, The Free Press, New York
17 Hartley J (1992) Vehicle Manufacturing Technology: A Worldwide Review of Trends for the Future. The Economist Intelligence Unit, London
18 Jaikumar R (1986) Postindustrial Manufacturing. Harvard Business Review No. 6 November-December
19 Jürgens U, Dohse K, Malsch T (1986). New Production Concepts in West German Car Plants. in: Tolliday and Zeitlin (1986)
20 Kern H, Schumann M (1992) New Concepts of Production and the Emergence of the Systems Controller. in: Adler (1992)
21 Kogut B, Zander U (1992) Knowledge of the Firm, Combinative Capabilities, and the Replication of Technology. Organization Science Vol 3 No 3
22 Locke R M, Negrelli S (1989) Il caso Fiat Auto. in: Regini M, Sabel C: Strategie di riaggiustamento industriale. Il Mulino, Bologna
23 MacDuffie J P, Pil F (1996) International Trends in Assembly Plant Automation: Round II Findings of the International Assembly Plant Study. Chapter 4.2 in this book
24 Migliarese P, Romano P (1984) Strategie di progettazione e organizzazione del lavoro: due casi di realizzazione di impianti innovativi in una grande azienda automobilistica. in: Ciborra and Lanzara (1984)
25 Nelson R R (1991) Why do Firms differ, and how does it matter? Strategic Management Journal Vol 12
26 Prahalad C K, Hamel G (1990) The Core Competence of the Corporation. Harvard Business Review May-June
27 Rosenbloom R S, Burgelman R A (eds) (1989) Research on Technological Innovation, Management and Policy. JAI Press, London
28 Sahal D (1985) Technological Guideposts and Innovation Avenues. Research Policy No 14
29 Studio Giano (1989) L'organizzazione del lavoro nella produzione della Tipo allo stabilimento Fiat di Cassino. Research Report January
30 Teece D, Pisano G (1994) The Dynamic Capabilities of Firms: An Introduction. Industrial and Corporate Change Vol 3 No 3
31 Tolliday S, Zeitlin J (eds) (1986) The Automobile Industry and its Workers: between Fordism and Flexibility. Polity Press, Cambridge
32 Tushman M L, Anderson P (1986) Technological Discontinuities and Organizational Environments. Administrative Science Quarterly
33 Tushman M L, Rosenkopf L (1992) On Coevolution of Technology and Organization. in: Baum and Singh (1992)
34 Volpato G (1996) Una strategia di riorganizzazione e rilancio; La Fiat negli anni'80 e '90. Utet, Turin
35 Womack J P, Jones D, Roos D (1990) The Machine that changed the World. MacMillan, New York

3.10 The Development of a Reflective Production System Layout at Volvo's Uddevalla Car Assembly Plant

K. Ellegård

3.10.1 Introduction

When you hear the expression "automobile production" I am sure that a mental picture of an assembly line immediately springs to mind. The assembly line in the automobile industry is probably one of the most wide-spread, industry-related concepts in the world, and hence an important factor of inertia to be taken into consideration when changes in production are about to be introduced.

The aim of this paper is, on one hand, to show that radical changes, such as abandoning the assembly line, are possible and, on the other hand, to indicate both the problems and the satisfaction that may arise when a radically new production system is introduced in the automobile industry. I shall use the example of the planning of a new production system at Volvo's Uddevalla plant[1], inaugurated in 1989. The new production system introduced there has been called the *Reflective Production System*[2]. It is based upon two basic principles. One principle concerns the technical system, i.e. a highly parallelized product flow with a new materials handling system; and the other principle concerns the build-up of competence, i.e. holistic learning. These principles are fundamentally different from those underlying the assembly line[3].

The planning group found that it was a time-consuming and difficult task to pave the way for the Reflective Production System. The new production system did not stand out as a clear idea from the start. It is true that members of the planning group said that they wanted to create something new. But they did not know right away how to tackle the job, and there were, of course, different opinions in the group regarding ways and means; „traditionalists“ confronting „radicals“.

However, many new ideas came up for discussion, and, step by step, a new logic took shape for the relatively labor-intensive assembly work required in large-scale production. These continually occurring, step by step changes in the planning process are shown in figure 3.10.1. The diagram indicates that initially there was a wide gap between the views of the proponents for different production

---

[1] The Volvo Uddevalla assembly plant is now shut down because of the haevy fall in demand for passenger cars in the early 1990's. One reason for the closure was that there was no paint shop and no body shop in the Uddevalla factory. Within Volvo, Reflective Production Systems are now being introduced in other factories.

[2] The Reflective Production System was named by a group of researchers in a book by Ellegård et al. [3].

[3] The principles are defined in more detail in [2].

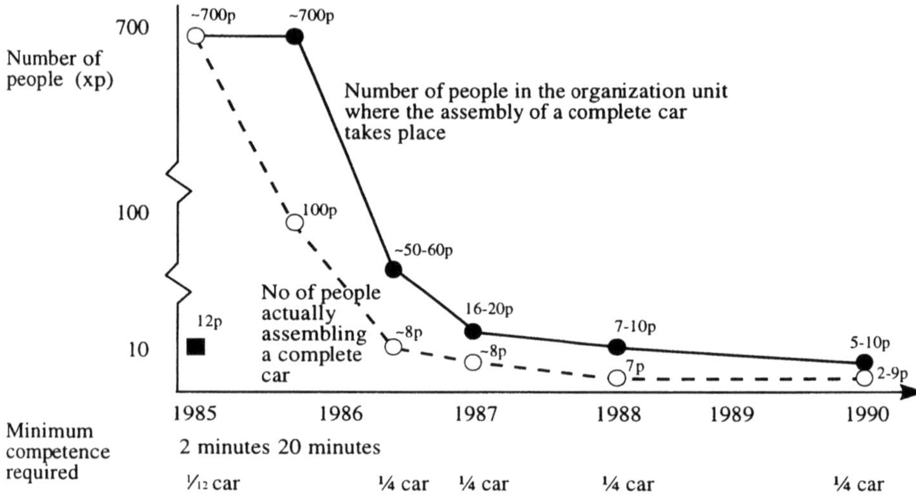

■ *The provocative idea, 12 people assemble the complete car together.*

○- - -○ *Proposed number of people assembling a complete car together.*

●——● *Proposed number of people working in the organizational unit assembling complete cars, i.e in 1985 the entire plant, in 1986 the product shop, in 1987 the team zone and the team from 1988 and later .*

Fig. 3.10.1. A diagram showing 1) the number of people involved in the organizational unit which is responsible for the assembly of the complete car, and 2) the number of people actually assembling a complete, individual car. There is a process of closing the gap between 1) and 2) during the planning period. - The proposed number of people assembling a complete car together equals the proposed number working in the organizational unit assembling complete cars, i. e. in 1985 the entire plant, in 1986 the product shop, in 1987 the team zone and the team from 1988 and later

systems within the planning group. The „radicals" favored the idea that a small, permanent team should be responsible for the total assembly of the complete car. The „traditionalists", who initially dominated the planning group, looked upon this idea as totally unrealistic and utopian. They favored an assembly-line style of production layout, with short work cycles and a high degree of division of labor. In the end, however, it was the radical idea that was actually implemented.

How did the different phases in the planning process evolve? What did the ideas concerning the factory layout look like? What happened during the planning process in order to make such big changes possible as indicated by the diagram in figure 3.10.1.? Why didn't the planning group immediately start planning the factory layout that was finally adopted, even though a proposal, which resembled it very much, existed from the very beginning?

3.10.2
Planning a new Factory

Planning an automobile factory requires a great deal of imagination, especially when the overall goal is to achieve better results than the competitors. There are, in principle, three ways to success:

1) To be better than everybody else in the industry, within the production system known and common to all competitors.
2) To be the most successful producer in implementing improvements within the framework of the existing production system.
3) To develop a radically new production system, not yet known to other competitors.

Most automobile companies try the first and the second way. When Volvo planned the Uddevalla factory, the third way was chosen, a high-risk choice which turned out to be successful.[4]

Of course, all companies want to increase profits from their operations. This ultimate goal can be broken down into more concrete subgoals, irrespective of which production system the company has chosen, e.g.

- high productivity,
- high flexibility,
- high overall efficiency and
- high quality.

All companies try to achieve these goals, and many efforts have been successful. There is, however, one fundamental factor influencing all four subgoals in industrial operations which has not been explicitly taken into consideration, i.e. the human factor. Focusing exclusively on the four above-mentioned goals will, in most social environments, sooner or later lead to recruitment problems and other problems related to the workforce. This is the situation which the Japanese automobile industry is facing today, and it is what Volvo, like many other companies in Sweden, were facing in the 1970s and 1980s.

When Volvo started planning the factory in Uddevalla in 1985, the planning group hence added a fifth concrete goal: *good working conditions*.

How did these five concrete goals influence the planning process at Uddevalla? How did all the different proposed plant layouts meet the human-factor goal? How did the plant actually meet the concrete goals set up for its operations?[5]

---

[4] Initially the Uddevalla plant was planned to be a complete automobile factory with a body shop, a paint shop and an assembly shop. However, early in 1986, the Volvo Car Corporation decided to build only an assembly shop in Uddevalla.

[5] The Volvo Uddevalla factory went into production in 1989. Due to falling demand in the market the total capacity of the plant was never utilized and in spite of very good results the plant was shut down in 1993. One reason for the closure was the falling demand in the car market. Another central reason was that the Uddevalla plant lacked body and paint shop.

3.10.3 Efforts to Achieve the Goals: The Evolution of a New Production System Reflected by the Layouts

The prevailing view of change and improvements in Sweden is, paradoxically, fairly static and often only two points in time are identified. A starting point is described as t0 and a goal is set up at t1. The road between the two points of time is often regarded as a highway with no alternative routes. The process itself, the movement from t0 to t1, is therefore neither seen as an instrument for achieving the goal, nor as a means of learning during the planning stage.

When Volvo's Uddevalla plant was about to be created, a fundamentally new orientation towards the process of change and improvement was developed. The focus moved away from the concept of just two distinct points in time (t0 and t1), with only one road between them, and subsequently concentrated on the movement at the crossroads as a means of fruitful learning, in order to find the most profitable road between the starting point and goal. This led to a radically new situation for the planning group which could now make the best choice step by step. This certainly meant extra work for the planning group. But at the same time, important factors were identified and could be taken into consideration. They were not simply left aside.

This process explains why so many different layouts were sketched during the planning of the Uddevalla factory. All but the very last two of these layouts were, however, drawings on paper which never left the writing desk. In this section, I will present the complete sequence of layouts developed by the planning group between 1985 and 1990.

3.10.3.1 A Provocative Idea

Volvo carried out many experiments during the 1970s in order to find an alternative production flow principle to replace the assembly line. One of the experiments was made at the Volvo Truck Corporation, where a small team of workers assembled complete trucks at a stationary dock station. The results were impressive, far better than expected in the wildest dreams of the project group. But the group and its engineers could not explain why the experiment was successful. Hence, no in-depth changes were introduced in the company's other factories and the assembly line continued to dominate.[4]

When planning of the passenger car factory in Uddevalla started, one of the engineers from the Volvo Truck experiment was recruited for the planning group. He found this to be an excellent opportunity to test the idea of a small team building complete cars. He was the father of one of the very first layout ideas presented in the planning group in April 1985: A small team of 12 people was to assemble the complete car, with the car standing still on one single station during the entire assembly process. This brief sketch of a layout indicates that each worker had to master at least 1/12th of the work necessary to assemble the complete car (compare with fig. 3.10.1.).

This idea was presented, but traditionalists immediately dismissed it as provocative, unrealistic and naive. They presented two main objections against implementing such a revolutionary idea: first, they maintained that workers were unable to learn more than a maximum of 20 minutes of assembly work without losing working speed and product quality. Second, should workers somehow really prove to be capable of learning more than 20 minutes of work, then tremendous problems would arise in terms of materials handling. It is not possible, they argued, for a small team to simultaneously handle the very large number of components necessary for the assembly operations at the work stations.

The idea, which was set forth within the planning group, was not, however, presented as the whole group's recommendation for a factory layout. The more traditionally minded members of the planning group did not take any notice at all of the new idea in their work on the group's recommendation. But the father of the idea commissioned a researcher with the development of a solution to the materials-handling problems. There must, he argued, be a way to arrange and structure the great number of components to be sent to the assembly team in order to support and facilitate their job.

The planning group also commissioned a researcher who had specialized in so-called holistic learning. His task was to show that it is not only theoretically possible for a person to learn how to assemble a complete car, but that enlarged competence is favorable for productivity, quality, flexibility and production-related improvements. The two researchers, with very different background knowledge, one from engineering, the other from education and psychology, very soon started cooperating as their ideas coincided.

3.10.3.2
An Initial Orientation Towards a Traditional Layout

In May 1985, the planning group presented its first official recommendation for the layout of the assembly shop in the Uddevalla plant. It showed a serial flow, based on assembly line principles. The layout did not take the goal to create good working conditions - i.e. the human-oriented goal - into serious consideration, but the document included a discussion on the human factor.

According to this layout, each member of the workforce of around 700 in the assembly shop had to perform a work cycle of about 2 minutes on each car. There were hundreds of serially linked work stations. Everybody had to make his own little contribution to every car completed. The degree of division of labor was hence very high. (fig. 3.10.2.)

This layout obviously did not meet the human-factor goal: 1) repetitive tasks due to short work cycles were very intensive, 2) the new holistic learning theory was not taken into consideration and 3) the new materials handling system was not introduced. Volvo's top management and union representatives all drew the conclusion that the layout was not good enough. The planning group was asked to create a more radical layout, meeting the human-factor goal more explicitly.

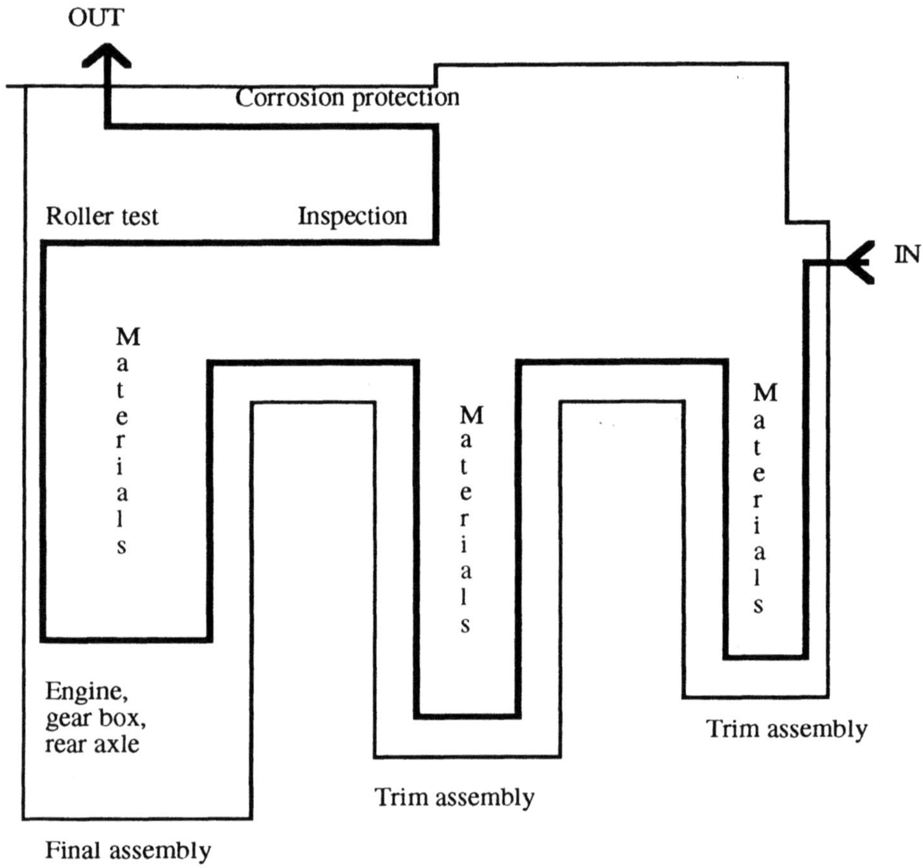

Fig. 3.10.2. The first layout presented by the planning group in May 1985. It was a traditional layout with a serially linked product flow through the plant and with short work cycles (about 2 minutes). Each of the around 700 workers had to perform his small part in the assembly for each individual car

3.10.3.3
A Big Change Within the Well-Known Logic of Automobile Production

It was not easy for the planning group to find a good alternative to its first layout. A solution to this problem was developed when traditional principles for materials handling were reappraised. After the engineering researchers in the group had placed all the components of a single car – a de-assembled car – on the floor in a big hall, the planning group realized that the workers would easily be able to visualize the completed car. Starting from a view of the whole car, continuing with the bigger, easily identified components and by going more and more into detail in this manner, the process of understanding the position of each component became much easier. If the components were given to the assembly teams in the form of previously prepared materials kits (with the components arranged exactly for the specific tasks of the assembly team), then the extended assembly work became easy to learn and perform. Several - previously traditionalist - engineers in the planning group realized that this was a good idea, and they prepared kits of 1/8th of the car, corresponding to a cycle time of around 20 minutes.[6]

The second official recommendation for a layout was presented in December 1985. This layout showed a semi-parallel flow in the assembly plant, i.e. the product flow was partly serial and partly parallel. In a serial flow, the work stations follow one another in sequence, and the product has to pass through all stations in the flow for its completion. In a parallel flow, the work stations are parallel to each other, which means that a product has to pass through only one of all the parallel stations for its completion. The work cycle time is much longer in a parallel flow than in a serial flow.

In the semi-parallel flow in the layout now recommended by the planning group, the car was sequentially assembled in eight shops, serially linked, each shop being called a product shop. Within each product shop, however, the assembly work required for each individual car, had to be carried out at only one of the total number of all the parallel work stations. Consequently, 1/8th of the car was assembled at each station. Because of the parallelization within the product shops, fewer than 100 workers (of a total workforce of 700 in the entire assembly plant) were directly involved in the completion of each individual car. This meant that the degree of division of labor was much lower than in the previous layout. The minimum skill requirement for each worker corresponded to 20 minutes of assembly work for the car. (fig. 3.10.3.)

The coordination of assembly and materials handling was untraditional. Assembly workers themselves were expected to prepare their own materials kits in the materials warehouse. There was also a certain amount of pre-assembly to be performed in each team.

By the time this layout was presented, it had become clear that the cost of the buildings required for this layout was too high. One reason was that the plant was located in a former shipyard on the shipyard dock. In addition, the building consisted of several floors.

Therefore, the planning group had to find a new and cheaper assembly plant location and layout. When this happened, the two layouts so far presented by the planning group had already been discarded. A creativity crisis emerged for many members of the planning group.

---

[6] According to the holistic learning principles, 1/8 of the car is too limited to reach the potential generative effects of these principles. The minimum competence level was thought to be at least ¼ of the Volvo car. Only then could each worker relate his own part of the work to the whole - the essential idea behind holistic learning.

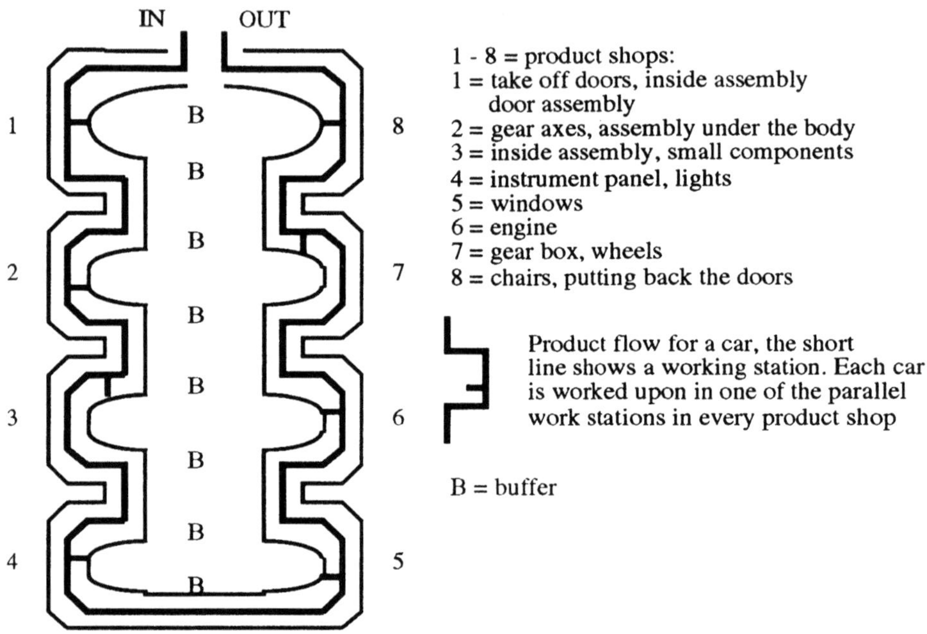

Fig. 3.10.3. The second layout presented by the planning group in December 1985. One eighth of each car had to be assembled at one station (of a larger number of identical, parallel stations) in every one of the eight product shops. The cycle time was about 20 minutes, and around 100 people were involved in the assembly of a complete, individual car

In order to find a way out of this deadlock, the project leader appointed a small, special subgroup, where the radicals within the planning group held a strong position. The mission was to create ideas for a new and non-traditional layout.

3.10.3.4
A Fundamental Change Forming the Base for a New Production System

The special group presented a brief sketch of a layout in January 1986. The main idea behind the layout was that there should be no assembly line at all in the assembly factory and in addition, the design was meant to make it physically impossible to arrange the product flow along an assembly line. Therefore, the assembly plant was now divided into eight separate buildings called product shops. The new materials handling principle of making component kits for each individual car was fully put into practice. Each product shop was to be supplied with component kits from one central materials handling shop. The crucial point was that each of the product shops now had to assemble its own complete cars (fig. 3.10.4.). In the previous organization, each car could pass from one work

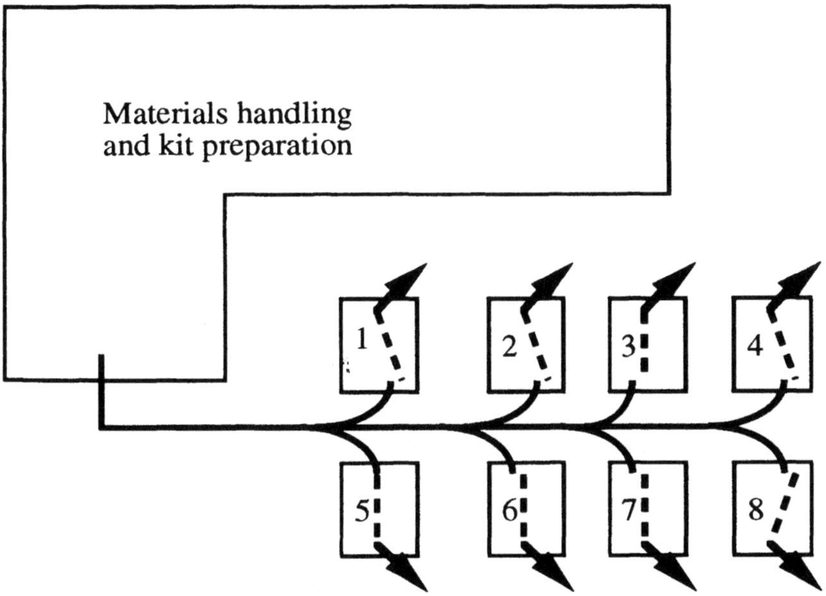

Fig. 3.10.4. Brief sketch of an idea for further development, submitted by a small, special subgroup within the planning group in January 1986. In each of the eight parallel product shops (located in eight separate buildings), workers had to assemble complete cars. (1-8 = Buildings, one building for each product shop, the product shops are arranged parallel. — — — —> = Product flow, indicating that complete cars are assembled in each product shop)

station in one product shop to a subsequent one in the next shop. With this organization, the finished car would hence have been the product of all the eight product shops.

The special subgroup never submitted a detailed layout of the plant interior, but the verbal presentation of the product flow made the group's intentions clear.

One important factor behind the group's idea was the favorable results from the experiments with materials structuring and kitting and from empirical investigations of the human potential for the learning of long work cycles. One worker, previously untrained in automobile assembly, learnt to assemble a complete car using the principles of holistic learning and materials arrangements. The results were good, and doubts about the human learning capacity in automobile assembly faded away.

By now, the planning group knew - theoretically and empirically - the potentials of the holistic learning principles and of the new materials handling principles, as well as the close relationship between them. The planning group also knew that they were expected to implement the new principles in a full-scale factory. Once again, they started developing a layout, this time in accordance with the new experience and knowledge.

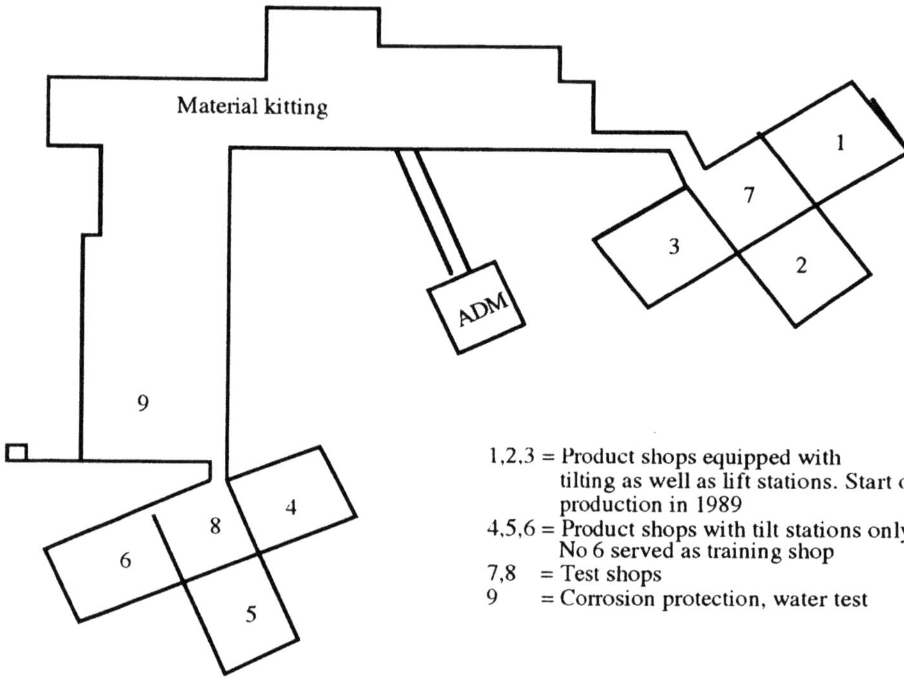

Fig. 3.10.5. The plant layout submitted by the planning group in June 1986. Complete cars had to be assembled in each of the six parallel product shops

3.10.3.5
Assembly of the Complete Car in a Product Shop

The concept of assembling complete cars in separate, parallel product shops had by now been adopted by the planning group as a whole, and Volvo's top management and unions supported its further development. The planning group was asked to develop a detailed layout. The result was submitted in June 1986.

The central materials handling shop was located in an existing building on the industrial site where the factory had to be built. The number of product shops was reduced to six. Attached to two corners of the materials handling building, six product shops had to be built. The three product shops at each corner were connected to each other and to the materials handling shop by means of a passage through which car bodies and components were sent for assembly and where the complete cars were returned for delivery. (fig. 3.10.5.)

The layout of all six product shops was meant to be identical. Thus, the product shops were arranged parallel, and complete cars had to be assembled in each of these product shops.

Each product shop was divided into four so-called "team zones". In each team zone, two teams performed assembly work. The team zones were serially linked

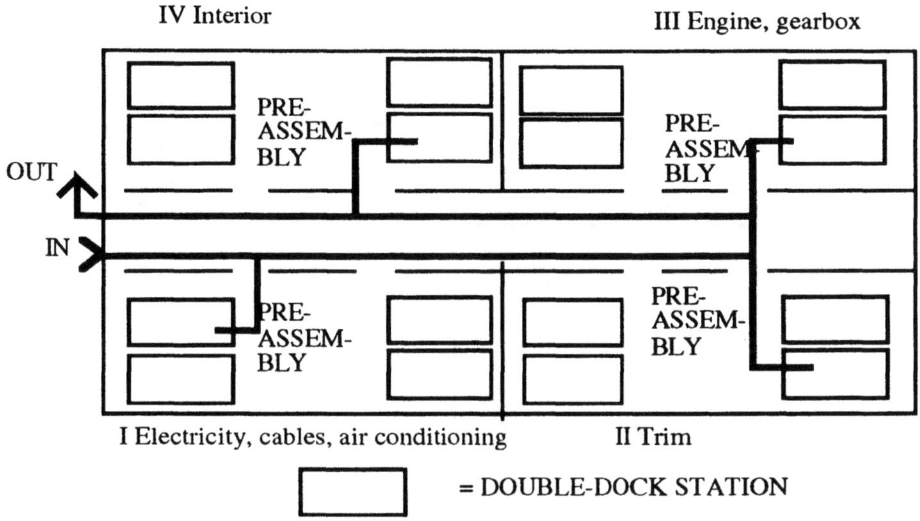

Fig 3.10.6. The layout of one product shop in June 1986 (compare with fig 3.10.5.). Within the product shops, there were four serially linked team zones in which a 1/4 of the car was assembled (the work contents are related to their respective functions in the car). Within the team zones, there were four stations parallel to each other. Eight workers, two in each team zone, were responsible for the entire car assembly work. (Phase in production: I = electricity, cables, air conditioning; II = trim; III = engine, gearbox; IV = interior. ——> = flow of one car)

to each other, each team zone completing one quarter of the car. Figure 3.10.6. shows the product flow through the product shop in the layout from January 1987.

The members of the first team zone assembled the first quarter of the car which was then sent to the next team zone. In the second team zone, the second quarter of the car was assembled and so on, until the car was completed in the fourth team zone, from where it was eventually sent to the roller test and then on to the undercoating and water test.

The components for each car had to be sent to each product shop in four different materials kits, one kit for each team zone. Thus, the first kit had to be sent, together with the body, to the team zone responsible for the assembly of the first quarter of the car. The next kit was sent to the second team zone, and so on.

Therefore, the minimum skills required for the two teams in each team zone were initially fixed at one quarter of a car, with the possibility of increasing the requirements at a later stage.

One quarter of each individual car was assembled while standing still at one of the stations within the area of each team zone. The need for a buffer area between the team zones arose in order to compensate variations in assembly time and to eliminate time losses due to operational interruptions. Once the car had left the first two workers in the first team zone, these workers had no further

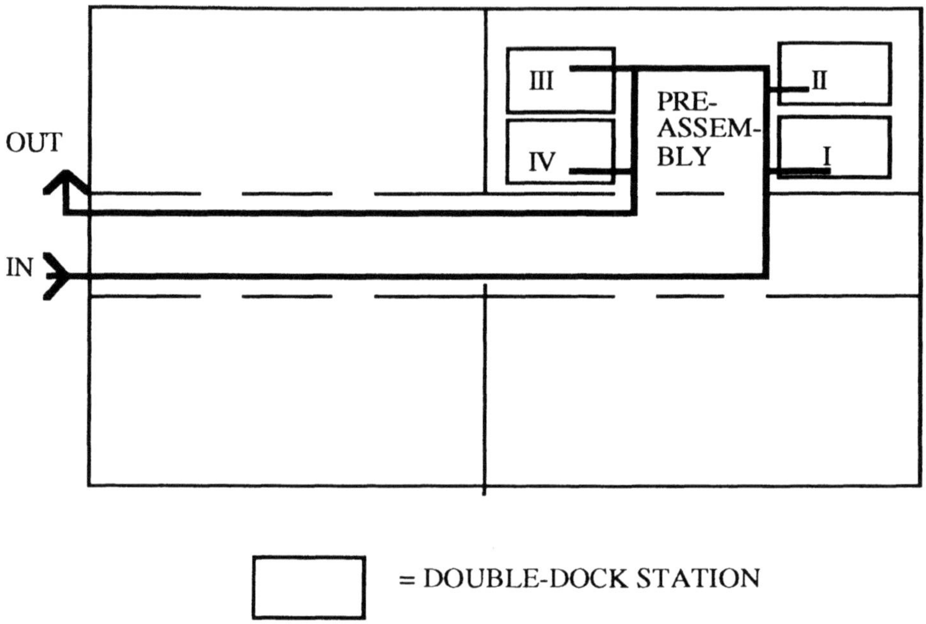

Fig. 3.10.7. The layout for the product shops as submitted in January 1987. The layout from June 1986 was scaled down, i.e. the four team zones were now arranged parallel to each other, and within each team zone, there were four serially linked work stations where every complete car had to be assembled by approximately eight workers. (Phase in production: I = electricity, cables, air conditioning; II = trim; III = engine, gearbox; IV = interior. ——> = flow of one car)

contact with this car. Therefore, the danger of information loss problems between the team zones arose. Quality standards might not always be met, for example.

The car had to be transported quite a long distance between the four different team zones, and transport problems could also arise even within the team zones.

3.10.3.6
Scaling Down: Assembly of the Complete Car in a Team Zone

During the subsequent six months, one question was eagerly discussed. It concerned the transport of the materials racks (carrying the materials kits), the bodies and the complete cars. Why were all these transport operations considered to be necessary? Why must the car under construction be transported within the product shop over such long distances and so many times?

One possibility to reduce transport operations, it was argued, would be to geographically concentrate the assembly within the product shop. The planning group therefore adjusted the layout inside each product shop for this purpose. The assembly of the complete car was now concentrated geographically. The shop floor area used for the assembly of one individual car was "scaled down", from the product shop scale to the team zone scale.

The degree of parallelization hence increased. Now not only the product shops, but also the team zones within the product shops were arranged in parallel. Each individual car was now transported to just one team zone area where it was completed. Within each parallel team zone, there were now serially linked work stations for the assembly of the individual car. The car had to be moved three times within the team zone area between the four work stations where the entire assembly work was performed, one quarter of the entire work at each station.

By now, many changes have been introduced since the start of planning in early 1985. Many ideas were taken into consideration, and the spirit and determination of the planning group were sometimes so high that detailed calculations of the consequences were not considered carefully enough. By spring 1987 (fig. 3.10.7.), when the new buildings were erected, and the first product shop was to be equipped, it became obvious that there was not enough space in the product shops to reach the capacity goal (40,000 units a year). Something had to be done, and very quickly. The solution to this problem was to reduce the space reserved for buffer stocks and transport within the product shops. The result was an assembly which was geographically even more compact.

3.10.3.7 Scaling Down: Assembly of the Complete Car in a Team: Stage I

The number of transport operations between work stations had to be reduced. Until now, the car had to be completed at four different stations in a sequence, a 1/4 of the car at every separate station. Hence, the car had to be moved three times between the assembly stations, and once more into and out of the team zone area. Now it was decided to assemble half of the car at each one of two different stations. This step had great implications.

In the layout submitted now, each team had to take the entire responsibility for the assembly of complete cars. The team had four work stations at its disposal. Each individual car, however, had to be assembled at only two of these stations. Therefore, one consequence of this geographically concentrated assembly was that assembly skills were even more concentrated in the organization, as complete cars had to be assembled by each team. In the previous plant layout, called "complete car in team zone" and discussed in section 3.6, the knowledge and skills to assemble only half of the car were expected from every team. The layout had been called "complete car in the team zone", as there were two teams in every team zone.

According to the new layout, all skills for the assembly of a complete car were required from each team. Hence, each team had the technical equipment necessary to complete each individual car at only two different stations. First, the body (and the relevant materials kit) was sent to a station equipped with a tilting device where the body could be turned 90 degrees. Second, the semi-assembled car was transported to a station with a lifting device, and the appropriate materials kit was sent there from the materials handling department. Half of the assembly

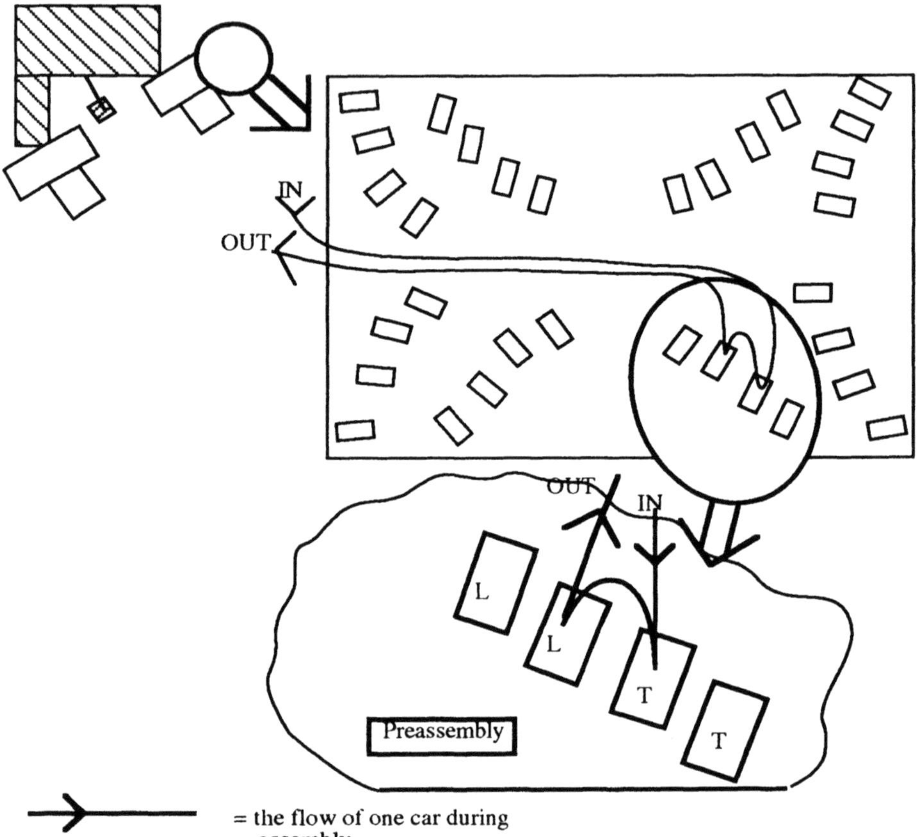

Fig. 3.10.8. The layout submitted in January 1988. This layout was implemented in the product shops Nos. 1, 2 and 3. Each team in the product shops had four stations at its disposal respectively, two lifting stations and two tilting stations. Each car was moved to one of the tilting stations where half the car was assembled, and then to one of the lifting stations where the car was completed. Seven workers jointly assembled the complete car

work was hence performed at the tilting station and the other half at the lifting station. (fig. 3.10.8.)

This layout was actually introduced in the first three product shops where production started in June 1989. In each product shop, there was a team organization consisting of eight teams, each with 8-10 members . The minimum skill requirement was still a 1/4 of the car for each team member, but the division of labor was initially fairly high: seven workers had to assemble the complete car at two of the four stations to which they had access in the team area (one tilting station (T) and one lifting station (L), see fig. 3.10.8.).

One advantage with the small area in which complete cars were assembled is that it was not necessary to start all teams or to open all product shops simultaneously. The product shops can be opened sequentially.[7] This advantage was used in the starting up of the plant. After some time in production however, it became clear to many people that the layout implemented in the first three product shops had some drawbacks. First, an invisible borderline developed between the workers involved in the initial assembly and the workers who performed the second half of the assembly. The borderline emanated from the geographical division of the assembly work into two different stations. Second, and as a consequence of the first drawback, the development of skills did not proceed as expected. Workers who had once started performing assembly work at the first station (the first half of the car), continued doing so, and the same happened at the second station. In other words, by reasons emanating from the arrangement of the technical equipment, the teams were divided into stable subgroups. None of the subgroups was in control of the full completion of the car. Hence, the third drawback: improvements initiated by workers were not as frequent as desired and expected.

At an early stage, the planning group focused on ergonomics and especially the movements and positions of the human body during work. Tilting equipment was installed in order to eliminate many dangerous body positions, but in the first three product shops only fifty percent of the assembly work could be done at a tilting station. Experiments were carried out in order to improve the layout from the ergonomic point of view.

3.10.3.8
Scaling Down: Assembly of the Complete Car in a Team: Stage II

There are ergonomic reasons for the repeated lifting and tilting of the car body during the entire assembly process rather than during the first half only. Fewer bends and under-up positions during work reduces the risk of work-related injury. In the Uddevalla plant, a small project group of assembly workers was formed, assisted by engineers and researchers, in order to develop a better assembly order using only a combined tilting and lifting device. They introduced a new assembly order, and they achieved substantial ergonomics and productivity gains from this alternative way of work organization.

One consequence of the work in this small group was that the product shop layout could be made even more compact: now, just one single station was needed for the complete assembly of a car. We note that this was the original idea proposed by one of the radicals in the planning group as early as in spring 1985, see section 3.10.3.1.

The result from the efforts in this small group of assembly workers and engineers was the basis for equipment planning in the three remaining product shops

[7] Conversely, if there is a need for reduced production, then one or two product shops could be closed, and the others could go on working with no disturbance.

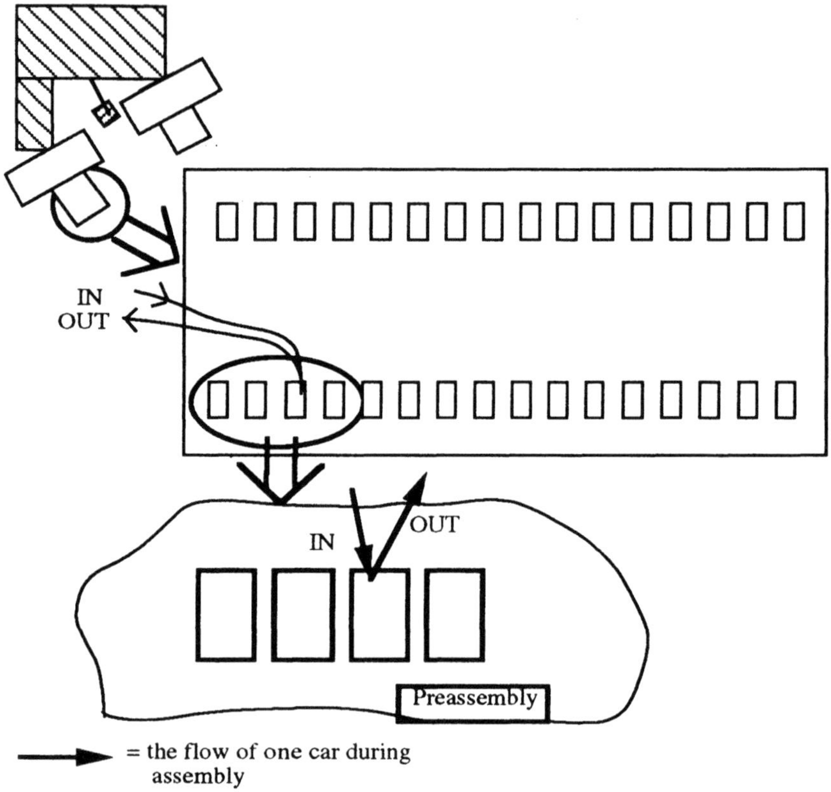

Fig. 3.10.9. The layout submitted in January 1990. This layout was implemented in the product shop Nos. 4, 5 and 6. Each team had four work stations (with a tilting device) at its disposal, but each individual car was completely assembled at just one station. There was a total number of eight teams in each product shop. The number of workers involved in the assembly of one car varied between 2 and 9

at Volvo's Uddevalla plant. This was the second layout which was actually implemented (fig. 3.10.9.).

In the three product shops equipped according to this so-called "tilt layout", each team still had four work stations at its disposal. Now, however, every car had to be completely assembled at one of these stations. Consequently, the car was not moved at all during the assembly process. The geographical stability of the car in conjunction with a minimum assembly skill requirement of a 1/4 of the car led to a flexible pattern in the division of labor. This layout opened up many possibilities for the use of different levels of the division of labor, between and within different teams. Instead of moving the car, workers could move from car to car, from station to station within the team area, in order to perform their work tasks according to the skills level of the team and the division of labor in the team as it had been agreed upon. [3]

3.10.3.9 Materials Handling

As of early 1986, materials handling was centrally organized in all the layouts. The big change from the former materials handling systems was that individual materials kits had to be prepared for each individual car. Therefore, a lot of materials picking and preparing could be done in the materials handling department. Materials handling was intimately related to information technology and computers. The assembly teams requested component kits via computer terminals. The terminals were used by the materials handlers to obtain lists of all the components necessary to prepare the kits. The lists were instructions for their work, so that they could find the proper components for the individual car demanded by a customer.

Different kinds of components were handled in different ways. Large components were placed directly on the materials racks. Middle-sized components were put into small boxes, together with components related to the same sequence in the assembly process. Very small components - screws and bolts, etc. - were automatically pre-kitted into plastic bags by a machine. The plastic bags were put into the small boxes, together with middle-sized components: for instance, a plastic bag with the special screws and washers for the assembly of the car mirrors were put into the same small box as the mirrors, and the box was finally put on the materials rack to be sent to an assembly team.

However, a central materials handling department should not be necessary in a Reflective Production System. It will probably be helpful to decentralize materials handling operations by having the assembly teams prepare their materials kits themselves. One main reason is that the quality of the materials kits (i.e. reducing the number of wrongly picked components in a kit) can be improved when the person who puts the materials kit together knows the position and the function of the component in the complete car. As a rule, workers in the materials handling shop do not know the position and functions of the components in the car, they think of the components simply as article numbers. This way of thinking differs vastly from assembly worker thinking and is far from the idea of holistic learning. Therefore, materials handlers and the assembly workers should at least know each others' tasks and ways of reasoning in order to maintain a constant and high quality level for component kits.

3.10.4 Results

Finding the best road among all the alternative routes that appeared during the planning process took a long time. Many, many changes in direction were made as new experience and new knowledge developed in the planning group. It might seem as a haphazard process, but from the very beginning, the group received a feedback regarding the correctness of the direction they had chosen, towards the layout favoring "complete car in the team". During the development process, it was tested both theoretically and empirically, and proved to be successful.

An interesting reflection is that – despite the long drawn-out planning process – the performance of the Uddevalla factory after just three years in operation showed better results than conventional assembly line plants in Sweden. Comparing the five goals set up for the Uddevalla plant with the eventual performance of the plant, the result was favorable:

Productivity:

Measured as the number of hours needed to assemble a car (direct and indirect time included). Productivity was very good in the Uddevalla plant. It was not satisfactory at the beginning, and many improvements were introduced during the first five years of operation. Thus, by late autumn 1992, the factory needed only about 32 hours to assemble a car. In the conventional assembly line plant in Sweden, 42 hours were needed for the same job at that time.

Quality:

The quality measurements at the Uddevalla plant showed favorable results. There was a steady improvement, and Volvo's goal for quality standards was met with a good margin. An interesting fact is that, even after the decision to shut down the plant, quality continued to improve rapidly, and before the closure in May 1993, it was far above the official Volvo goal.

Efficiency:

One measure of efficiency is the cost for model changes. Three years in succession, the Uddevalla plant made the annual model change cheaper than the big assembly line factory in Göteborg, as measured per car. This applies to both investment costs per car and education costs per car. Considering the logic of the Reflective Production System, this was an expected result. First, the product shops used only a few expensive devices. Second, employee skills were generative, i.e. the existing knowledge creates a very quick understanding of the changes which are required in the assembly work already mastered.

Flexibility:

The Uddevalla factory was flexible in several ways; here are just a few examples: 1) it was possible to simultaneously produce many different car models and many variants in the plant, with extremely limited balance losses. In fact, the number of car models which could be produced at one and the same time could equal the number of teams. 2) New models could be introduced efficiently and quickly. One team might invent a good working order which could then become the recommended working order for the other teams. 3) Variations in demand could be met by shutting down just one product shop rather than the entire factory. 4) It is possible to work according to different forms of division of labor in the different teams, depending on the skills level and the individual preferences by the team members.

Good working conditions:

In the Uddevalla factory, there were two important principles for recruiting workforce. One was the aim that at least 40% of the workers should be women. The other was to employ different age groups, young, middle-aged and older people. The overall intention was to make use of individual differences in production, by a suitable gender and age mix and by looking upon the people's different skills and experience as important driving forces which could be displayed in the different situations facing each team.

The workers in the teams could decide themselves when the next car to be assembled had to arrive at their work stations. This is a consequence of abandoning the assembly line. Therefore, workers were, on the one hand, able to achieve the desired product quality from the beginning, even if problems arose during the assembly process, as there was no external force transporting the car away. On the other hand, they could order a car earlier than expected if assembly work was done smoothly and if the previous car was ready ahead of schedule.

The team was made up of about eight people and each member of the team (not always the same person) performed an indirect work task related to the assembly work. Such indirect tasks included quality assurance, improvements in production techniques, educational efforts, administrative tasks for the team, etc. Assembly workers at the Uddevalla factory hence enjoyed a flexible work environment.

Another important human factor at the Uddevalla factory was that work cycles were long, so that many ergonomic problems were reduced or even avoided. Repetitive tasks are a central problem in ergonomics, as the human body cannot tolerate highly repetitive and short cycle work without a high risk of work-related injury. Assembling a 1/4 of a Volvo car means a cycle time of about two hours. Therefore, no worker had to repeat the same operation more than four times a day, where the division of labor principle implied that four workers jointly assemble the complete car.

3.10.5 Final Remarks

However, due to the heavy fall in the world market demand for passenger cars, Volvo's Uddevalla assembly plant was shut down in 1993. One reason for the closure was that there was no paint shop and no body shop in the Uddevalla factory. One explanation - in addition to the official one given by the Volvo Car Corporation management (see also footnote No. 5) - is that many people in Göteborg, where the parent company is located, were convinced from the very beginning that the Uddevalla factory could not become successful, and hence did not care to follow its development. Many people in Göteborg expected failure. Therefore, it was difficult for many of them, who were not directly involved in the planning of the plant, to really understand its prerequisites and its future potential. It also prevented them from seeing that the plant was, in fact, successful.

In early 1995, however, Volvo decided to re-open the plant. This time, a separate subsidiary was founded, named Autonova, as a joint venture where TWR

(Tom Walkingshaw Racing) owns 51% and Volvo 49%. Autonova will operate according to the reflective productions system principles, and the layout differs somewhat from before. It is too early yet to make a detailed description of this layout. One major change, however, is that materials handling is now integrated into the assembly work. This might result in substantial improvements in the overall productivity. Autonova will be a complete car plant, so that a body and a paint shop will be built. Production of sports cars and convertibles will start in late 1996, even though the first cars were already built there within the framework of a training program for employees in June 1995.

Within the Volvo organization, the Reflective Production System thinking is being introduced in other factories too, as the understanding of the logic of the Reflective Production System grows. In Göteborg, trucks are assembled according to the Reflective Production System principles, and in Skövde, the same applies to engine assembly. Other plants in the Volvo Corporation are in the initial phase of putting a Reflective Production System into practice. An understanding of this alternative to the assembly line has developed internally after the initial inspiration from the Uddevalla plant.

3.10.6 References

1 Ellegård K (1989) Akrobatik i tidens väv. En dokumentation av projekteringen av Volvos bilfabrik i Uddevalla. Choros 1989:2. Dept of Human and Economic Geography, Göteborg University

2 Ellegård K, Engström T, Nilsson L (1990) Reforming Industrial Work -Principles and Realities in the planning of Volvo's car assembly plant in Uddevalla. The Swedish Work Environment Fund

3 Ellegård K, Engström T, Johansson B, Medbo L, Nilsson L (1992) Reflektiv Produktion. Industriell verksamhet i förändring. AB Volvo

4 Engström T, Jonsson D, Johansson B (1994) Design Assumptions and Empirical Evidence Concerning Parallelized Long Work-Cycle Assembly. In: Eds Bradley G E, Hendrick H W (eds) Human factors in organizational design and management-IV. North-Holland

4

Issues and Dynamics

4.1 Strategies for Assembly Automation in the Automobile Industry

T. Fujimoto

4.1.1 Introduction

The purpose of the present chapter is to identify and analyze some of the alternative strategies for assembly automation, from both a historical and a total system point of view. In order for automobile manufacturers to form and implement assembly automation systems for the future, such strategies may serve as useful guidelines. The types of strategies that the present paper explores are, in a sense, ideal types, in that the pure form of each strategy may be hard to find in actual auto makers: in practice, they normally adopt hybrid or mixed strategies. However, it may be useful to clarify the direction and shifts in emphasis in terms of basic objectives, philosophies and policies for constructing future assembly systems. This paper will describe and analyze four types of assembly automation strategy[1].

4.1.2 Types of Assembly Automation Strategy

Let us now explore some types of strategies for assembly automation in the automobile industry. By assembly automation strategy we mean a coherent set of decisions on building and utilizing capabilities of assembly automation in order to improve manufacturing performance. It is a subsystem of overall manufacturing strategy that is associated with performance and capability of the total production system [e.g. 19]

Figure 4.4.1. presents four major types of assembly automation strategies that the author has identified, based on historical and comparative analyses of the assembly process of the auto makers. The classification is based on differences in the focus of performance that the automation systems are expected to improve[2].

The first criterion for the classification is what to improve. This means, the automation strategies may be classified according to whether it is targeted toward

---

[1] The strategies for assembly automation are affected by the overall trends of total production systems in the automobile industry. For the evolution of automobile production systems, see the introduction to this book.

[2] Note that the four strategies do not logically exclude against each other. That means that an actual company can adopt a mix of multiple strategies.

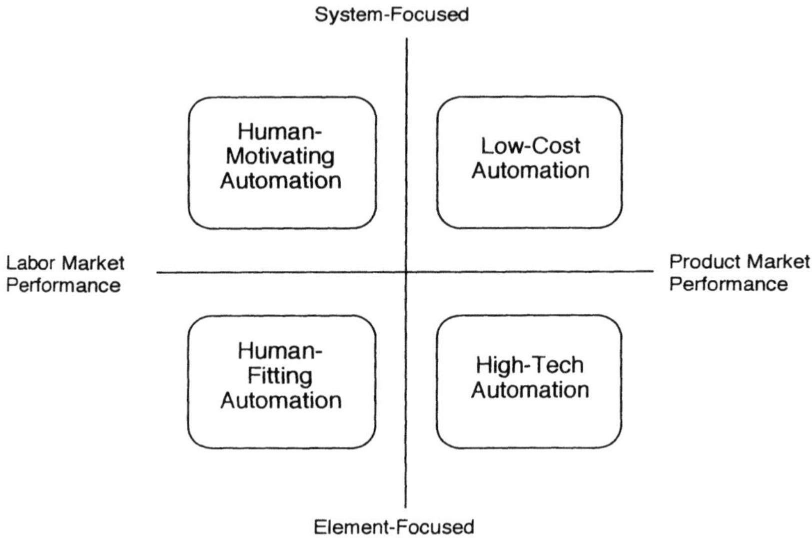

Fig. 4.1.1. Four types of assembly automation strategies

product market performance or labor market performance. Product market performance, or competitiveness, measures how a firm's products attract potential customers and satisfy existing customers. Quality, cost and delivery are often referred to as major competitive indicators in manufacturing. Labor market performance, on the other hand, means job attractiveness for potential employees and job satisfaction for existing workers. At the same time, companies need to satisfy shareholders and, in the long run, they may focus on performance improvements to a particular shareholder group at a given time.

The second criterion of classification is how to achieve improvements in a given performance. One approach is *element-focused;* this assumes that superior performance of each element of the automation system, such as each individual piece of automation equipment or work station, should add up to enhancement in the total system performance. The other approach is *system-focused;* here it is argued that total system performance is more than a simple sum of element performance, and that basic design or conceptualization on the total system should precede element designs.

Based on the two-dimensional scheme, we can classify the basic strategies for assembly automation into four categories:

1. *High-tech Automation Strategy*: This approach focuses on technological improvements of individual automation equipment. It is assumed that this will contribute eventually to certain factors of product market performance, such as cost and quality competitiveness.

2. *Low-cost Automation Strategy*: This strategy directly targets total system performance. It aims at improving productivity, quality, delivery and flexibility with simple automation equipment in order to save total investment cost. If an automation advancement introduction is expected to cause a decrease in the total system performance, such automation will be deliberately avoided under this strategy.
3. *Human-Fitting Automation Strategy*: This strategy focuses on enhancing the attractiveness of each work station by replacing so called 3-D (dangerous, dirty and demanding) tasks with automated or semi-automated equipment. In other words, it tries to improve job satisfaction by eliminating dissatisfaction of individual workers at a physical or physiological level.
4. *Human-Motivating Automation Strategy*: The last approach is oriented to total system designs for employee satisfaction and job attractiveness. It pays particular attention to the problem of job alienation, inherent in the traditional Ford-style moving assembly lines, and tries to modify or radically change the overall production systems and work organizations. Or, alternatively, it may start with a highly automated assembly line and try to blend a repetitive task (i.e. residual work) and a non-repetitive task (i.e. maintenance and engineering) into one job in order to alleviate stress and boredom resulting from the former tasks.

The above automation strategies do not exclude one another. Moreover, it should be noted that these strategic types are ideal types: each of the actual firms is likely to adopt a mix of multiple strategies. However, the firms may focus, in relative terms, on a particular strategy at a given time, and the focus may shift in the course of their competitive pressures, labor situations and company performance changes.

The patterns of the strategic focus may also differ within firms or regional groups (fig. 4.1.2.). The choice of assembly automation strategies would be more or less linked to the transformation of the total production system concepts discussed in the introduction to this book. For example, many of the Western mass-producers, such as GM, VW and FIAT appear to have pursued the *high-tech automation* strategy during the 1970s and 80s, but, in terms of its overall competitiveness, this strategy has been challenged by the *lean production system*. On the other hand, the prevalent strategy among the major Japanese auto firms, such as Toyota, has been the low cost strategy, which seems to be an important component of Toyota-style production systems. This strategy contributed to the competitive advantages held by Japanese firms during the 1970s and 1980s, but, due to the domestic problem of labor shortage in the late 1980s to early 1990s, these companies shifted their emphasis to the human-fitting automation strategy. More recently, Toyota has started to incorporate certain human-motivating elements into its design of assembly processes and assembly automation equipment [32]. On the other hand, some European companies, pursuing the Volvoism production concept, have apparently attempted to introduce human-motivating automation (e.g. automation in the parts picking area at Volvo Uddevalla plant), but they have not yet been able to demonstrate reasonably high performance in the market place[3].

Thus, although the auto companies in different regions tended to follow different evolutionary paths, none of the above strategies for assembly automation, in its pure form, seems to be dominant in the global market at this point. In the author's opinion, again, the recent trend in this regard seems to be that of convergence and hybridization of automation strategies, rather than competition for survival among different approaches. Before examining this trend, however, the subsequent sections will first look at each of the strategies in more detail.

4.1.3 High-tech Automation Strategy

4.1.3.1 Characteristics and Background

High-tech automation strategy has been a prevalent approach by many Western auto makers, and has also influenced some Japanese auto makers[4]. The main (internally consistent) elements of the high-tech automation strategy seem to be as follows[5].

1. Contribution to advancement of automation technology. Motivation to production engineers.
2. Technological optimism, or "automation for the sake of automation" mind set.
3. Tendency to rely on expensive equipment that may have excessive functions from a total system point of view.
4. Tendency to de-emphasize robust equipment design which takes future improvements into account.
5. Tendency to buy equipment from outside specialist suppliers rather than making it in-house.
6. Top-down equipment design and improvements by specialist production engineers.
7. Emphasis on technological "great leap forward" by introducing big automation systems.

While the intended goal of the *high-tech strategy* is to enhance product competitiveness in the market place through advanced technologies, the backbone philosophy behind this strategy seems to be a technological optimism or progressiveness, or the notion that advancement of automation technology will al-

---

[3] It has been reported by some researchers, however, that the productivity and quality level of the Uddevalla plant, before it was temporarily shut down in the mid 1990s, were significantly improved See, for example, the papers by Ellegård and Berggren in this book.

[4] Toyota's Tahara #4 assembly plant, for example, is said to have been influenced, at least indirectly, by such automated assembly plants as the Fiat's Cassino plant and VW's Hall 54.

[5] Note that, in this context, the descriptions of this strategy may be biased toward the negative side.

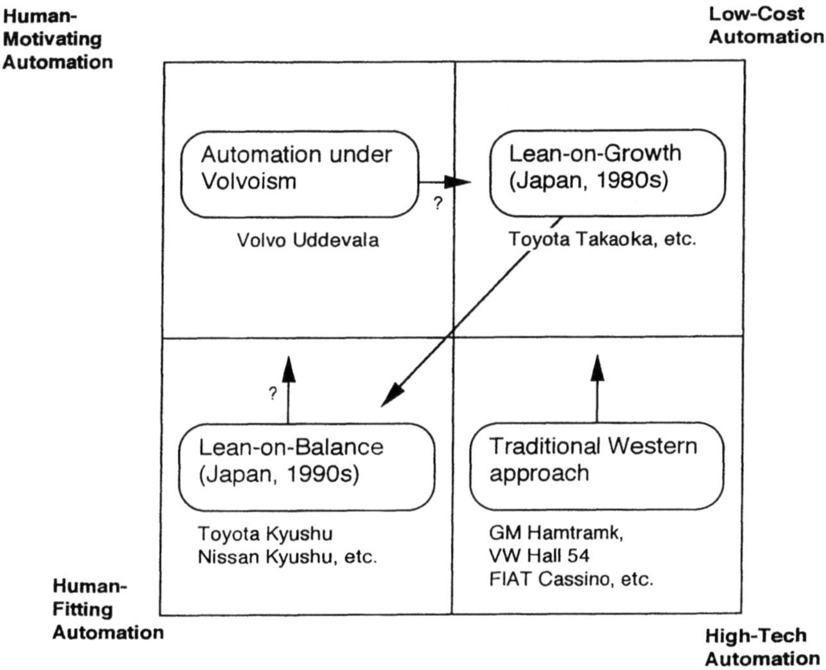

Fig. 4.1.2. Typical examples of assembly automation strategy most automatically result in improvements in the overall competitiveness of the production systems. Thus, although the firms pursuing this strategy certainly contributed to the advancement of automation technology itself, they tended to end up pursuing *automation for the sake of automation*, regardless of its overall competitive performance. They also tended to introduce individual high-tech equipment to the assembly lines without changing the basic concept and design of conventional mass production processes.

The internal logic of technological advancement is also emphasized. Technology-oriented notions, such as *the higher the automation ratio, the better*, *the more intelligent the robots, the better* or *the closer to unmanned operations, the better*, tend to be taken for granted, regardless of their competitive consequences. Therefore, this strategy tends to overshoot to such an extent that excessive automation is pursued beyond the optimal level in terms of overall productivity or quality performance.

There seems to be at least a few philosophical reasons why the high-tech automation strategy has become prevalent in Western auto makers. Firstly, Western mass production factories have tended to pursue a higher degree of functional-horizontal specialization than their Japanese counterparts. This may have created a certain divisional parochialism: The production engineering group may have pursued its divisional goals for technological advancement, even at the cost of company-wide goals for competitiveness. Secondly, vertical specialization between engineers (i.e. system builders-improvers) and workers (i.e. system operators) was more prevalent in Western auto makers, which tended to nurture the former's preference for technological big jumps by big systems rather than company-wide efforts for continuous equipment improvements. Thirdly, the lack of trust between workers and managers in some of the conventional Western factories might have fostered the notion that workers are inherent sources of defects and line stops and thus, have to be eliminated altogether by way of automation. Conversely, the postwar Japanese auto makers, de-emphasizing horizontal and vertical specialization while trusting the potential of workers as human resources, for philosophical reasons did not pursue the high-tech automation approach to an extreme. However, even Japanese auto makers seem to have had a tendency toward high-tech automation, especially during the "bubble economy" era, from the late 1980s to early 1990s, when most of the Japanese auto makers were enjoying an abundant cash flow but were faced with the problem of labor shortages.

4.1.3.2
Performance

As far as assembly operations are concerned, competitive performance by factories pursuing high-tech automation has not been impressive. According to the report by IMVP (International Motor Vehicle Program), for example, European auto makers, some of which seem to have adopted the high-tech approach (e.g. VW's Hall 54 and FIAT's Cassino plant), showed a somewhat higher ratio of final assembly automation on average than the Japanese, but their average assembly productivity was much lower than that of their Japanese counterparts [40].

In the USA, GM invested about 40 billion dollars during the first half of the 1980s to construct a new generation of high-tech assembly plants, culminating in its Hamtramck plant; this plant started up in the mid 1980s. Although the Hamtramk plant contained an ambitious level of factory automation, it suffered from a very high level of down time, due to frequent machine stops and slow recovery times. Although GM toned down its automation strategy in its assembly plants during the late 1980s (e.g. Wilmington and Linden plants), since the early 1990s, the company is still said to be the lowest in assembly productivity among the big 3. Ironically, some of the Ford assembly plants (e.g. Chicago and Atlanta plants producing Taurus), as well as the better performing "transplants" of the Japanese auto makers in the US, are said to have demonstrated productivity levels nearly as high as the better Japanese assembly plants, despite lower automation ratios.

Finally, let us take a brief look at the comparative study of Japanese and British assembly automation systems in the automobile and consumer electronics industries by Tidd [36][6].According to this study, the sample British factories tended, on average, to use more expensive and sophisticated robots, but the Japanese demonstrated a higher level of flexibility in the overall automation system

6 See also Chapter 2.3 of this book (by Tidd).

[see, also, 20,21]. As shown later, the Japanese firms tended to improve other factors of the manufacturing system, such as product design (i.e., design for automation), jigs, and work design before introducing robots and other automated equipment [see, also, 38]. Consequently, the Japanese system achieved a high level of flexibility with relatively simple and inexpensive robots. British factories, by contrast, tended to introduce relatively sophisticated robots without redesigning the overall production system. The British case seems to be a typical example of high-tech automation.

4.1.4 Low-Cost Automation Strategy

4.1.4.1 Characteristics and Background

Low cost automation was an important subsystem of the "lean" production system in the 1970s and 80s. This strategy had the following characteristics[7]:

1. *Focus on Overall Competitiveness*: Automation is recognized as means to achieve improvement in the competitiveness of the total production system. The problem of *automation for the sake of automation* in the high-tech approach is thus carefully avoided. This approach tries to achieve a given level of total system performance with the simplest, most reliable, and least expensive automation equipment.
2. *Total System Optimization*: Automation is regarded as just one component of the total manufacturing system which includes product design, jigs and fixtures, materials, work design, process flow design and so on. These factors are simultaneously optimized from a total system point of view, as opposed to designing automation equipment alone without changing the other factors.
3. *Simple Automation*: In order to reduce investment costs, the design of automated equipment is oriented toward having "just enough" functions (e.g., flexibility) for the target operations. If semi-automation or power-assisted devices are estimated to be more cost-effective, advanced automation technology is deliberately avoided. Thus, as opposed to the notion that *the higher the automation ratio, the better*, the concept of *optimal automation ratio* is widely accepted in the low cost strategy.
4. *Robust Design*: Although the automated equipment may have just enough functions for current operations, it also adopts robust design, in that it is easy to modify or to add functions for future changes or improvements.

---

[7] Contrary to the case of high-tech automation, the following description of the low cost strategy may be biased toward the positive side of the system, because the author emphasizes the link between the low cost strategy and the high competitive performance by the effective Japanese auto makers of the 1980s.

5. *In-house Production of Automation*: The low cost automation strategy tends to result in a higher ratio of in-house design/fabrication of equipment. Standard equipment purchased from outside specialist vendors tends to have excessive functions for target operations. The user-company may estimate cost saving, by eliminating excessive functions, to be larger than the economy of scale effect of the outside vendor. Equipment designed and made in-house may also be easier to improve and maintain for the in-house engineers, supervisors, maintenance workers, as well as operators.
6. *Incrementalism*: Rather than trying to introduce a big and advanced automation system at one time, the low cost strategy tends to emphasize incremental approaches towards making islands of automation and gradually expanding or connecting them.
7. *Compatibility to Continuous Improvements*: Low cost automation must be compatible with the core elements of the Toyota-style production system or the organizational problem solving mechanism. The equipment may deliberately be designed to automatically reveal manufacturing problems and to allow human intervention in response to the contingency. The concept of *Jidoka* (semi-automated equipment that automatically detects defects and stops operations) is a typical example. The equipment is jointly maintained and improved by plant engineers, maintenance workers, supervisors, team leaders and operators, with the supervisors playing a pivotal role. Total Productive Maintenance (TPM) is also built into the system.

Although the above list is, again, a description of an ideal type, many actual automobile and auto parts makers seem to have adopted automation strategies that were more or less similar to the above model. The description of the Japa nese auto makers, reported by Tidd [36], as well as the case of Nippondenso by Whitney [39], seem to be generally consistent with the above model. Also, according to the author's survey, Japanese auto makers in the mid 1980s had set a very conservative upper limit to automation investment: 5 to 10 million yen for automation equipment equivalent of one person per shift, depending upon the company [11]. Other empirical surveys on automation at Japanese auto makers during this period seem to be generally consistent with the above description of the strategy.

4.1.4.2
Performance

The low cost automation strategy which Japanese auto makers adopted during the 1970s and 80s, is believed to be one of the contributors to the competitive advantage that the main Japanese auto makers enjoyed during the same period. Although the Japanese automated welding operations more aggressively than their Western counterparts, they were rather conservative in automating final assembly operations, as the IMVP report indicated [40]. Thus, ironically, the low cost automation strategy, applied to the final assembly area, meant keeping automation ratio low and avoiding excessive automation, both quantitatively and qualitatively in order to maintain overall competitiveness. In other words, Japan's world class auto makers had apparently estimated the optimal assembly automaton ratio to be near zero.

According to the author's survey, Japanese auto makers pointed out that (i) many product variations, (ii) limit of space, (iii) shapes and materials of parts, (iv) existence of tasks inside the body and price of the automated equipment were the main constraints against automation in final assembly [11]. It is also believed that, unlike certain consumer electronic goods, the aggressive pursuit of design for automation tends to result in a deterioration of the product's design quality. Thus, in the past, those companies focusing on overall competitive performance tended to be rather conservative in automating final assembly.

Overall, empirical researchers have not found any correlation between automation ratio and overall competitiveness in the final assembly area. For example, Toyota's Takaoka plant, which industrial observers believed to be one of the most productive assembly plants in the world during the 1980s, has had virtually no robot in the final assembly area. The NUMMI plant (Toyota-GM joint venture), which is believed to be one of the most productive in North America for the same period, had also adopted a very low level of automation [25]. By contrast, none of the highly automated assembly plants of the day, such as VW Hall 54, the FIAT Cassino plant and the GM Hamtramck plant, came close to the productivity levels of the above low-tech assembly plants.

4.1.5 Human-Fitting Automation

4.1.5.1 Characteristics and Background

Human-fitting automation strategy attracted attention of Japanese auto makers in the late 1980s to early 1990s, as labor shortages had become a serious limit to the continued growth of Japanese auto makers. This strategy is characterized as follows:

1. *Worker-Oriented*: Performance in the labor market (i.e., ability to attract and satisfy workers) is given higher priority than that in the product market as criteria for automation decisions. For example, if the introduction of a robot assembly cell significantly improves the attractiveness of the work place but, at the same time, increases the unit product cost, the company adopting this strategy will still introduce this cell.
2. *Element-Focused*: Improving the attractiveness of assembly operations is oriented towards individual work stations, rather than the entire assembly system. While the traditional Ford-style moving assembly lines remain basically unchanged, those firms following the human-fitting automation strategy break down assembly tasks, evaluate them individually and try to automate the most unattractive work stations.
3. *Physical Improvements*: The definition, *attractiveness of the work place*, is physical or physiological, rather than psychological or philosophical. That

> means that the human-fitting approach emphasizes the elimination of 3-D (dangerous, dirty and demanding) tasks by automating them. Another big issue of assembly work, or work alienation and boredom on the traditional assembly line, is generally outside the scope of this strategy. In other words, the human-fitting approach aims at motivating workers to join the assembly lines by satisfying them at the bottom levels of Maslow's hierarchy of needs (i.e. physiological and safety needs) [27].

There are a number of reasons why the human-fitting strategy of assembly automation became a focal point among Japanese auto makers in the early 1990s. Firstly, direct assembly work in the Toyota-style (i.e., lean) production system has essentially been the same as that of traditional Ford moving assembly lines, although the former trained and deployed multi-job (multi-skilled) workers rather than Ford-type single-skilled workers. Typical tact times (1 to 2 minutes) are also the same for the two systems. Although there is a debate on whether the Toyota-style system is more demanding for workers than the Ford style mass production lines of the same production capacity, it would be hard to prove that Toyota's assembly lines are significantly less demanding physically than typical American or European assembly lines. It is true that the Toyota-style system has had various policies, emphasizing the potential of workers as a human resource, such as worker participation in continuous improvements, job enrichment and enlargement by way of multi-skilling programs and corporate welfare programs. However, such human-oriented policies, usually found off the assembly line, coexisted with highly stressful work on the line. In other words, direct assembly work had remained physically demanding, even in the early 1990s.

Secondly, as the shortage of labor became a serious problem for Japanese auto makers in the early 1990s, the lack of popularity of the automobile assembly lines was recognized as a serious constraint for domestic automobile production. Although the subsequent recession since 1992 has alleviated the labor supply crunch, the problem still seems to exist in the long run[8]. Thus, in the late 1980s and 1990s, Japanese companies started to see assembly automation as a means to make assembly work less demanding and more attractive, rather than as a means to improve competitiveness.

Thirdly, the period in which Japanese auto makers were faced with the labor shortage problem was also the time when perceived capital cost was very low under the bubble economy. As companies believed that they would be able to use equity finance, such as convertible bonds for building highly automated plants, they became less conservative about investing in automated equipment.

In short, the nature of assembly lines under the Toyota-style system, the emerging labor shortage and the low level of perceived capital cost were all factors which led Japanese auto makers to shift their strategic focus from *low cost strategy* to *human-fitting strategy*. In a sense, this shift was linked to the change from lean-on-growth to the lean-on-balance system at the total system level [13,14,18].

---

[8] For example, there is certain evidence that Toyota still considers improvements in the attractiveness of work places to be its long term goal in the mid 1990s.

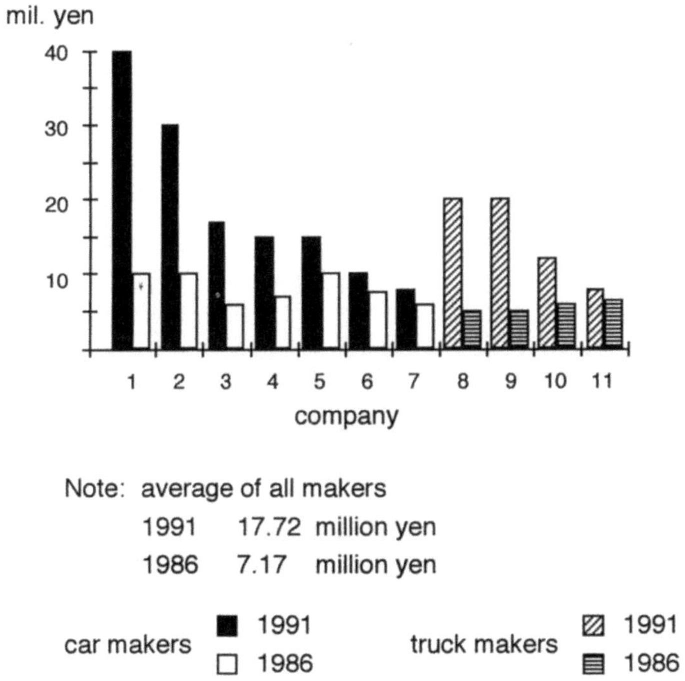

Fig. 4.1.3. Upper limit of automation investment

However, it should be noted that the human-fitting approach, in its pure form, does not answer the question of how to solve the long-discussed problem of work alienation, inherent in the Ford-Toyota-style moving assembly lines. Also, the Japanese assembly lines that incorporated the human-fitting automation concept (e.g., Toyota's Tahara #4 plant, Nissan's Kyushu #2 plant, Mazda's Hofu #2 plant, Honda's Suzuka #3 plant, etc.) tended to be criticized because of their high capital spending, especially after 1992, when the post-bubble recession started to squeeze their cash flow.

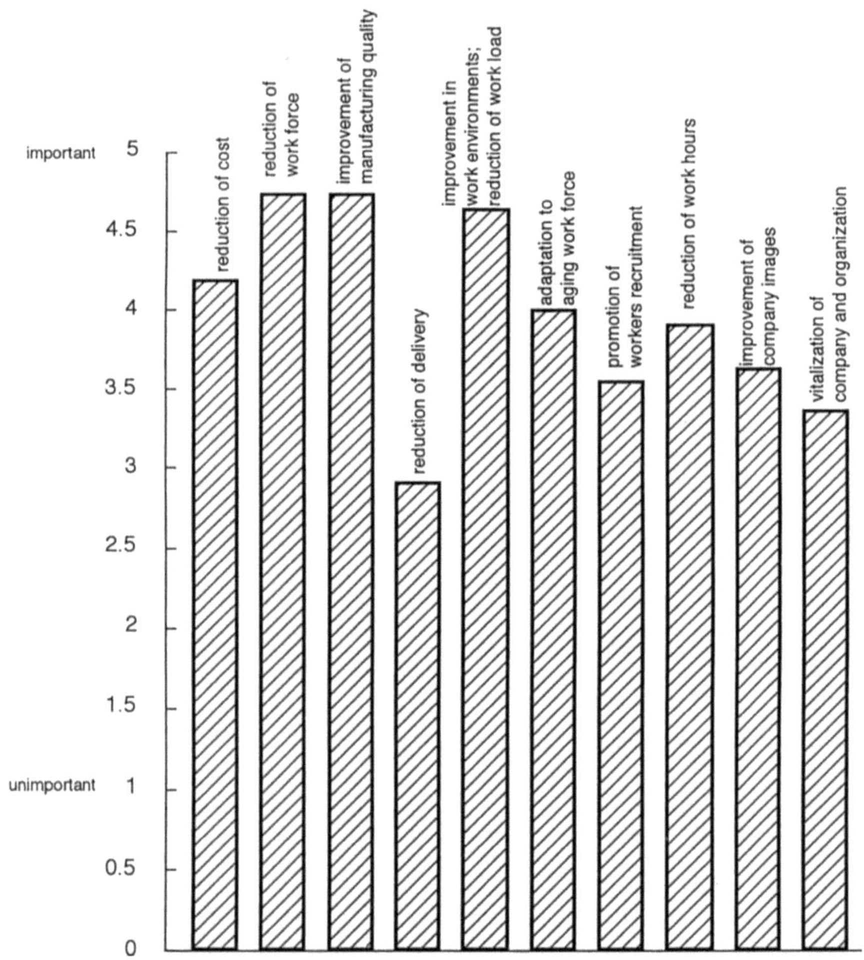

Fig. 4.1.4. Objectives for assembly automation (survey results in 1991)

4.1.5.2
Empirical Evidences

A survey, conducted by a study group together with the author in 1991, seems to indicate some circumstantial evidence that is consistent with the Japanese shift in strategic emphasis from low cost to human-fitting automation[9].

[9] The three empirical studies described here were all conducted in a study group called The Research Committee on Optimal Automation System in the Automobile Industry, chaired by Professor Koichi Shimokawa and sponsored by ten Japanese motor vehicle manufacturers.

1. *Upper Limit of Investment*: A survey on the upper limit of investment on assembly automation equivalent to one worker (fig. 4.1.3.) shows that about half of the Japanese auto makers studied increased the ceiling significantly. Thus, at least some of the Japanese auto makers may have departed somewhat from the low cost strategy between 1986 and 1991 [11].
2. *Motivation for Automation*: The same survey in 1991 also asked about the relative importance of objectives for, or expectations from, automating final assembly lines toward the year 2000 (fig. 4.1.4. See, also, [11][10]). Of the ten potential objectives investigated, quality improvements and reduction of workers received the highest score (4.7 points). What was more striking, however, was the score for improvement of work environments and reduction of workload (4.6 points). This was recognized to be almost as important as the first two factors, and was significantly higher than cost reduction (4.2 points). This result seems to indicate that, although competitiveness (particularly that of quality) remained an important motivator for automation, improving the physical condition of work places became more important as a motivator, and that by the early 1990s, the companies had shifted their focus somewhat from the low cost automation strategy to the human-fitting strategy.
3. *Task Characteristics and Automation Installation*: Another survey by the author's study group in 1992 indicated that actual patterns of robot installation on the relatively automated assembly lines in Japan was generally consistent with the above results on the motivation for automation.
   Firstly, the survey selected 35 major work stations in the final assembly process, and questioned the respondents (production engineers representing ten Japanese auto makers) on work stations with installed robots and automated equipment in the case of their most automated line. The responses were averaged in order to construct an automation installation ratio. The results are summarized in fig. 4.1.5. [For further details, see 15].
   Secondly, the survey asked for the respondents' subjective judgement on the characteristics of each work station. For each of the 16 task characteristics, such as *parts are heavy*, *parts are complex in shape*, *work environment is bad*, *dangerous task*, they were asked whether each description fitted each of the 35 work stations (1 or 0). Then, by averaging their responses by work stations and characteristics, *task characteristics indicators* were constructed (table 4.1.1.). Then, using the above indicators in a multivariate regression analysis (N = 35), an examination was carried out to see whether the pattern of actual automation installation by work stations could be explained by certain task characteristics. Because putting all the independent variables (i.e. task characteristic indicators) in the regression model creates multicollinearity problems, the author tried, as the first cut, to find some models that may indicate why automation installation ratios at some work stations were higher

The author appreciates the cooperation of the participating companies, administrative efforts of Seigo Onishi and Akimasa Kawata, as well as the assistance in data analysis from Professor Hisanaga Amikura (Chiba University) and Takashi Matsuo (doctoral program, University of Tokyo).

[10] The respondents answered by selecting the degree of importance on a 5 point scale (1 = unimportant; 5 = important) for ten potential objectives for assembly automation

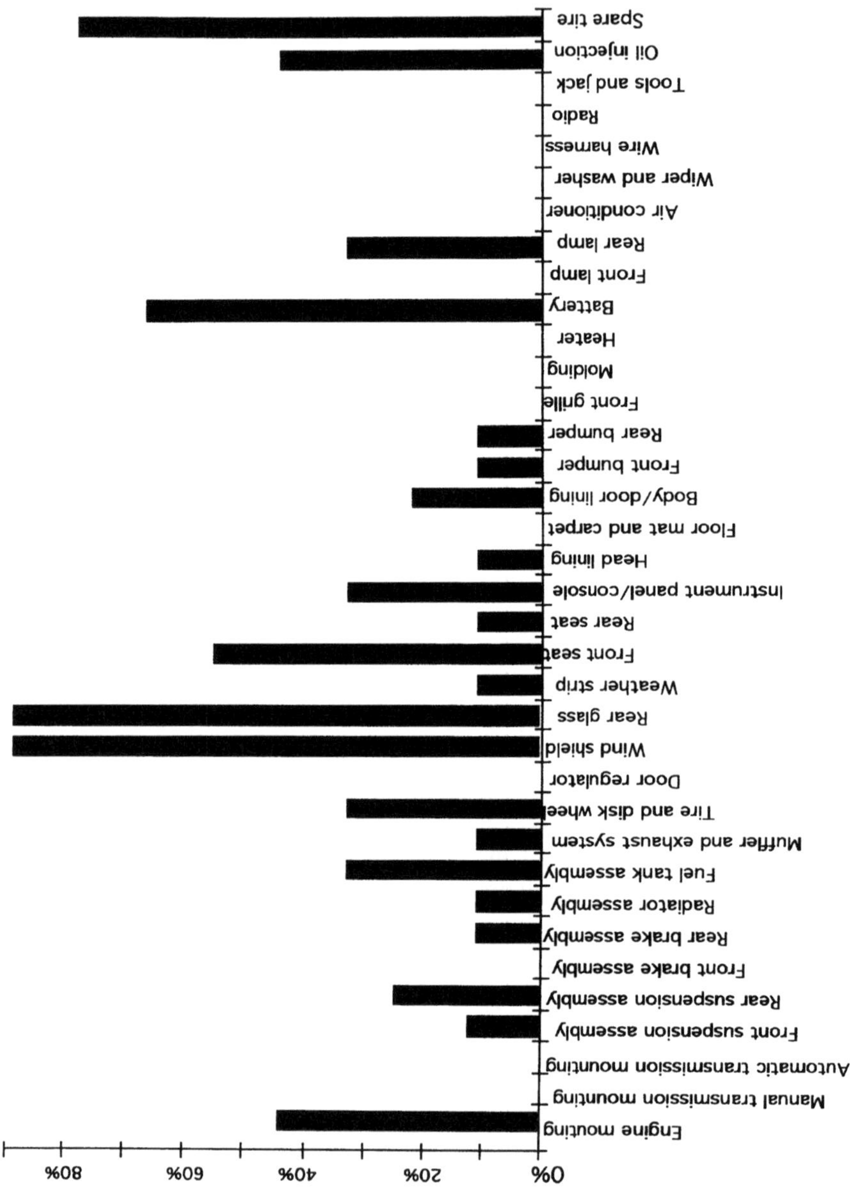

Fig. 4.1.5. Robotization index by process steps (The index measures the fraction of the companies)

Table 4.1.1. Task characteristics indicators by work station (The indicators at each cell show affirmative response ratios by the respondents at Japanese auto makers.)

task characteristics \ main assembly work stations	engine mounting	manual transmission	automatic transmission	front suspension	rear suspension	front brake	rear brake	radiator	fuel tank	muffler and exhaust	tire and disk wheel	door regulator	wind shield	rear glass	weather strip	front seat	rear seat	instrument panel/control	head lining	floor mat and carpet	body-door lining	front bumper	rear bumper	front grille	molding	heater	battery	front lamp	rear lamp	air conditioner	wiper and washer	wire harness	radio	tool and jack	oil injection	spare tire
heaviness of parts	100	100	100	100	78	33	33	67	100	78	100	11	89	89	0	100	78	78	14	56	0	89	89	0	0	44	100	0	0	44	0	22	0	0	0	100
bulkiness of parts	67	50	50	63	56	56	67	56	67	100	56	0	67	67	33	67	78	67	100	100	56	100	100	14	38	67	11	11	13	44	0	56	0	0	0	56
complexity of parts' shape	78	38	38	63	67	56	56	11	33	78	11	56	0	0	22	33	33	78	0	44	33	56	56	0	38	56	0	33	50	44	56	78	0	11	0	0
softness of parts	0	0	0	0	0	11	11	11	11	11	11	0	0	0	89	56	67	56	89	100	78	44	44	14	63	0	0	11	13	0	44	89	0	0	0	11
number of sub-parts	67	0	0	25	33	22	11	11	22	11	11	0	0	0	0	33	22	67	0	0	33	22	22	0	0	22	0	0	0	33	22	22	0	0	0	0
variety of parts	56	25	13	38	33	67	67	44	56	44	67	0	44	44	0	56	56	78	33	67	78	56	44	29	13	33	56	56	38	22	11	44	63	0	25	22
assembly inside cabins	33	0	0	13	11	11	11	44	11	11	0	67	0	0	13	67	78	67	100	89	33	0	0	0	0	89	33	11	13	89	33	78	63	56	13	22
difficult body-parts alignment	78	13	13	38	44	22	22	67	67	78	22	33	33	33	67	44	56	56	78	67	22	56	56	14	50	22	22	11	13	22	33	56	13	22	13	33
demanding posture for assembly	89	50	50	75	78	33	33	89	89	100	67	22	22	22	56	100	100	89	100	89	33	78	78	14	13	89	44	22	38	78	44	56	38	56	13	56
poor work environment	33	13	13	13	11	0	0	22	11	22	11	11	44	44	22	11	11	0	11	33	0	0	0	0	0	0	11	0	0	0	0	11	0	0	63	11
dangerous jobs	56	38	38	25	22	0	11	0	22	11	22	0	22	22	0	11	0	0	0	0	0	0	0	0	0	11	22	0	0	0	0	0	0	0	0	11
high skill requirements	78	38	38	25	33	22	22	22	44	44	33	78	44	44	44	11	11	56	22	11	33	22	22	0	63	22	0	0	0	33	0	22	13	0	0	0
many person hours needed	22	13	13	38	33	11	22	0	22	11	0	22	0	0	33	0	0	33	0	11	11	0	0	0	13	11	0	0	0	22	11	44	0	0	0	0
critical parts for vehicle safety	78	63	63	88	89	100	100	11	67	0	67	0	22	11	0	22	0	0	0	0	0	0	0	0	0	0	0	0	0	0	0	0	0	11	25	11
risk of mis-assembly if manual	0	0	0	0	0	11	11	0	0	0	11	0	11	11	0	0	0	0	0	0	0	0	0	0	0	0	0	0	0	0	0	0	0	0	13	0
risk of low quality if manual	22	13	13	25	22	11	11	0	0	0	11	11	78	78	11	0	0	0	0	0	0	0	0	0	25	0	0	0	0	0	0	13	0	0	13	0

than those at others[11].Table 4.1.2. shows some models with a reasonably good fit (standard errors in parenthesis).
As the table shows, the automation and robot installation ratio tended to be positively correlated with both competition-related indicators, particularly in quality (e.g. *defects occur if manual operation, quality assurance is difficult if manual operation*), and job-related indicators (e.g. *parts are heavy, dangerous job*). The automation ratio was also negatively correlated with the complexity of the parts' shape.

Thus, a preliminary analysis indicates that actual patterns of automation installation in the recent assembly lines in Japan (e.g., table 4.1.2.) were generally consistent with their intention in the motivation study (e.g., fig. 4.1.4.). This means that concerns about quality and those about the demanding nature of assembly work were identified as two main motivators for assembly automation in both cases. The above three studies combined seem to indicate that the Japanese firms of the early 1990s shifted their focus to the human-fitting approach, at least partially, while quality has remained the most important criterion for automation investment.

4.1.6 Human-Motivating Automation

4.1.6.1 Background: From Human-Fitting to Human-Motivating

As mentioned earlier, the concept of automation that is "friendly" to human beings became a focal point at Japanese automobile assembly plants in the late 1980s and early 1990s. Assuming that the shortage of labor is a problem which Japanese auto makers have to face on a long-term basis, despite the recession in and after 1992, the future automation concepts that they pursue will have to continue to take job attractiveness and employee satisfaction into account[12]. The human-fitting automation of their new generation assembly plants was the first step along this line, although initial experiments may have cost them too much, judging from the fixed cost burden of the new factories which Japanese auto makers were faced with in the recession of the early 1990s.

However, the human-fitting approach is not the only way in which automation makes work places more attractive. Human-motivating automation, which assists alternative assembly systems to the Ford-Toyota-style assembly lines, may also deserve serious consideration.

---

[11] For future investigation, a combination of factor analysis and multiple regression analysis may be applied here.

[12] The author's interviews with Toyota's production managers, production engineers and union leaders in 1995 indicate that the company is still committed to improving work conditions and work attractiveness at the shop floor, as a long-term objective as of the mid 1990s.

Table 4.1.2. Selected regression results - depent variable: automation equipment installation ratio by work stations (standard errors in paranthesis)

regression models: / variables:	model 1	model 2	model 3	model 4	model 5
constant	0.8	0.3	1.2	1.5	2.2
The component is heavy	3.5 (0.8)	2.9 (0.8)	–	2.6 (0.8)	–
Dangerous task	–	–	7.6 (2.7)	–	7.0 (2.5)
Frequent misassembly in manual operations	27.6 (7.8)	–	26.5 (8.8)	6.0 (1.7)	22.8 (8.3)
Difficult to assure quality in manual operations	–	6.5 (1.9)	–	–	–
The component consists of many piece parts	–	–	–	5.3 (2.3)	6.2 (2.5)
The shape of the component is complex	–	–	–	-5.0 (1.5)	-5.0 (1.7)
R^2	0.50	0.49	0.37	0.63	0.51
Degree of freedom	33	33	33	31	31
F-value	16.4	16.1	9.6	13.3	8.1

In this regard, the existing approach by Japanese auto makers tends to focus only on physical or physiological aspects of the existing Ford-style assembly lines, or on alleviating the so called *3D problems* (dangerous, dirty and demanding jobs). Consequently, the existing human-fitting approach tends to be element-focused, i.e. it tries to automate 3D tasks at each individual work station while keeping the Ford-style moving assembly lines basically unchanged[13].

In retrospect, the traditional lean production system was, in a sense, human-oriented outside the direct assembly jobs, as it promoted worker participation in continuous improvements by way of suggestion systems and small group activities, job enlargement through multi-skilling training, the stable employment

[13] Robotized work stations often stop the bodies and make the body transfer intermittent by disconnecting them from the moving conveyers in order to maintain accuracy of body alignment, but such automated stations tend to be "islands" in the middle of traditional moving assembly lines.

policy, relatively egalitarian wage systems, as well as other company welfare policies; its direct assembly jobs, however, were still essentially based on the concept of the Fordistic moving assembly lines. In fact, critics of Toyota-style factories often concentrated on the high-stress nature of their assembly lines [e.g 24][14].

4.1.6.2
Characteristics of Human-Motivating Automation

Against this background, some companies have started to seek alternative modes to the Ford-Toyota style assembly line concept. As mentioned earlier, Volvo and some other European auto makers started experiments along this line (e.g., Uddevalla plant), but so far they have not proven sufficient competitiveness vis-a-vis the lean assembly lines. The main idea of human-motivating approach is to use automation in order to alleviate work alienation and psychological stress, inherent in moving assembly lines, while maintaining competitiveness in the international market. Rather than tackling the problem of physiological stress at each work station, this approach tries to improve the psychological, social or philosophical aspects of assembly work by focusing first on the assembly system as a whole (i.e., system-focused). In other words, human-motivating automation attempts to make the work place more attractive by motivating people at the higher level of Maslow's hierarchy: socialization, self-esteem, and self actualization [27].

As the idea of using automation for new forms of assembly systems is relatively new, we have not yet identified the full characteristics of this approach. A basic guideline may, however, include the following:

1) *System-Focused*: While the human-fitting approach tries to make each individual work station more human-friendly through automated equipment, this approach focuses first on the overall system level and tries to identify an alternative assembly work design that can attract and satisfy workers better than the existing systems.
2) *Motivation at Higher Levels*: While human-fitting automation tries to make the work place more attractive by eliminating physical work conditions that may create workers' dissatisfaction, the human-motivating approach emphasizes certain assembly automation methods that help the work organization motivate workers by overcoming work alienation, boredom due to repetitive tasks and psychological stress inherent in traditional assembly line work.
3) *Automation as By-Player*: In the long run, workers are still expected to be the main players in the alternative assembly system. Thus, this approach does not regard unmanned assembly operations as the ultimate goal of assembly automation. Automation is introduced only when it is consistent with the human-motivating nature of the assembly system as a whole.

---

[14] Toyota and its union recognized the problem of unattractive jobs on the assembly lines in the late 1980s and organized joint management-union committees which discussed this issue, as well as other issues for improving work attractiveness.

4.1.6.3 Three Paths of Human-Motivating Strategy

There seem to be at least three paths in which automation is applied to the problem of work alienation and boredom in repetitive assembly work. The first approach is closely related to Volvoism (Uddevallaism), discussed earlier: applying advanced automation technology to non-Fordistic assembly systems and making their productivity reasonably high compared with that of the Fordistic ("lean" style, in particular) assembly lines.

The second path is a more moderate approach: modifying the assembly line configuration so that the work on the lines becomes less stressful and more meaningful.

The third approach begins with a highly automated assembly line (e.g., VW's Hall 54) and involves changing training programs, modifying work designs by way of job enrichment and blending repetitive and non-repetitive tasks into one job for each worker in the hope that the problem of repetitive work may be alleviated by the nature and pace of the non-repetitive, non-routine tasks. Automation tends to be applied to indirect tasks (e.g. material handling and parts picking) in the first case, while automation of direct assembly tasks is assumed in the third approach.

Automation and Volvoism: Although we have not seen many concrete examples of human-motivation automation, the experiments by Volvo at its Uddevalla assembly plant may be an exceptional case [4]. At the Uddevalla type plant, each car is assembled by a team of workers in a booth, as opposed to the assembly line, while a kit of parts for each vehicle may be picked up and delivered by automatic guided vehicles (AGVs) or manual carts[15]. Thus, whereas the Uddevalla system is, in a sense, a return to the pre-Ford stationary assembly method, it may use automation not for direct assembly work but for indirect material-handling or parts-picking jobs, which was a major bottleneck of the pre-Ford assembly system. Also, the same kind of ideas may be applied to the case of the neo-craft production system.

Although Volvo decided to close its experimental plants (Uddevalla and Kalmar) in the mid 1990s, the company decided to reopen the former as a plant for certain niche models. So, the Uddevalla experiment continues. In any case, the challenge of such alternative methods is how to overcome the problem of relatively low productivity. Some forms of advanced automation technologies may be applicable for this purpose.

Human-fitting Automation for Modified Assembly Lines: The second approach is a less radical one: to modify the assembly lines in order to make them more attractive, more meaningful and less stressful, and to apply assembly automation systems that are consistent with the modified line. Such assembly lines try to compromise the high efficiency of traditional assembly lines and the job attractiveness of the alternative assembly concept.

One recent example of this approach is the new assembly line concept at the Toyota Kyushu plant, as well as Toyota's more recently renovated lines (see chapter 3.2 of this book). In such plants, a final assembly line is divided into five to

[15] In the actual Uddevalla plant in the early 1990s, manual carts were used for parts delivery.

twelve line segments, each of which is about 100 meters long. These segments are de-coupled by some buffer bodies, a meaningfully related set of assembly jobs are assigned to each segment and a work group (typically about twenty workers) often takes charge of one segment. By belonging to a particular work group (i.e., line segment) and by rotating jobs within it, the workers may find their jobs more meaningful, although the cycle time for each worker is still short (typically one to two minutes), unlike the Uddevalla system. Also, the buffers between the segments are expected to alleviate the psychological pressures of stopping the entire assembly line when problems arise. In short, while maintaining the assembly line concept, the new Toyota approach tries to make each work group more autonomous, both physically and organizationally [30]. Accordingly, the assembly automation system is designed consistently with this autonomous line concept: automated equipment is designed to be compact, simple and safe, so that it can be operated close to the workers within each segment, as opposed to being isolated from them.

Compared to the Uddevalla case, this is perhaps a much less radical departure from the traditional Fordist concept, however, Toyota's recent approach may be regarded as an attempt toward human-motivating assembly operations in which the role of automation is rather limited. While such modified assembly lines may enjoy a similar level of efficiency as traditional assembly lines, we have to keep in mind that they may not be completely free from the stress and work alienation that is inherent in the assembly line configuration. Further empirical researches need to be carried out on this issue.

Job Enrichment at Highly Automated Assembly Lines: The third possibility is to start with a relatively highly automated assembly line, which itself may be a traditional Fordist line, and to try to alleviate the problem of work alienation. This type of job redesign for automated lines may be first implemented in areas other than final assembly, such as body welding and machining, which from a historical point of view, have been more automated than the final assembly area.

Generally speaking, when direct assembly jobs are robotized, two types of tasks may be created: the tasks of monitoring, teaching, maintaining and improving the automated equipment on the one hand, and on the other hand, "residual" tasks that handle what the robots cannot do [22]. The former tend to be non-repetitive and require new types of judgmental skills [2] or engineering knowledge; the latter tend to be repetitive, fragmented and de-skilled. If the two types of tasks are carried out by different groups of workers, automation may cause polarization of workers in terms of work alienation: those who control robots and those who are controlled by robots. For example, while job classifications in Japanese assembly plants have been relatively simple with just two main categories (i.e., direct / semi-direct workers, and maintenance workers), the introduction of robot automation may further divide the former category: *operators* doing teaching, monitoring, minor maintenance and so on, and other direct unskilled workers doing residual activities[16].

---

[16] In one Japanese company, operators' skill level is regarded as equivalent to that of team leaders in manual assembly areas. The operators have been formally trained through Off-JT programs and pilot plant operations at this company.

A potential danger of the work organization described above is that it is difficult to make the residual job meaningful for the workers doing this job. This is particularly the case when the robotized zone is clearly separated from the manual zone, since it means isolating residual workers in the automated zone. The problem is solved when the residual work for automation is eliminated altogether, but this is both technically and economically difficult in the foreseeable future.

One way to avoid the work alienation of residual work might be to let the robot operators do the peripheral jobs by way of job enlargement and job rotations. Another possibility may be to combine direct assembly tasks and the residual jobs by locating robots and manual assembly workers adjacently to each other. In any case, it should be noted that assembly automation, if carelessly introduced, may aggravate the problem of work alienation for a certain group of workers. New types of training programs, job enlargement/ enrichment and production process redesign are necessary in order to avoid this.

To sum up, there seems to be at least three potential paths toward human-motivating automation: the first is to start with non-traditional, stationary assembly systems that aim at job attractiveness and make it more competitive by using automation; the second is a modified assembly line with more autonomous line segments and a moderate use of automation within each of these segments; the third is to start with relatively highly automated assembly lines and to try to make them more attractive by way of job redesign and training programs. In any case, however, so far, there have been relatively few explicit attempts to identify optimal automation systems for the purpose of human motivation. Further experiments are needed along this line.

4.1.7
Future Prospect: Convergence, Hybridization, and Diversity

4.1.7.1
The Trend of Convergence and Diversity

We have examined four types of assembly automation strategies, which are summarized in table 4.1.3. It should be noted again that each type of strategy is essentially an ideal type, and that actual auto makers may choose mixed strategies rather than pure ones. Nevertheless, the present paper also indicated that a certain group of companies, normally clustered around a certain geographical region, have tended to emphasize a certain strategic type at a certain point in time (fig. 4.1.2.). This tendency seems to be natural, as a group of companies facing similar competitive and labor environments will try to adapt their assembly automation capabilities to the challenges posed by the environments. For example, during the 1970s and 1980s, some of the main Japanese companies had apparently tended more toward *low cost automation* and less toward *high-tech automation* than average Western auto makers, however, around 1990, they then shifted their emphasis to *human-fitting automation*.

Table 4.1.3. Summary of the four types of assembly automation strategies

	High-Tech Automation	Low Cost Automation	Human-Fitting Automation	Human-Motivating Automation
Main Objectives	Competitiveness through advancement of automation technology	Competitiveness through total system improvements	Improvements of physical work conditions at each work station	Elimination of work alienation
Key Measures	Advanced automation equipment for each work station	Total system approach to automation with low cost and limited functions	Automation of „3D" tasks despite increase in manufacturing cost	Automation supporting alternative systems to traditional assembly lines
Strength	Contribution to advancement of automation technology Reputation as high-tech company	Contribution to total performance in cost, quality and flexibility Compatible with continous improvements of the lean system	Contribution to attractiveness of work place Reputation as high-touch company	Contribution to attractiveness of work place Reputation as high-humanization company
Weakness	Advanced automation may not contribute to quality and productivity Lack of trust between labor and management may be worsened	Assembly work place may not be attractive enough to workers	Investment on difficult automation may result in loss of competitiveness through high fixed cost	Utopian pursuit of humanization may result in loss of competitiveness through low productivity
Typical Examples	US and European mass producers of 70s and 80s: *GM Hamtrank* *VW Hall 54* *FIAT Cassino*	Japanese mass producers of 70s and 80s: *Toyota Takaoda* *Honda Suzuka #1*	Some Japanese producers around 1990: *Nissan Kyushu #2* *Toyota Tahara #4* *Honda Suzuka #3* *Masda Hofu #2*	Some European makers under Volvoism influence: *Volvo Uddevala* *Mercedes Rastat* *Toyota Kyushu (to some extent)*

What about the future of assembly automation? Are automation strategies going to converge to the extent that there will be only one best strategy in the world auto industry? Or are they going diverge, so that each auto maker will choose a certain pure strategy rather than being "stuck in the middle"? None of this seems to be happening. Judging from the general trend of the total production systems, in which assembly automation is an indispensable subsystem, the global trend for assembly automation strategies in the auto industry appears to be a combination of both convergence and diversity.

First of all, as both product competition and inter-firm cooperation are becoming more global in the auto industry, it is only natural for the companies in different regions to start learning from each other's production systems. This will be the case, both at the total production system level and the assembly system level. For example, while Western auto makers introduce some elements of low-cost automation in the process of learning the lean production system, the Japanese auto makers may find the automation strategies adopted by some European car makers useful for solving their problems in the labor market.

Secondly, convergence in the automation strategy does not mean that all the auto makers in the world will start to follow one best way to automate assembly operations. Inter-firm and inter-regional differences will still remain. In fact, the early 1990s have shown that many auto companies in the world started to blend certain elements, adopted from other firms and regions, with their existing capability, accumulated over the years. A natural result of such "hybridization" experiments could be a diversity at detailed levels of the production system and manufacturing strategy. Compared with Toyota's main assembly factories of the 1960s (i.e., Takaoka and Tsutsumi), for example, we tend to find more diversity among its new generation plants: process designs and manufacturing strategies at the Tahara, Kentucky, and Kyushu (Miyata) plants appear to differ considerably from one another, although they still share the core philosophy of the Toyota Production System. Above all, hybridization of the traditional Toyota System and certain elements from European factories seems to make the Miyata plant somewhat unique among the Toyota factories[17].

To sum up, a general trend for future production systems and assembly automation strategies in the world auto industry seems to be a combination of *convergence* at the basic level, *mutual learning* throughout firms, *hybridization*, *experiments* and growing *diversity* at the detailed level.

4.1.7.2
Future Research: Inter-Firm Comparison of Strategic Profiles

If the basic trend is hybridization, then a relevant question for future research in this field would be not so much how to classify each auto firm into a certain pure strategic type (i.e., into a particular cell in fig. 4.1.1.), but rather to compare and analyze the *profiles of automation strategy* of the auto firms. This means that each of the auto firms can be evaluated in terms of its efforts and achievements in the four automation strategies. It is theoretically possible for each firm to pursue more than one strategy at the same time, but actual firms may emphasize one individual strategy at a certain point in time. In order to explore this issue of convergence, hybridization and diversity in assembly automation strategies on international scale, a series of comparative studies may be needed along this line.

[17] The *Kyushu model* has nevertheless been prevailing rapidly among Toyota's assembly factories since 1994.

		plant A	plant B	plant C	plant D
high-tech automation indices	Automation ratio >15%	●			●
	Many assembly steps (>150)	●			
	AGVs for body transfer	●			●
	Intelligent robots used		●		
	Average axes of robots > 5	●		●	
	Vision sensors used for alignment	●		●	
	Investment/capacity >1M yen/unit-month	●		●	
	total point (max. = 100)	86	14	43	29
low-cost automation indices	Upper limit of investment < 10M yen / person				●
	# of semi-automation equipment> # of robots		●	●	
	Mechanical methods used for alignment		●		●
	Average axes of robots < 4				●
	Inhouse robot development				●
	Investment/capacity <1M yen/unit-month		●		●
	total point (max. = 100)	0	50	17	83
human-fitting automation indices	Work environment more emphasized than cost	◍ even	●	●	◍ even
	Human fitting clearly stated in plant charter	●	●	◍	●
	Demanding (heavy parts) stations robotized	◍ 7/15	◍ 7/15	◍ 8/15	◍ 8/15
	Dirty (oil injection) station robotized	●			●
	Dangerous (engine mount) station robotized	●	●	●	
	total point (max. = 100)	80	70	60	60
human-motivating automation indices	Cycle time is long (> 10 minutes)				
	Small # of work stations (< 30)				
	Stationary assembly system adopted				
	Maintenance and assembly tasks merged			◍	
	total point (max. = 100)	0	0	13	0

● = affirmative (one point) ◍ = partially affirmative (half point)

Fig. 4.1.6. Preliminary results of strategic profile analysis for assembly automation - plants A, B, C, D are all Japanese final assembly lines that were installed in the late 1980s to early 1990s. In each of the four indices, equal weighting was used as the first cut. The data were collected by th JTTAS group (Fujimoto, Amikura, Matsuo)

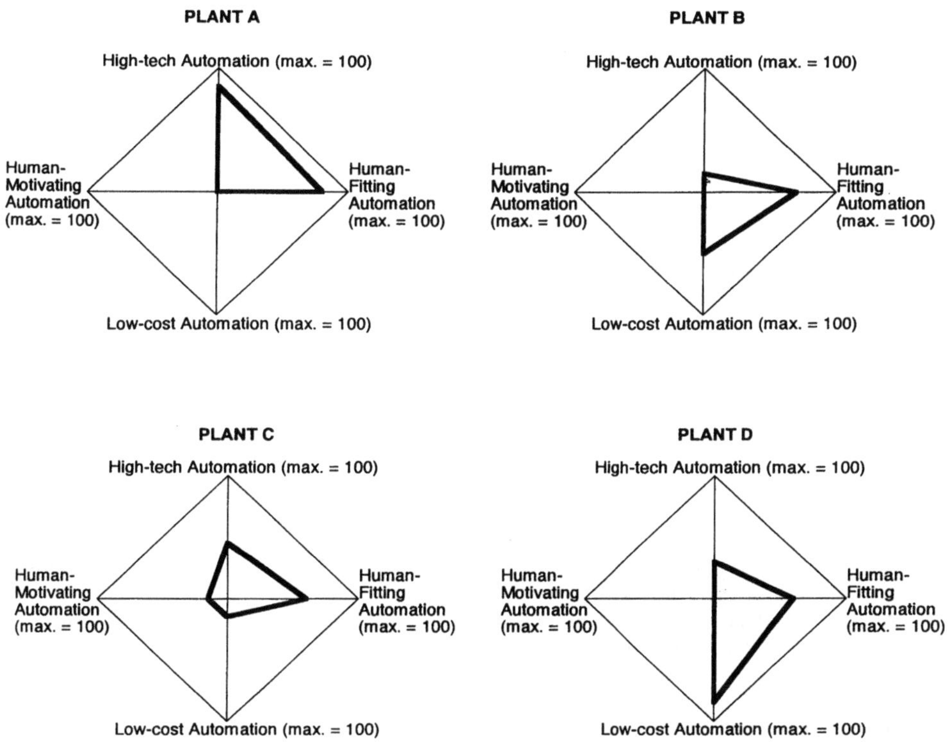

Fig. 4.1.7. Strategic profiles of four Japanese final assembly lines

For example, auto companies and their assembly plants may be compared and analyzed by using a set of variables that collectively form "profile indices" for the four strategic types. A preliminary example of such data for some of the new generation Japanese assembly plants is shown in figure 4.1.6. By aggregating the data with a certain weighting system, we can create the indices for assembly automation strategies and compare them across firms and plants. This may show the practitioners where they are and where they are going in the context of global and dynamic evolution of the world auto industry into the 21st. century[18]. Figure 4.1.7. shows examples of the strategic profiles derived from figure 4.1.6., which indicates that the strategy mixes by the recent Japanese assembly plants have been, in fact, significantly different from one another, despite the fact that they all emphasized the human-fitting automation strategy.

The measurement system for analyzing the strategic profile of automation is by no means perfect. The next step would be to refine the set of variables that

[18] A simplified version of indicators measuring system profiles is shown, for example, in [7] Chapter 9 and Appendix.

represent the strategic pattern of each assembly plants better. Making the analysis international is another agenda.
The world auto industry of the 1990s is certainly in a period of transition. Strategies for assembly automation must be evaluated in this context. Each company will have to decide how to accumulate manufacturing capability toward a desirable mix of high-tech, low-cost, human-fitting and human motivating automation, in order to cope with the dynamics of product and labor markets, while maintaining an internal consistency for the total system. To the extent that the nature of this challenge is pre-competitive, there will be growing opportunities for auto firms to learn from each other's experiences, visions and strategies.

4.1.8 References

1 Abernathy W J, Clark K B, Kantrow A M (1983) Industrial Renaissance. Basic Books, New York

2 Adler P (1983) Rethinking the Skill Requirements of New Technologies. Harvard Business School Working Paper #84-27

3 Amikura H (1991) Seisan System no Gakushu Mechanism - The Learning Mechanism for Production Systems. Business Review, Hitotsubashi University Vol 37 No 1: 54-76 (in Japanese)

4 Berggren C (1993) Volvo Uddevalla - A Dream Plant for Dealers?. Working Paper, Royal Institute of Technology, Department of Work Science

5 Bright J R (1958) Automation And Management

6 Camuffo A, Volpato G (1994) The Evolution of Manufacturing Automation in the Italian Automobile Industry: A Study of Fiat Auto Plants. Presented at the International Conference on Assembly Automation and Future Outlook of Production Systems, Hosei University, November, 1993. Cf. chapter 3.9 of this book

7 Clark K B, Fujimoto T (1991) Product Development Performance. Harvard Business School Press, Boston

8 Cole R E, Adler P S (1993) Designed for Learning: A Tale of Two Auto Factories. Sloan Management Review Spring: 85-94

9 Cusumano M A (1985) The Japanese Automobile Industry. Harvard University Press, Cambridge

10 Ellegard K (1995) The Development of a Reflective Production System Layout of the Volvo Uddevalla Car Assembly Plant. Chapter 3.10 of this book

11 Fujimoto T (1992) Why Do Japanese Auto Companies Automate Assembly Operations? Presented at the Berlin Workshop on Assembly Automation. November, 1992. Research Institute for the Japanese Economy Discussion Paper 92-F-15, University of Tokyo

12 Fujimoto T (1992) What Do You Mean by Automation Ratio? Presented at the Berlin Workshop on Assembly Automation. November, 1992. Research Institute for the Japanese Economy Discussion Paper 92-F-16, University of Tokyo. (Chapter 2.4 of this book)

13 Fujimoto T (1993) At a Crossroads. Look Japan September 1993: 14-15

14 Fujimoto T (1994) The Limits of Lean Production. Politik und Gesellschaft, Friedrich-Ebert-Stiftung, Germany January: 40-46

15 Fujimoto T, Matsuo T (1993) Note on the Findings of the Assembly Automation Study (Second Report) - A Survey of the Japanese Auto Makers. Presented at the International Conference on Assembly Automation and Future Outlook of Production Systems. Hosei University, Tokyo, Japan November, 1993

16 Fujimoto T, Takeishi A (1993) Jidosha Sangyo no Seisansei -Productivity of the Automobile Industry. Soshiki Kagaku (Organizational Science) Vol 26, No 4: 36-43 (in Japanese)

17 Fujimoto T and Tidd J (1993) The UK and Japanese Auto Industry: Adoption and Adaptation of Fordism. A Paper Presented at the Conference on Entrepreneurial Activities and Enterprise Systems, University of Tokyo Research Institute for the Japanese Economy, Gotenba City, January 1993. Japanese Translation: Keizaigaku Ronshu (The Journal of Economics) Vol 59 No 2 and 3, The Society of Economics, University of Tokyo

18 Fujimoto T, Takeishi A (1994) Jidosha Sangyo 21 Seiki he no Shinario (The Automobile Industry: A Scenario toward the 21-st Century]. Seisansei Shuppan, Tokyo

19 Hayes R H, Wheelwright S C (1984) Restoring Our Competitive Edge. John Wiley & Sons, New York

20 Jaikumar R (1984) Flexible Manufacturing Systems: A Managerial Perspective. Harvard Business School Working Paper # 1-784-078

21 Jaikumar R (1986) Postindustrial Manufacturing. Harvard Business Review November/December: 69-76

22 Jürgens U, Dohse K, Malsch T (1986) New Production Concepts in West German Car Plants. In: Tolliday S, Zeitlin J (ed) The Automobile Industry and Its Workers: Between Fordism and Flexibility. Polity Press, Cambridge: 258-281

23 Jürgens U (1992) Presentation at the Berlin Workshop on Assembly Automation. November 1992

24 Kamata S (1973) Jidoha Zetsubo Kojo. Gendaishi Shuppankai

25 Krafcik J (1988) Triumph of the Lean Production System. Sloan Management Review Fall: 41-52

26 Kumasaka H (1988) Jidosha no Kumitate Gijutu no Genjo to Shorai - Current Status and Future of assembly Techniques of Automobiles. Jidosha Gijutu (Automotive Technology) Vol 42 No 1: 72-78 (in Japanese)

27 Maslow A H (1954) Motivation and PersonalityHarper and Row, New York

28 Monden Y (1983) Toyota Production System. Institute of Industrial Engineers, Atlanta

29 Nevins J L, Whitney D E (1989) Concurrent Design of Products & Processes. McGraw-Hill, New York

30 Niimi A, Miyoshi K, Ishii T, Araki T, Uchida K, Ota I (1994) Jidoka Kumitate Rain ni Okeru Jiritsu Kanketsu Kotei no Kakuritsu. (Establishment of Autonomous Complete Process Planning for Automobile Assembly Line). Toyota Technical Review Vol 44 No 2: 86-91 (in Japanese)

31 Penrose E T (1959) The Theory of the Growth of the Firm. Oxford: Basil Blackwell

32 Shibata F, Imayoshi K, Eri Y, Ogata S (1993) Kumitate Sagyo Futan no Teiryo Hyokaho (TVAL) no Kaihatsu" (Development of Assembly Load Verification) Toyota Technical Review Vol 43 No 1: 84-89 (in Japanese)

33 Shimokawa K (1992) Japanese Production System and the Factory Automation. Discussion paper for the Berlin Workshop on Assembly Automation. November 1992

34 Schonberger R J (1982) Japanese Manufacturing Techniques. Free Press, New York

35 Teece D J, Pisano G, Shuen A (1992) Dynamic Capabilities and Strategic Management. Revised, June 1992. University of California at Berkeley Working Paper

36 Tidd J (1989) Next Steps in Assembly Automation. Presented at International Policy Forum, International Motor Vehicle Program, MIT, May

37 Tidd J (1991) Flexible Manufacturing Technologies and Industrial Competitiveness. Pinter Publisher, London

38 Whitney D E (1986) Real Robots Do Need Jigs. Harvard Business Review May-June: 110-115

39 Whitney D E (1993) Nippondenso Co Ltd - A Case Study of Strategic Product Design. C.S. Draper Laboratory Working Paper, CADL-P 3225

40 Womack J P, Jones D T, Roos D (1990) The Machine That Changed the World. Rawson Associates, New York

4.2 From Fixed to Flexible: Automation and Work Organization Trends from the International Assembly Plant Study

J. P. MacDuffie · F. K. Pil

The considerable attention given by researchers and managers alike to the diffusion of "lean" or "flexible" production concepts throughout the world automobile industry carries with it the assumption - often implicit - that flexible automation is an integral part of this alternative production paradigm. The ample anecdotal evidence available about automation trends in automotive manufacturing is typically based on the newest and most advanced plants of various companies, focusing on their most "cutting-edge" technological installations. What is lacking is more systematic data about how automation use differs across assembly plants around the world, in terms of the relative capital intensity of different parts of the assembly process, the types of automated equipment that are used, and the specific tasks to which automation is applied.

In this paper, we provide a data-driven overview of all these issues, coupled with an exploratory analysis of the factors underlying the adoption of flexible automation. We draw on data that we have collected as part of the International Assembly Plant Study, sponsored by M.I.T.'s International Motor Vehicle Program. These data were collected in two phases, with Round 1 in 1989 and Round 2 in 1993/94. Table 4.2.1. contains the regional distribution of plants in the Round 1 and Round 2 samples; 43 plants participated in both phases of data collection.
Previous papers have provided a description of automation trends for the Round 1 data [3], have used automation as a control variable in cross-sectional and longitudinal analyses that investigate the impact of flexible production systems on productivity, quality, and product variety outcomes [4,7,8,10], or have treated automation as a factor that potentially influences the adoption of "high involvement" work practices [6,10,11]. In this paper, we will emphasize changes in the utilization of automation over time, using data from both phases of the assembly plant study, both within and across plants and companies in different regions. Our interpretation of these trends is heavily influenced by our interviews with managers and engineers, conducted during the plant visits we carry out as an integral part of the assembly plant study [9].

The paper is organized into seven sections. First, we define how we measure automation in the assembly plant study. Second, we describe the overall regional trends in the use of automation from 1989 to 1993/94. Third, we explore the patterns of usage for robotic equipment across regions, emphasizing in particular the significant move by many companies towards the replacement of fixed or "hard" automation with flexible, programmable automation. Fourth, we explore

Table 4.2.1. Distribution of participants in the international assembly plant study

Regional Group[1]	Round 1 (1989)	Round 2 (1993/94)
Japan/Japan	9	12
Japan/N.A.	4	8
Europe	24	21
U.S./N.A.	16	25
New Entrants	11	18
Australia	6	4
total:		
Plants	70	88
Companies	24	21
Countries	17	20

[1] JP/JP = Japanese plants in Japan, JP/NA = Japanese plants in North America, Europe = U.S. and European plants in Europe, U.S./NA = U.S. plants in North America, and NE = New Entrant Plants (Korea, Mexico, Brazil, Taiwan)

departmental differences in the use of automation, emphasizing the evolution in thinking about the most effective way to automate various tasks in the body, paint, and assembly shops. While automation levels continue to rise in the body and paint shops, a different approach is being taken in the assembly department, the most labor-intensive area of the plant and yet the place where total automation solutions have been most elusive. Fifth, we describe how trends in the adoption of flexible automation are linked to the adoption of flexible work practices that seek to boost worker involvement in production-related problem-solving. Sixth, we summarize what we have learned about the performance implications (in terms of productivity and quality) of the automation trends described here. The seventh section presents our conclusions from these analyses and our speculation about future trends in automotive manufacturing automation.

4.2.1 Measuring Automation

Our definitions of automation vary by department. We use automation measures developed by Krafcik [2,3] that in some cases overlap heavily with measures commonly used in the industry and in other cases are unique to the assembly plant study. The body shop is the area of greatest consensus on how automation should be measured. Virtually every plant we have seen measures automation in the body shop as some variant of the percent of spot welds that are placed automatically, although some plants also add measures of the accuracy of aperture fitting, or the extent to which material selection and transport is done automatically. We take the percentage of spot welds that are automated as our baseline measure but we add to it, for plants that do arc welding, the fraction of arc welds that are done automatically. Arc welds are converted to an approximate equivalent number of spot welds (i.e. 1 meter of arc welding equals 150 spot welds, with the average plant using just over 2 meters of arc welding per vehicle).

In the paint shop, there are five main activities that can be automated: electro-coating, primer/surfacer application, interior painting, top coat painting, and application of joint sealer. Electro-coating is inherently automatic and is generally not considered in measures of paint-shop automation. The remaining activities are included by most plants in their measures of automation, although the placement of joint sealer is sometimes considered separately from the paint process itself. We measure the automation level for each activity by looking at the percentage of total surface area that is painted by automated equipment rather than manually - or for sealer, the percentage of total sealer length applied by automated equipment. Our departmental measure of paint automation weights each activity by its approximate labor content (in percentage terms) in the least automated plants in the sample: a 50% weight for sealer automation, and a 16.67% weight for top coat, interior, and primer/surfacer. Because some paint shop activities tend to be either fully automated or not at all (100% or 0%), the total paint automation measure can increase dramatically if one of these activities becomes automated.

Definitions for automation in the assembly department (comprising trim, chassis, and final assembly lines) vary across different automobile companies. Fujimoto [1] provides a good overview of some of the definitions used in Japan. Some plants measure the percent of production steps automated, others look at labor savings, and still others look at the fraction of overall production time that vehicles spend in automated stations. In order to develop a consistent measure across manufacturers, we estimate the labor required to perform each automated assembly task in relation to the total labor content in the assembly area.

For the overall measure of assembly plant automation, combining all departments, we weight the automation level in the body, paint, and assembly areas by the percentage of total labor content applied to those activities in the least automated plants in the world: 31% for the body shop, 19% for the paint shop, and 50% for the assembly shop. This measure is labeled "Total Automation" and is expressed as the percentage of total production steps that are automated.

The departmental measures described above do not necessarily make a distinction between "fixed" (or "hard") automation and "flexible" (or "programmable") automation. So we also have measures that capture the extent to which flexible automation is being used. In the body shop, we distinguish spot welds applied by "fixed" or "hard" automation from those applied by "flexible" automation. We do the same in the paint shop for various activities. Also, since robots are the most prominent form of flexible automation and can be found in all three departments, we measure the total number of robots used in the plant, adjusted for the plant's capacity. This measure is labeled the Robotic Index and is expressed as the number of robots per vehicle produced per hour. Robots are defined as having at least three axes of motion, although for some analyses, we further differentiate simple robots (3-4 axes of motion) from more sophisticated equipment with 6 or more axes of motion.

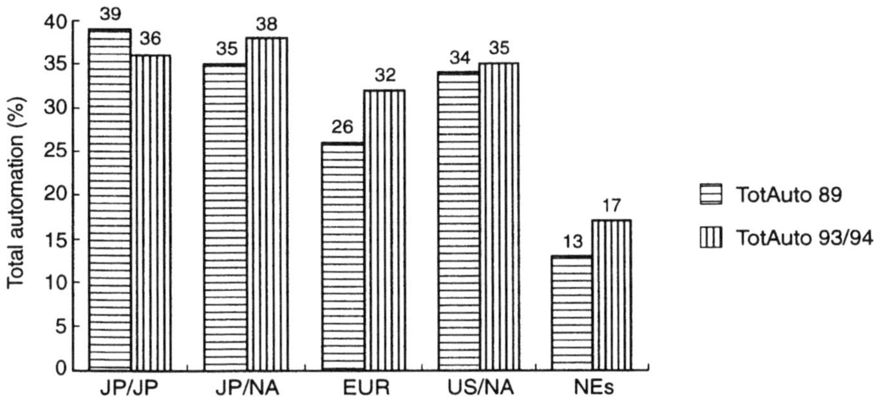

Fig. 4.2.1. Change in total automation in matched sample of plants (n=43) JP/JP= Japanese plants in Japan, JP/NA= Japanese plants in North America, Europe=U.S. and European plants in Europe, U.S./NA = U.S. plants in North America, and NE = New Entrant Plants (Korea, Mexico, Brazil, Taiwan)

4.2.2 International Trends in Automation Usage, 1989 - 1993/94

By the late-1980's, when the Round 1 data were collected, there was a significant disparity in the use of automation across regional groups of plants. Japanese companies were by far the greatest users of automation, followed by the American "Big Three" companies. In Europe, there were extraordinary efforts at some individual plants to achieve dramatic increases in automation (particularly in the assembly area) by some manufacturers; the Wolfsburg plant of Volkswagen and the Cassino plant of Fiat stand out as well-known examples. But overall, European companies tended to have lower automation levels than Japanese and American companies. Plants in New Entrant countries (e.g. Mexico, Brazil, Taiwan, Korea) tended to use minimal automation, most of it concentrated in the placement of welds that were difficult to place manually, although some individual plants (particularly in Korea) were highly automated.

The sample average for Total Automation increased roughly 10% for the matched sample and 20% for the unmatched sample from 1989 to 1993/94. However, the broader trend is one of convergence in automation levels across regional groups (fig. 4.2.1.). European plants experienced significant increases in automation, while plants in Japan report slight reductions in their average automation levels. Several New Entrant plants also increased their automation levels. Particularly in Korea, the boost in automation reflects the fact that new plants are being built with very high levels of automation, possibly as a reaction to rising labor costs. Moving towards a more capital-intensive production system may also reflect an attempt to limit union influence by some Korean companies.

As table 4.2.2. reveals, automation levels have been on the rise across all departments since 1989. However, the most striking change has been the shift in the type of automation used. The last decade has seen a dramatic replacement of fixed automation with robotics. Indeed, the average number of robots per vehicle/hour increased 60%, from 2.3 in 1989 to 3.7 in 1993/94 for the matched sample of plants. For a plant with an annual production volume of 250,000, this is equivalent to an increase from 144 robots to 231 robots. This trend is most apparent in body shops and, to a lesser extent, in many paint shops. There, the switch from solvent-based to water-based paints has also provided the impetus to increase the use of robotic interior painting. We will review trends in the use of robots in more detail below.

Table 4.2.2. Change in automation levels by type in matched plants (n=43)

	1989	1993/94
Total Automation	27 %	29 %
Weld Automation	63 %	69 %
Paint Automation	36 %	39 %
Assembly Automation	1 %	2 %
Robotics (robots/vehicle/hour)	2.3	3.7

While many European, American, and Japanese plants have added new automation, we have also observed plants in all parts of the world pulling out automation that was either unreliable, underutilized, or exceedingly complex. Problems with equipment reliability appear to be particularly strong in the assembly area or in subassembly areas of the body shop. We have observed instances where plants work overtime, and sometimes even a third shift, because some automated equipment has failure rates as high as 30%. Equipment we have seen removed ranges from tire placement and bolting equipment in assembly to door fitting and welding equipment in the body shop. That is not to say that these operations cannot be automated. However, poor equipment designs, coupled with some plants' inexperience with preventive maintenance and on-line repair activities, can render some of this equipment more of a liability than a benefit.

There are some plants where robotic equipment sits idle during large portions of the cycle time because there are too many robots operating in too small a space. As a result, several plants have actively scrapped "dead-wood" equipment. Not just excess robots and unwieldy assembly automation are receiving the ax, but so are Automated Guided Vehicles (AGV's), which require that a tremendous amount of space be left empty as they provide unattended transport and variable routing of vehicles between successive operations. The reduction of automation that is underutilized or exceedingly complex in relation to its benefits actually has little impact on the measure of automation in the plants, in part because equipment like AGVs is not measured by most standard automation measures

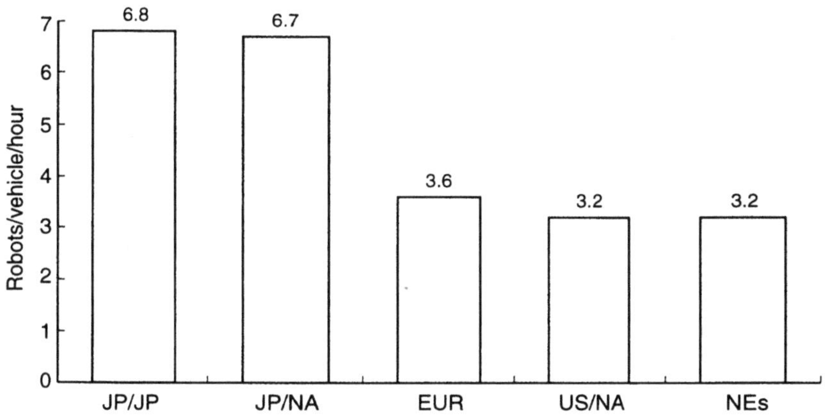

Fig. 4.2.2. Robot use by region, 1993/94

(including ours), and in part because pulling out equipment such as robots that are causing interference can actually enhance the utilization rate of the remaining equipment.

4.2.3
Use of Robotics

Although the overall levels of robotics have increased since 1989, large regional variation remains in the extent to which robots are used in 1993/94, as shown in figure 4.2.2.

Japanese plants in Japan as well as in North America make the most extensive use of robotics, while U.S. plants use robotics the least. The average New Entrant plant has more robots than the average U.S. plant, but the New Entrant average obscures significant variance among plants in this regional group. Some of the Korean plants utilize more robots than plants in Europe and the U.S., and even some of the plants in Japan. On the other hand, plants in countries like Taiwan, Mexico, India, and Brazil, use very few robots.

While there is variance in the number of robots used by plants in different regions of the world, the distribution of robots across departments within the plants is much more similar, with the preponderance of robots used in the weld area (fig. 4.2.3.). Japanese plants tend to use somewhat more robots in their paint departments, in part because they are more likely to automate interior painting - an activity that needs robots both for painting and for opening and closing doors and front hood and rear trunk lids.

Although U.S. and European plants generally have the fewest robots, their robots tend to be the most complex. Indeed, nearly 100% of the robots used in many

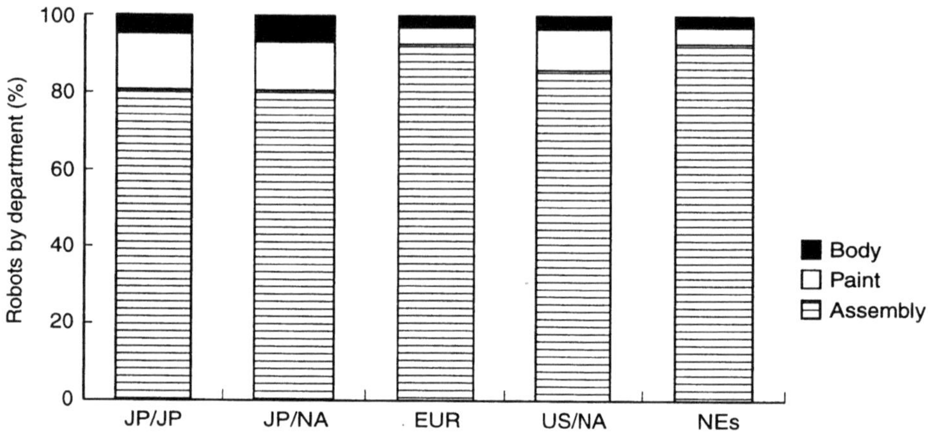

Fig. 4.2.3. Robot use by department, 1993/94

European plants have six or more axes of motion, compared to only 70-75% in average Japanese plants. Some Japanese manufacturers, particularly Honda and Suzuki, have been quite adept at minimizing robotics investment by identifying the cheapest robotics for each task. As a result, many of the robots used by these companies are lighter, smaller, and have fewer axes of motion.

Not only can robots differ in their basic capabilities for movement but they can also be deployed quite differently from plant to plant. For example, in the welding of subassemblies in the body shop, robots can maneuver weld guns around panels or assemblies being welded, or they can move the objects being welded through stationary weld guns. Robot placement also differs across plants. In the body shops of some plants, robots work in pairs or quads on a vehicle as it comes through, whereas in other plants, as many as 8 or 10 robots can be working at a single station. The advantage of the latter is that the vehicle moves through fewer stations, with more welds applied each time the body is fixed in position by jigs, increasing dimensional accuracy. More welds per station can also mean that robots spend less time being idle. However, as the number of robots working in a given station increases, robots are more likely to get in each other's way (interference), and idle time can increase. Several Japanese manufacturers have become extremely skilled at designing and placing robots to minimize such interference and to increase the number of welds placed at each station.

Not all robots perform tasks that would be captured by commonly-used measures of automation. For example, over half the plants in our sample use robots for parts transfers and placement in the body shop. Another increasingly popular use is in-line inspection of dimensional accuracy in the body shop, using vision systems fixed to sophisticated robots capable of several axes of motion. Currently,

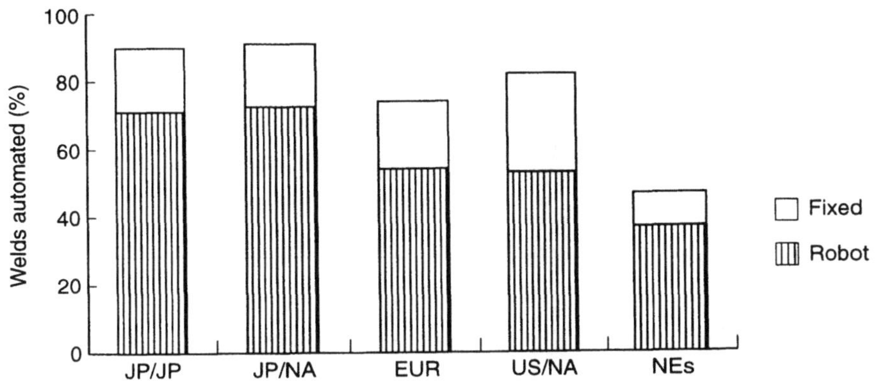

Fig. 4.2.4. Type of automation used for spot welds, 1993/94

only about 15% of the plants in our sample use robots for that purpose, but we expect that number to increase over the coming years.

4.2.4
Automation use by Department

Here we review automation trends in the body, paint, and assembly departments.

4.2.4.1
Body Shop

In the body shop, levels of automation are very similar across the different regions of the world, as shown in figure 4.2.4. Most plants now exceed 80% automation of total spot and arc welds, and many are approaching 100% automation of spot welds. The exception are some of the plants in New Entrant countries, where funds for capital investment may be scarce and low labor costs reduce the incentive to invest in labor-displacing automation. While most Korean plants have weld automation levels on par with those of other plants in the world, plants in other New Entrant countries automate welds sparingly. Generally, the welds that are placed automatically at these plants are those that require heavy weld guns, and those where accuracy of placement is important because the welds are visible to the customer (e.g. the welding of the roof to body sides).

Despite similarity in the number of welds placed automatically, U.S. plants make much greater use of fixed automation than plants in any other regions. This is feasible only because U.S. plants generally produce only one or two models in their plants and use dedicated body lines for each model. Similarly, both U.S. and European plants are less likely than their Japanese counterparts to have framing

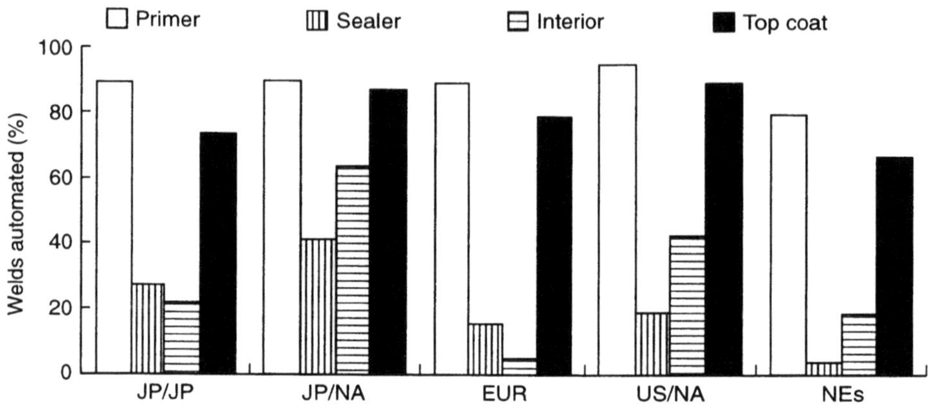

Fig. 4.2.5. Paintshop automation

stations where the sides, roof, and underbody are married at once. However, most plants in the U.S. and Europe that have made recent investments in their body shops are likely to have such framing stations, and generally also have the capability to handle multiple body styles and models, although that potential is not always used.

One of the most flexible weld systems is the Intelligent Body Assembly System (IBAS) used by Nissan all over the world. This system uses robots not just for welding, but also for holding the bodies. The framing station contains special robots that can be instantly reconfigured to hold body parts for any number of models or body styles. Indeed, IBAS is capable of handling all Nissan models and body styles, with the exception of truck models. Toyota's Flexible Body Line (FBL) and Mazda's Circulation Body Assembly Line (C-BAL) are similar.

There are also important regional differences in the automation of arc/seam welding in the body shop. U.S. plants generally do this manually, whereas the average Japanese plant automates over half of these welds.

4.2.4.2 Paint Shop

As mentioned above, the electro-coating process, in which the vehicle body receives its first coats of protection against rust, is always automated.. After that point, there is greater variance in what plants automate, as figure 4.2.5. shows. Most plants fully automate the application of the primer/surfacer coat, as well as the top coat. Because of the way the automated equipment is set up, either these activities are fully automated or not at all. However, some plants have multiple paint booths, and while top coat may be fully automated in one of those booths, it may be manual in another, thus permitting a plant's overall automation level for top coat painting to be between 0 and 100%. Also, the addition of a two-tone (a second paint color used to provide an accent on certain parts of the body) may be done manually.

There is much greater variance in the extent to which interior painting and the application of joint sealer is done automatically. Interior painting is almost never fully automated, with the exception of plants using a space frame design (where the absence of panels provides easier access to the vehicle's interior). In all but a few cases, production workers provide some degree of finishing touch to the interior painting to cover up areas unreachable or poorly covered by robotic spray.

Joint sealer is applied to seams and joints to waterproof the vehicle and reduce wind noise. The application process is one of the most labor intensive activities in plants that do not automate it. Some plants have turned to automating most of the joint sealer application process, while others have looked for ways to improve tolerances and change body panel designs to reduce sealer content. Sealer spray via robots has become popular at a few plants, particularly those belonging to European and Japanese companies. However, it is not that widespread as yet, with less than half of all plants doing it at all, and less than 15% automating more than half of the sealer application.

The newest plants, and in particular the Japanese transplants in North America, have the greatest amount of automation in their paint shops and are most likely to automate interior painting and sealer application. Thus the likely trend into the future will be steady increases in the overall level of paint shop automation as new investments are made, either in new plants or retrofitting older plants.

There are also other innovations taking place in paint shops that do not alter the level of automation but do fundamentally alter the process of paint application. Perhaps the most significant is the switch to water-based paints from solvent-based paints to reduce Volatile Organic Compound (VOC) emissions. The switch to water-based paints has started in many American, European, and Japanese plants. However, many of the plants upgrading equipment have not yet made the full switch to water-based paints. Most are trying it in one of their paint booths with solid colors. Although the technology exists to apply water-based metallic paints, the paint quality is inferior to that achieved with solid colors, so most plants continue to use solvent-based metallic paints.

4.2.4.3 Assembly Area

The assembly area of the plant is the most labor intensive portion of any car factory, containing, on average, 60% of all production workers. Very few assembly operations are automated. Indeed, in the average plant only 6 assembly steps are automated and most plants automate less than that. Although each region has its outlier plants, the average automation level in assembly by region ranges from 1.1 to 2.2%.

Table 4.2.3. Assembly operations most frequently automated (percent of plants with automated process)

1. Windshield Sealer Apply/Install	67 %
2. Rear Glass Sealer Apply/Install	56 %
3. Tire Disk Wheel	50 %
4. Engine Mounting	43 %
5. Suspension Assembly	38 %
6. Spare Tire Insert	34 %

As shown in table 4.2.3., the majority of plants use automated equipment to apply sealer to the windshield and rear glass, and to place them on the vehicle. These two are tasks that are prone to quality problems when performed manually and they are relatively easy to automate. The other operations listed in table 4.2.3. are often fully automated because they are strenuous tasks if performed manually and the robots needed to automate these tasks are relatively simple.

Overall, however, it must be said that progress towards automating the assembly process has been slow, sporadic, and not all that successful. After highly publicized campaigns in the 1980s to automate assembly tasks - carried out by companies ranging from Volkswagen to Fiat to General Motors - failed to produce desired results despite massive investments, most companies are approaching assembly automation more cautiously, even in their greenfield plants. Fiat, whose Cassino plant was advertised as being one of the most automated in the world, has taken a step back from automation in the assembly area at its new Melfi plant, and has focused instead on introducing flexible work practices. At Nissan's new Kyushu site, the assembly area was designed to accommodate tremendous amounts of automation, yet it now seems unlikely that Nissan will ever follow through with all of its original automation plans. Toyota, which experimented with highly advanced automation in its Plant #4 at Tahara in the mid-1980s, has since reduced its emphasis on automation at its new Kyushu plant and its newly retrofitted RAV4 line at the Motomachi plant.

Automation in the assembly area is not limited to the automation of particular tasks on the line. It is exhibited most clearly in the assembly line itself. Gone in most plants are the old chain conveyors that run through the whole assembly area. Instead, newer assembly lines have multiple conveyance systems, with the primary distinction on the main line being between underbody and other work. Off the main line, many plants now have feeder lines for subassembly of dashboards, engine/transmission/strut assemblies, and doors (most plants have moved, or are moving to doors-off assembly, where the vehicle doors are removed at the start of assembly and reattached downstream, after most interior installation tasks have been completed.) In over two-thirds of the plants, overhead conveyors are used to permit underbody work to take place outside of the dreaded "pit", and most plants not using such a system are converting to it. More modern lines have "tilt" overhead conveyors to permit easier access to the underbody.

For work on the upper body of the vehicle, there are two main types of conveyors: platform conveyors that only carry the vehicle and conveyors that move both the vehicle and the person working on the vehicle. The latter is found primarily in newer plants. Also found in some plants are accordion hydraulic systems that raise and lower the car as needed. Some companies in the 1980's utilized AGV's to transport the vehicles, but these have fallen out of favor because they are too expensive to purchase and maintain and, as mentioned above, they require too much space. The most dramatic departure from traditional assembly lines, with their synchronous cycle times at each station, is the variable speed conveyance system at Nissan Kyushu, which permits individual cars to move asynchronously through the assembly process, advancing more quickly, more slowly, or stopping, dependent on the requirements of the task performed at a particular station. We see no sign that this asynchronous approach is being adopted by other companies.

We noted earlier that plants have been relatively unsuccessful in attempting to automate most assembly tasks fully. As a result, we now see an alternate trend - the expansion of what we term "automation assist." These are forms of automation that support the production worker, but do not necessarily replace him or her. There are two primary reasons for automation assist: enhancing ergonomics and reducing extraneous production worker movement. To eliminate the ergonomic strain from installing heavy parts, tools that lift and place parts like seats, tires, and doors are increasingly common. Also important from an ergonomic standpoint are tools that reduce the pressure placed on the wrists when applying the full torque during a bolting operation. Some plants, like Toyota Kyushu for example, let production workers place bolts, screw them on lightly, and then utilize a fixture which automatically applies full torque. Unlike full automation, automation assist tools do not actually install the parts - they only place the parts in a position where the production worker can more easily install them, or finish off a task started by the production worker.

We have observed various forms of automation assist directed at reducing excess movement by workers. In some plants there are platforms that move under the raised body on which the worker can stand while doing certain underbody tasks. After work on one vehicle is finished, the platform automatically moves the worker back to the next vehicle. There are simple robots that deliver a tray of tools to the worker inside the vehicle for the installation of interior parts, so the worker doesn't have to carry the heavy tools while maneuvering into position. In one instance, we saw a device that physically moves the worker into and back out of the vehicle on a chair at the end of a mechanical arm.

Perhaps the most prominent and popular form of automation assist are carts that carry parts and tools. They move along beside the vehicle from the time it enters a work station to when it leaves and then are returned, either automatically or manually, to the starting position. By having the parts and tools move with the vehicle, the production worker saves considerable amounts of walk time, thus reducing both the physical demands of the job and wasted motion. These are quite popular with workers and often have acquired nicknames - for example, "line side limo's" and “dollies” in some U.S. plants, "Viking ships" in some Japanese plants, and "les servantes" in some European plants. The most interesting aspect of most forms of automation assist, and in particular, the line-side limo's, is that they are low-cost forms of automation, generally developed in-house by teams of engineers and production workers.

4.2.5 The Role of Flexible Workers

While we mostly focus on trends in the use of automation, we will also briefly discuss the relationship between the use of flexible automation and the need for flexible workers. To consider this relationship fully, we must first take stock of the strategic goals that firms need flexible automation to pursue. Flexible automation allows for multiple products to be built in a single plant and/or for rapid model changes (both major and minor) over time. Investing in flexible automation thus facilitates strategies of more product variety and shorter product life cycles. Robotic weld and paint equipment can also be adjusted more easily to accommodate incremental process improvements or engineering changes.

How do these strategic goals relate to the firm's approach to work organization and workforce flexibility? The link between flexible work organization and flexible automation is hardly technologically determined. Robots do not require teams to operate effectively, nor multiskilled workers. But the decision to invest in flexible automation and the decision to invest in new forms of work organization are increasingly interconnected.

There are many ways in which flexibly-deployed workers capable of effective problem-solving are critical to achieving the strategic goals associated with flex-ible automation. In plants building many different models, workers have heightened responsibility for accommodating greater product complexity without productivity or quality penalties, mastering a higher variety of tasks, making sure the right parts go on the right vehicle, working with team members to find the most efficient layout for parts and tools, and identifying product-specific quality problems.

Then, to accomplish rapid changeovers from one model to another, work methods must be revised and well-tested in advance to avoid quality problems during product launch. Workers who are accustomed to job rotation within and across teams and to involvement in *kaizen* activities that refine work methods over time are critical resources in achieving an effective changeover. Programmable automation also lends itself more readily to worker involvement in making incremental process changes. The ease of minor reprogramming by workers (which can often be done by physically "teaching" the robot where the new weld spot should go by moving the weld tip to the exact spot) removes the technical barriers to incremental change - unlike fixed automation, for which any changes require engineering involvement and substantial cost.

Analysis of regional trends for both flexible automation and flexible work organization [5] reveals that plants using flexible forms of work organization (e.g. teams and job rotation) and emphasizing high levels of workforce training are likely to implement flexible automation most heavily. (Note that this relationship does not necessarily hold in the reverse, i.e. plants that implement flexible automation are not necessarily more likely to implement, subsequently, flexible work organization [11]) In contrast, there remains a strong association between plants that rely heavily on fixed automation (e.g. U.S. "Big Three" plants) and more traditional approaches to work organization. The importance of "fit" between technological and organizational capabilities is particularly well-supported by the extremely high level of both kinds of flexibility at the Japanese plants in Japan and North America. While the presence of good "fit" does not conclusively indicate the direction of the causality, it is worth noting that Japanese plants have had flexible work organization much longer than they have had programmable automation. This suggests, as previously mentioned, that the presence of flexible work organization can help make investments in flexible automation more feasible and/or cost-effective.

4.2.6 Performance Implications

Here we briefly summarize our findings, reported elsewhere [7,8], on the relationship between our automation measures and economic outcomes of productivity and quality. Automation in both the body and paint shops bears a strong relationship to plant efficiency in terms of labor productivity. As automation increases, labor requirements clearly decrease. This is not true for full automation of production steps in the assembly area. Full automation of assembly steps is expensive and the equipment often requires significant maintenance labor which offsets savings in direct labor. As a result, automation in the assembly area is generally only done when there are significant quality or ergonomics gains. However, automation assist appears to offer significant performance improvement in the assembly area for relatively little capital investment.

In terms of quality, while Round 1 analyses showed no relationship between the level of automation and the level of defects, in Round 2 we find that increases in automation are consistently associated with improved quality (measured using the J.D. Power Initial Quality Survey data), even controlling for work practices, scale, product complexity, and vehicle design age. We surmise that this differ ence between Round 1 and Round 2 is the result of companies learning how to use their new automation more effectively, to eliminate automation that is overly complex or plagued by too much downtime, to avoid automation that offers relatively little benefit in relation to its costs, and to support their use of flexible automation with investments in flexible work organization and human resource policies.

We find some empirical support for the idea that some companies are learning how to gain more fundamental performance advantages from their use of new technology. When flexible automation, as captured by our Robotics Index, is used in conjunction with flexible work organization and human resource practices such as work teams, problem-solving groups, high levels of training, there is a positive interaction effect - i.e. productivity and quality improve by more than the sum of the impact of the automation or the new work practices on their own.

4.2.7 Conclusion

The common perception that the diffusion of flexible production concepts is linked to the increased utilization of flexible automation is mostly accurate, although data from Rounds 1 and 2 of the International Assembly Plant Study reveal a more nuanced story across departments within assembly plants and across regions. Worldwide, automotive assembly plants experienced a moderate increase in automation from 1989 to 1993/94, with increasing convergence in average automation levels across regional groups. Important regional differences remain, however, not only in the amount of automation found in the body, paint, and assembly departments but in the relative use of flexible versus "fixed" or "hard" automation.

Japanese plants in Japan and North America continue to install the most automation in their plants, on average, followed by plants of the U.S. "Big Three", European, and Korean companies. Automation levels in the weld shop often exceed 90% in the advanced industrialized countries, while automation in the assembly area remains consistently low, between 1% and 2%, in these same plants. Paint shops show the most change over these five years and the most regional variation.

While overall automation has not increased dramatically since 1989, the more striking trend is the substitution of flexible automation (primarily robotic equipment) for fixed automation. This change in the makeup of the tool stock, even where overall automation levels remain relatively stable, reflects the high levels of product complexity that many plants now handle (due to a combination of the company's product strategy and its export activity) as well as steady improvements in the price/performance ratio for flexible equipment. This helps explain some of the regional variation in the use of flexible automation. Japanese and European plants use robotics the most and also have the highest levels of product variety and export activity. U.S. plants of the "Big Three" companies have worked to reduce their (already relatively low) levels of product variety and still do little exporting, and accordingly continue to rely much more heavily on fixed automation than plants in any other region.

Innovations in assembly plant automation have also emerged in ways not easily captured by traditional measures of automation. Perhaps the most important of these is the rapid proliferation of different types of automation assist in the labor-intensive assembly departments of plants around the world. In contrast with previous efforts to automate assembly tasks fully, which proved to be expensive, subject to frequent downtime, and difficult to adapt to different products, the automation assist approach minimizes investment costs, relies on simple and reliable technologies, and can be customized to different work stations. Automation assist is also often welcomed by workers because it does not aim to replace their jobs but to support them, it can improve ergonomics and reduce fatigue associated with excess walking time and workers can often be involved in the design and implementation process for these installations.

We consider briefly the links between flexible automation and a flexible workforce and the performance consequences of the increased use of flexible automation. Achieving the full benefits from flexible automation with respect to various strategic goals appears to require a skilled, motivated, and flexible workforce, and the assembly plant data reveal an increasingly close correspondence between flexible automation utilization and the adoption of new work practices. While automation levels have consistently been strong predictors of labor productivity at the plant level, the relationship between automation and quality outcomes has changed over time. In Round 1, there was almost no association between automation levels and quality, but in Round 2, the relationship is strong, suggesting that many companies have learned - mostly through adjustments in their mix of automation to achieve the best match of tool to task - how to gain improved quality as well as productivity. Finally, we find some synergistic interactions between investments in flexible automation and investments in a flexible workforce. Performance improvements are greater when both kinds of investment are made than could be predicted by summing the individual contributions of each type of investment.

In summary, we find that the conception from the early 1980's that full automation would sweep the automobile industry and make production workers obsolete is long-gone. While a few infamous examples of rash over-investment in automation are best-known, we found that company after company has concluded through their own initiatives that massive increases in assembly automation are not cost-effective. Instead, the industry has witnessed a tempered investment in automation overall, the steady replacement of fixed automation by flexible automation, the elimination of unreliable or overly complex technology installations, and the development of low-cost initiatives like automation assist.

We expect that over the coming years, companies will increasingly recognize the variety of ways in which flexible workers and flexible automation can serve as important complements to each other. The strong emergence of automation assist in diverse plants around the world is just one of many indicators in this direction. It may well be that the strategic goal of flexible production is best achieved when companies recognize the complementarities between these two factors and allow their investment plans for automation and corresponding investments in employee training and new forms of work organization to evolve accordingly.

4.2.8 References

1 Fujimoto T (1992) Measuring Automation in the Assembly Area. Working Paper, University of Tokyo
2 Krafcik J F (1988) Comparative Analysis of Performance Indicators at World Auto Assembly Plants. M S Thesis, Sloan School of Management, M I T
3 Krafcik J F (1989) A Comparative Analysis of Assembly Plant Automation. Working Paper, International Motor Vehicle Program, M I T
4 MacDuffie J P (1995) Human Resource Bundles and Manufacturing Performance: Organizational Logic and Flexible Production Systems in the World Auto Industry. Industrial and Labor Relations Review Vol 38: 199-221
5 MacDuffie J P (1996) International Trends in Work Organization in the Auto Industry: National-level vs. Company-level Perspectives. in Wever K, Turner L (eds) The Comparative Political Economy of Industrial Relations. Industrial Relations Research Association, Madison, WI: 71-113
6 MacDuffie J P, Kochan T A (1995) Do U.S. Firms Invest Less in Human Resources? Determinants of Training in the World Auto Industry. Industrial Relations Vol 34 No 2: 145-165
7 MacDuffie J P; Krafcik J F (1992) Integrating Taechnology and Human Resources for High-Performance Manufactoring. in Kochan T A, Useeva M (eds) Transforming Organizations. Oxford University Press, New York: 209-226
8 MacDuffie J P, Pil F K (1996) 'High-Involvement' Work Systems and Manufacturing Performance: The Diffusion of Lean Production in the World Auto Industry. Working Paper, Department of Management, Wharton School
9 MacDuffie J P, Pil F K (1995) The International Assembly Plant Study: Philosophical and Methodological Issues. in Babson S (ed) Lean Work: Empowerment and Exploitation in the Global Auto Industry, Wayne State University Press, Detroit: 181-198
10 MacDuffie J P, Sethuraman K, Fisher M L (1996) Product Variety and Manufacturing Performance: Evidence from the International Automotive Assembly Plant Study. Management Science Vol 42 No 3: 350-369
10 Pil F K (1996) Understanding the International and Temporal Diffusion of High-Involvement HR and Work Practices. Dissertation, Wharton Business School
11 Pil F K, MacDuffie J P (1996) The Adoption of High-Involvement Work Practices. Industrial Relations Vol 35 No 3

4.3 Rolling Back Cycle Times: The Renaissance of the Classic Assembly Line in Final Assembly

U. Jürgens

4.3.1 Introduction

The assembly line has been a controversial socio-technical innovation from its inception. While work physiologists and psychologists were divided about the effects on quality of work and workers' satisfaction before World War II [cf. 12: 163-172] the critics became more and more numerous during the 1950s and 1960s. "Of all occupations in modern industry none has attracted such controversial comment as that of the assembly worker, and especially of the auto assembly worker on the 'final line'" - Walker and Guest state in 1952. [33: 9] Countless scientific studies and socially critical discourses dealt with the negative effects of automobile work on the workers: highly repetitive work tasks tied to the speed of the assembly line, physically highly strenuous, but intellectually stupefying for the workers, stifling any initiative and sense of responsibility at the level where work is carried out. In the middle of 1960 Goldthorpe determined that "assembly-line production in the automobile industry is now generally regarded as the *locus classicus* of worker alienation". [14: 235] During the 1960s a wealth of arguments was produced by field studies which showed the adverse effects of paced line work, not only on the workers' health and work satisfaction but also on labor relations in the plant and even on efficiency. Hence, more and more arguments were raised against the economic superiority of assembly line production, and alternative forms of process organization were suggested. By the mid 1970s the feeling was widespread that the days of the classic assembly line were over. The controversy shifted to the question of the future direction: In any case, the old assembly line did not seem to belong into a modern auto factory. It could hardly have been imagined that precisely the assembly line would survive the age of Fordism after all.

The 1990s, however, see the renaissance of the classic assembly line. With the success of the Japanese manufacturers on the world market, Western manufacturers recognized that world class manufacturers had not questioned the assembly line principle to the same extent as had been done in the West and in particular in Europe. The example of Japan and Western experience with problems with mechanization and alternative forms of work organization came together. The closure of the Swedish plants which had pioneered alternative directions was a highly symbolic act, and confirmed for many the victory of the classic assembly line concept over its alternatives. This is not only true for more radical complete car assembly concepts but also for the new forms of work based on dock or cell assembly principles. These work structures, which had been most popular among Swedish and German companies during the 1980s are being scrapped to a large extent in the mid 1990s. And even earlier, the protagonists of avantgarde assembly automation projects had to give in and revise their strategies.

At the heart of the contentions surrounding the assembly line is the issue of cycle time. The length of the individual work cycle is one of the major planning parameters and determining for the quality of work on the line. With longer work tasks the number of operations and thus the variety of job content in general increases. Consequently, job enlargement has become a necessary prerequisite for job enrichment in most of the quality of work life/humanization of work literature. In practice, however, we observe a similar trend in reverse. Whereas process designers thought to lengthen the job cycle in the seventies and eighties they have in the mid-nineties reversed gears, and reoriented themselves toward the one-minute cycle characteristic of the classic assembly line. In this chapter we will discuss this reorientation. Is it due to an organizational learning process reflecting the fact that alternative forms did not match expectations or stand up to critical scrutiny, or have new factors come into play to explain the turnaround? The first section deals with the arguments brought forward by the critics since the 1950s, which had questioned the effectiveness. The second section deals with the experience manufacturers had with alternative forms of work organization in assembly they had introduced in the course of the 1980s. In conclusion we will discuss the implications for the quality of work on the assembly lines. In the 1990s, with the return to the short cycle time, will we experience the return of old conflicts about speed-up, repetitive work and degradation of human labor, or do alternatives exist to ensure the sustainability and attractiveness of assembly line work?

4.3.2
The Critics of the Assembly Line

Hardly any systematic research had been done on the problems of work on automobile assembly lines despite its practical and theoretical importance, precisely in this industry, and this was true, even up to the 1960s. Assembly line work in the automobile industry had, as Walker and Guest observed in their pioneering study on "The Man on the Assembly Line" with field studies made in the late 1940s, "been most discussed but least studied in a systematic fashion" [34: 4f.], and Chinoy, who carried out field studies for his book on "Automobile Workers and the American Dream" just after World War II, wrote in a postscript in 1975: "The evidence, unfortunately and, in view of the symbolic significance of the assembly line, surprisingly, is limited" [4: 140].

The pre-war research had been mainly concerned with the issue of industrial fatigue, which was investigated from the viewpoint of the natural laws of the "human motor" [26: 120]. This concern of the emerging European "science of work" was first applied to assembly line work during World War I, when the British Minister of Munitions appointed a Health of Munitions Workers' Committee, which exercised some influence over British munitions plants, reducing hours, introducing rest-pauses and holidays, and identifying health risks [26: 275]. The research carried out during the 1920s and 1930s, while drawing from investigations in various industries, expanded the knowledge about individual and inter-individual performance variation and various strain and stress factors of paced work while also increasingly recognizing the problem of monotony at work. One of the first studies of working conditions on assembly lines in auto manufacturing was made by Walker and Guest in the late 1940s. Walker and Guest listed as characteristics of mass production jobs in car assembly (1) the mechanical pacing of work, (2) the repetitiveness, (3) the minimum skill requirement, (4) the predetermination in the use of tools and techniques (5) the minute subdivision of product worked on and (6) the surface mental attention. They continue: "For the engineer all of the above characteristics are brought into focus in what is known as the job cycle. Each worker must perform a prescribed number of operations within a set time limit and, in the case of those working on moving conveyors, within a given distance along the line." [34: 12]

Regarding the central importance of the length of the job cycle, it could be expected that the cycle "would become a major point of controversy and contest when it came to the design of new assembly processes". Thus it could be expected that national and local differences would lead to a broad variety of solutions specifically with regard to the cycle time issue. The opposite holds true, however. Since the inception of car assembly on conveyor belts, a one-minute job cycle has almost continuously ruled and constrained work planning for almost a century. What is then the legitimization of the one-minute job cycle, which seems so firmly ingrained in the mind of process planners in almost all countries?

The influence of time and motion study has to be brought in at this point. The method of elemental time studies initiated by Taylor and refined by Bedaux seems to have established a firm mental grid of work times and motions scaled according to the one-minute logic. Thus, the rating survey films used for training industry engineers and process planners showed operators working at or near a 60 rating on the Bedaux scale, called the day work speed or the normal effort. (While a rating of 80 shows the work speed under incentive conditions.)

"Normal pace (on the 60/80 Bedaux scale) is achieved, when a worker is producing 60 units a work per hour. It has been agreed that this pace is demonstrated by 'a man of average stature walking at 3 miles per hour, unloaded, on level ground, and under normal atmospheric conditions'." [7: 52]

The one-minute standard seemed to be confirmed by studies on the costs of learning assembly operations. Kilbridge concluded from several field studies and laboratory experiments: "When the task length is very short a simple and restricted motion pattern is repeated continuously. Observations indicate that this induces cramping and excessive muscle fatigue which inhibit the worker's ability to maintain the uniformly fast pace. The reaction is known in industry and is commonly called 'short-task fatigue'. As the task length increases and the motion pattern becomes more diversified simple muscle fatigue declines. Beyond a point, however, a countervailing influence tends to reduce the pace attainable. Forgetting, fumbling, loss of motor skill and rechecking time increase, resulting in a slowing of pace called 'long-task delay'." [21: 519] According to his empirical research these negative effects come in already when the job cycle exceeds the one minute mark. Kilbridge found that, in order to reach the normal pace for a one minute MTM standard time, approximately 1,000 cycles were required to learn the job. When the job content was increased the number of repetitive cycles needed to reach the normal pace grew. The learning time seemed in fact to be a monotonous function of the standard time of the job. Thus, the learning cost argument seemed rather to support the shortening of cycle times. Kilbridge: "As work is subdivided and deskilled, the rate at which operators learn their tasks increases sharply and the learning costs associated with employee turnover and product or model changes decline. There is a point, of course, at which, although learning costs may continue to decline, other economic and psychological factors inhibit further division." [21: 516]

Shortening the task times has not proved to be a solution for labor problems like employee turnover. They rather made the situation worse because, as Baldamus had pointed out earlier, "... - on the whole - turnover rates are inversely related to the level of skill and the shortness of the work-cycle" [2: 51]. Labor turn-over, worker dissatisfaction and de-skilling were central topics for the critics of the assembly line with most of the pioneering empirical studies dating back to the late 1940s. This is true for the studies of Friedmann [11] and Touraine in France [31], Walker and Guest as well as Chinoy in the US, and for the study by Wyatt and Marriott in British car assembly plants [36].

Thus Walker and Guest observed a clear association between absenteeism and mass production characteristics and a certain turnover tendency related to these characteristics. To like or dislike a repetitive job was, as Walker and Guest showed, moreover strongly dependent on their experience with their former jobs. Still there was a clear correlation between the number of operations performed per job cycle and the degree of job interest. The most clear-cut results with regard to different process layout configurations and with regard to differences in work cycles were provided by the research of Wyatt and Marriott. The authors showed a drastic increase of their index of work satisfaction depending on the length of the cycle time among body shop workers of a British car plant [36: 23].

While experimental psycho-physical knowledge about the 'human factor' seemed to support the short job cycles of the classic line, real labor unrest and its most wide-spread form, labor turnover, forced production engineers to reconsider their assumptions. The cost for training obviously was not just a matter of the learning curve, but also of the leaving rates.

The revision in the learning curve debate reflects a shift in the factors influencing the cycle time decision and process design in many companies in the 1960s. Job enrichment was to become the new paradigm. This paradigm shift was supported by a new generation of studies on the stress and strain situation at work, which emphasized the relevance of factors such as decision latitude and social support in the work situation, thereby overcoming the narrow confinements of ergonomics in work science. Studies by Kornhauser [23] among American auto workers showed negative effects of machine paced jobs on psychological health. Caplan et al. [3] reported that the assembly line workers had the highest level of somatic complaints and dispensary visits among 23 occupational groups. Both studies showed that in the absence of coping mechanisms like individual

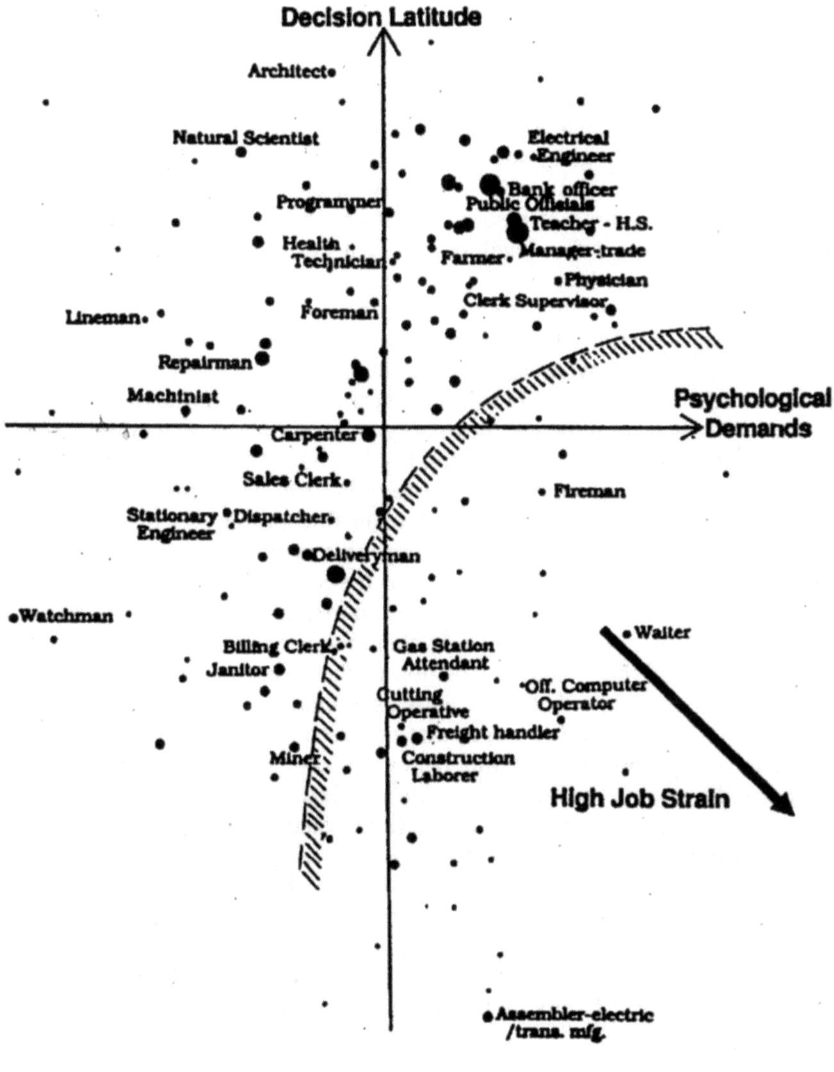

Fig. 4.3.1. The occupational distribution of decision latitude and psychological demands among men (n = 2,897) based on US Quality of Employment Surveys (1969, 1972, 1977). Source: [19: 177]

decision latitude and social support at work, the psychological strain and physiological stress increased the risk of coronary heart disease. And the assembly workers were seen to be particularly bad off in this regard. The negative character of assembly line work is shown by figure 4.3.1., which presents findings of

Karasek and associates [19, 20], who charted the job characteristics of 130 occupations in the US drawn from the US Quality of Assembly Service with regard to their decision latitude and psychological demands. The occupational group of assemblers (electric and transportation equipment manufacturing) were found to be at the top of the psychological strain risk scale, and, at the same time, lack the decision latitude and social support necessary to cope with this situation. Thus, they are among the "victims of restricted learning opportunities, social isolation, and restricted possibilities for development of a positive social identity because their work is so fragmented and distant from customer feedback." [19: 182]

Coetsier's study on the subjective assessment of line work by the line workers themselves, carried out in a Belgian assembly plant in 1966, confirmed the importance of social support whereas the assembly line was seen by the worker "as a strange and hostile being that forces him to work according to the pace it imposes on him and on which he depends inexorably in its capacity as worker." [5: 120] No personal act is permitted on the assembly line, the human personality is willingly and knowingly neglected as the assembly line has, among other things, been designed precisely to eliminate the human fluctuation. All the more important, according to Coetsier, become the compensatory elements in the work situation: "The worker looks for compensation for the loss of respect for his personality. As a general rule he finds this compensation in companionship with his fellow workers. This companionship is also furthered by the mutual interdependence of the workers and by the spatial disposition of the assembly line. A new element is found in companionship that is at least as important in the determination of the work climate as the elevated wages that go together with assembly line work." [5: 120] The study showed that the factor "relations with work companions" received a highly positive weight in workers' subjective assessments thus compensating for some of the negative estimations in the other items. With regard to task content, the study showed further that the items "meaning of work", "repetitive character of work", which were emphasized most by the critics were not evaluated as negatively as effort related to "data to be remembered" and "set-backs" i.e. having to struggle to keep up with the line due to faulty parts, inadequate equipment or other factors. This item obviously correlated closely with the "imposed pace", which scored most highly on the negative side among the work organization items.

While Coetsier did not investigate the influence of job cycle length on worker satisfaction and the level of labor conflicts on the shop floor, a German research project carried out between 1968 and 1971 in an assembly plant (of trucks, however) demonstrated a clear correlation between the length of the work cycle and the level of conflicts or work dissatisfaction.

Work dissatisfaction and conflict levels were highest, where job cycle times were the shortest (between 2,5 and 5 minutes). They reached their bottom with cycle times between 15 and 20 minutes, and showed a marked increase again for cycle times beyond 20 minutes. This curve is interpreted as the result of two opposing forces: While the level of attention required by the job may be low with short cycle time and only a few task elements, the high time pressure of the short cycle and the high pressure to conform with the forced pace creates a high poten-

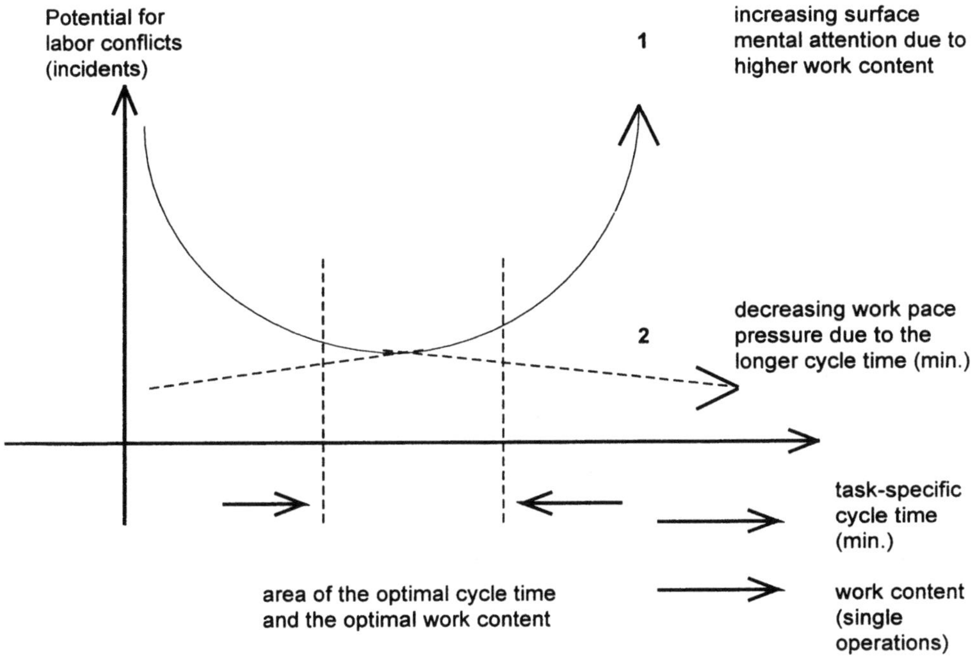

Fig 4.3.2 The influence of task specific job cycle length and number of elemental operations on the perceived conflict potential at work - schematic diagram. Source: [10: 202]

tial for conflicts. When job cycles become longer the pressure due to the pacing of the machinery is reduced, which leaves a greater margin to the individual for own decisions. With job cycles being further increased, and thus the labor content further enlarged, a high level of attention is required, which leads to higher levels of conflict again. "Thus the optimal length of the task-specific job cycle will be in the area, where conflict reduction due to an increased decision latitude corresponds with an increase of inconveniences due to the higher labor content which has to be coped with." [10: 203, transl. U.J.]

The conclusions about the effect of lengthening the work cycle is similar to those of the learning curve research (fig. 4.3.2.). With increasing cycle time the negative effects of speed and pressure diminish when the area of the optimal cycle time is reached until, with even longer cycle times, negative effects of complexity and increased attention become dominant. This view was also shared by Ulich [32: 271]. With regard to the seemingly linear relationship between cycle time and work satisfaction observed by Wyatt and Marriott [36] he expressed his doubts as to whether this linear relationship would continue. "We assume rather a curved linear, in this case a reversed U-shaped relation between complexity characteristics of work and human behavioral characteristics. [36]

The question remains open, however, where exactly the area of the optimum is situated.

In any case, the optimum was to be expected beyond the one-minute mark. The German Metal Workers Union (IG Metall) concluded from studies such as Euler's but more so from the events at Lordstown and the Swedish debate over new forms of work that it should exert its influence in order to make sure that assembly lines were designed according to the level of knowledge reached by work science. IG Metall demanded that job cycles should not fall short of 1,5 minutes of any new assembly line to be set up in the future and that there should generally be no further shortening of existing job cycles. The 1,5 minute rule became part of a regional wage framework agreement struck in 1973, and hence became a benchmark for works councils in all German auto companies. (cf. for a discussion of this agreement: [28].

At the beginning of the 1970s the pressure for work reform and the critique of assembly line work was felt by manufacturers worldwide. In the USA the need for improved job design was highlighted by the Work in America report of the Upjohn Institute in 1973. In response to the question, 'what type of work would you try to get into if you could start all over again', only 43 % of white collar and only 24 % of blue collar workers said, they would choose the same kind of work. Short cycle times, and hence repetitiveness resulting from increased specialization and mechanization introduced for greater efficiency, were regarded as a primary cause for this dissatisfaction. The labor rebellion at the General Motors assembly plant in Lordstown, where assembly operations had been laid out with a 36 second job cycle, was regarded as signal that this type of work was not accepted by workers any longer. (This interpretation, however, is controversial, see [16: 104 f.] In a less spectacular way the issue of worker discontent was put on the agenda in Japan by the book of Kamata, who reported from his experience as temporary worker on the line at Toyota. [18] Kamata's report even though dismissed as an outsider's view was regarded as embarrassing within management circles.

A survey carried out by Muramatsu and Miyazaki published in 1976 reflected a self critical view from the perspective of production engineering. While the assembly line was held up by the authors "as the most advanced application of the 3 S-principles "standardization, specialization and simplification", which "spell the principles in the development of modern industry" [24: 311] certain shortcomings were admitted. Among them in particular the vulnerability of the system of assembly line work with regard to labor are underlined:

- the system inherently requires flexibility for overtime and consequently its operation becomes almost impossible in circumstances where overtime work is refused or prohibited;
- the whole system may become unstable when a worker does not conform with the established standard procedure,
- if the average rate of absenteeism becomes larger than 25 % the conveyor-line may no longer pay off.

The research group around Muramatsu and Miyazaki was established in 1970 with the goal of designing new production systems aiming at the development of a "fusion system", which simultaneously accomplishes the improvement of productivity and the satisfaction of the "workers' desires". Several surveys were carried out in the course of the 1970s using a sample of mostly line workers from various industries including major car manufacturers and automotive supplier companies. The surveys show that the "implementation of job enlargement" and "line or non-line" work stood out as influencing factors in 1974. They lost in importance and became almost marginal in the following years [25: 302], however. A special survey made for Toyota showed that three factors: rest time, job rotation and the number of workers in the group had the greatest effect on motivation. As a response Toyota introduced a rotation system, which by increasing workers' versatility helped to meet the fluctuation in demand for different types of cars. The authors describe the case of Toyota's Tsutsumi Plant in 1974, where "in order to enhance the worker's desire 'to improve himself and fulfill his desire to overcome various frustrations'" several key factors for improving the workers' motivation were introduced. The new system was designed to:

- increase the opportunity for the worker to exhibit his ability,
- offer the opportunity to the worker to learn a wide variety of jobs without restriction,
- allow the worker to take breaks freely during his working hours,
- adjust the number of workers per group to 8 to 15 men depending on shop characteristics. [24: 6].

These measures explain, according to Muramatsu and his associates, the major performance improvements at Toyota in the following years. Thus, the ratio of versatile operators increased from 40 to 95 % between 1974 an 1980, the number of suggestions per worker per year increased from 14 to 50, the percentage of defects declined by 46 %, and man-hours per car were reduced from 10,9 to 7,6 in 1980 [25: 308].

The direction taken by Toyota in response to the concern of workers dissatisfaction left the process layout and the traditional one-minute cycle untouched. Also, work planners at General Motors, who had the opportunity to draw conclusions from the Lordstown conflicts in the design of their "Southern Strategy" assembly plant in Oklahoma City, the only new assembly plant planned in the 1970s, compromised on the ideal of a "high speed" plant only slightly, with a process layout of 75 units per hour, i.e., a job cycle of 45 seconds. At the same time the plant was intended to become the first real team-based assembly plant of the Big Three. [cf. 16: 106 f.]

The job enrichment (JE) literature viewing JE as a weapon to combat poor morale and motivational attitudes provided management with little guidance on economically optimal choices of JE parameters. Thus, by the mid of the 1960s a different approach developed, which treated JE as a managerial policy designed to lower cost. This approach underlined the inefficiencies and cost penalties related to certain types of process organization. In this sense Globerson et al. stressed the importance of training costs in relation to employee turnover as a major factor: "While poor morale due to job dissatisfaction may directly cause some loss of output, the increased voluntary leaving rate, with associated hiring and training costs, appears to be the dominant negative term in the overall cost equation." [13: 345] A higher labor turnover among line workers as compared to other workers was regarded as an established fact in this debate. Thus, Kilbridge had found in a case study of two companies that turnover was 21.9% for line workers, 12,0% for batch workers and only 6% for workers on non-repetitive jobs. From a JE experiment at SAAB it was reported that an increase of cycle time from 2 minutes to an average of 20 minutes led to a drop in the leaving rate by 50%. Drawing data from this experiment Globerson and Crossman established a curve of training costs per job as a function of cycle time and labor turnover. They concluded. "For each job there appears to be a cycle time minimizing total training cost. However, lacking an adequate empirical model of the cycle time/leaving rate relationship, this function can only be estimated. Also in doubt is the exact influence of cycle time on learning curve parameters. Making reasonable assumptions in both these respects and using the reported results of a JE experiment in automotive assembly, cycle times in the range of 10 - 30 minutes were calculated to minimize total training cost." [13: 354]

Despite the relevance of assumptions on, for instance, learning performance for designing assembly processes neither theoretical models based on work science assumptions nor practice reports have provided the necessary guidelines. Thus, Wild stated in 1975 that no adequate procedure was available for the estimation of time loss resulting from learning under conditions of differing task complexity and model mix condition. [35: 440]

The implementation of job enrichment programs, particularly in Sweden as the example showed, offered the opportunity to compare the relative merits and shortcomings of certain types of mass production systems since the mid-1970s under real world conditions and, as Wild stated, "to encourage more objective and thorough evaluation." [35: 443]

In regard to the advantages of work structures with stationary assembly, the proponents could draw from early work science results generated under the "human motor paradigm". The research on performance characteristics of repetitive work (summarized by Dudley [7]) had shown that an operator will not work at a constant rate. He or she would rather tend to work faster than could be habitually and consistently maintained (the normal effort). Dudley: "Approximately 66 % of the cycle time will be at or shorter than the mean. The variability is a perfectly normal human characteristic, and it cannot be prevented by an indi- vidual, however hard he tries ...." [7: 43].

This skewed appearance contrasts with a more scattered distribution of times achieved by a trainee or unskilled operator. However, evidence indicates that the skewed distribution of operation times of unpaced workers becomes less marked when the working pace is controlled as on assembly lines and that this tendency increases directly with the degree of control. [7: 44]

Another disadvantage of the serial assembly system results from the difficulty in attaining the optimum division of labor on assembly lines. This is also known as the balancing problem. When task structures and work assignments become highly specialized, a point will be reached at which the cost of non-productive work and line balance delay exceeds the savings resulting from this division of labor. On assembly lines it is difficult to divide work evenly among all operators, and operators having shorter assignments will have some idle time. This balance delay is usually attributable to restrictions imposed on the ordering of work elements by fixed stations and machines on the line. Wild found that - all other factors being equal - lower cycle times gave rise to higher balance losses and vice versa [35: 450].

In a systematic attempt to evaluate different types of assembly systems Wild differentiated between five sources of inefficiencies: 1. learning, 2. set-up, 3. balance loss, 4. division of labor loss, 5. system loss. Engström et al. have used this typology recently in order to demonstrate the superiority of "organic flow" or "reflective production" assembly systems such as at Uddevalla over more conventional systems in terms of efficiency [8,9].

Engström and Medbo compared three idealized systems: a serial flow system, an example would be the Volvo Torslanda plant, a semi-parallel flow, would be Saab's example Malmö plant, and an "organic flow" system, of which the Uddevalla plant of Volvo would be example. Building on Wild's typology they measured three types of loss situations: "(1) it is not possible to divide evenly between the work stations - balance loss, (2) a large amount of the work at the work station is required for materials handling - division of labor loss, and (3) the assembly worker cannot vary his work pace naturally - system loss". (ibid.) Empirical data were taken from Swedish plants controlling for differences in product design and thus in the "ease of assembly" of the product. The results are shown in table 4.3.1.:

Table 4.3.1. The losses for three types of assembly system.

	Serial flow	Semi-parallel flow	Organic flow
(1) Balance loss	30 %	15 %	5 %
(2) Division of labor loss	25 %	20 %	15 %
(3) System loss	80 %	65 %	40 %
Total loss	135 %	65 %	40 %

Source: [9]

The data show a striking superiority of the "organic flow" type of assembly processes over the conventional serial flow assembly line in the Swedish context. Regrettably, similar studies on a broader international scale do not exist, and the data of the International Motor Vehicle Program on assembly plant efficiencies [37] cannot be differentiated according to different assembly system configurations.

4.3.3
Learning from Experiences from Longer Cycle Work

A fundamental difficulty which the critics of the assembly line had to face were the uncertainties as to the real benefits and costs of alternatives. The evidence drawn from different industries (or from truck assembly) could not convince the skeptics, and the same held true with regard to the Swedish work reforms as long as they were confined to small satellite plants. The uncertainties linked with a change in paradigm caused the majority of manufacturers, when the pressure for change culminated in the 1970s, to choose either to seek for improvements on the basis of the given process design or to seek for compromises in the process design which take into account the demand for longer work cycles without changing the principle design configuration. Among such compromise approaches particularly the drift work approach and the modular approach became wide spread "alternatives" to the work structures on the classic assembly line.

a) Drift work implies a de-coupling of the individual job cycle from the production cycle of the line. The individual operator works over two and more stations on the same car while standing on the moving line or walking along with the line flow. This type of work has been practiced in many companies, particularly where certain options with high labor content items like air conditioning have to be installed in every other car. After the 1,5 minutes minimum standard for cycle times had been established by union demand, working over several cycles became almost general practice in German assembly plants. Thereby the traditional short cycle process layout could remain unchanged. At Volkswagen, for instance, where the line speed used to be changed according to the level of absenteeism it was agreed that, when the cycle times dropped below 1,5 minutes, work operation would be switched to double cycles. Where no technical restrictions existed, work was performed over three and four stations.

The practice was regarded as satisfactory both from the point of view of industrial engineering as well as from the workers' viewpoint. It helped to reduce cycle losses due to difficulties with line balancing, difficulties, which had multiplied with the increase in variants and options. For workers it provided a means to increase their range of communication with fellow workers and increased task variety.

When in 1988 VW took the first step toward group work and assembly operations it built on its drift work experience. Workers received a pay supplement for mastering tasks within their work systems. This multi-skill requirement was regarded as fulfilled by the practice of working over several work stations. At this time the average was four work stations with up to eight stations in individual cases at VW's Wolfsburg plant.

Drift-work helped to reduce cycle time loss and increased workers flexibility for different jobs on the line. It had its disadvantages, though: the distances to be walked back to the original station increased and the pedestrian traffic required broader aisles between the line and the parts storage area. Drift-work also made it difficult to trace back quality defaults to the individual workers, thus impeding the direct quality feedback.

b) The modular work approach required changes in the process layout. The approach was regarded as an answer both to increasing model mix problems and to demands for a humanization of work. Special work areas were created for specific sub-systems, particularly those where labor content varied highly between the different car models, variants and options. The reduced complexity of the tasks on the remaining main line helped to standardize jobs here and overcome line balancing problems. At the same time process planners in the modular areas could opt for alternative concepts with parallel work stations allowing for stationary work. The self-moving platforms developed by Fiat for its Robogate bodyshop and by Volvo for its Kalmar plant became the most popular solutions for between-station transport, even though some companies opted for the less expensive solution of installing sub-assembly lines or benches to perform these modular sub-assemblies (cf. chapter 4.9 by Mishina in this book).

The compromise strategy, with long cycle work operations in parallel assembly stations and a shortened main line (where drift work was practiced) became the most popular solution among German manufacturers in the 1980s. Thus modular assembly areas were created for, for instance, the assembly of doors, instrument panels or cockpits, seats, wiring harnesses, engine dressing, drive train etc. BMW, for example, installed an AGV course at its Regensburg plant, Opel introduced the "new production concepts" in most of its European plants. Volkswagen installed a flexible assembly line with parallel assembly stations at its Wolfsburg plant, parallel to the rigidly coupled mechanized assembly area in building 54, and Mercedes Benz, which had assigned a huge area in its Sindel-fingen plant for modular wiring harness installment, took stationary assembly a step further in its latest German assembly plant in Rastatt which opened in 1992 [17].

The process layout in modular production areas offers the opportunity to create work structures with meaningful tasks and more individual freedom in performing the job. With regard to the issue of the optimal length of the job cycle, process planners had to rely on the experiences with drift work and on hunches as there were no clear findings in work science to guide their decisions. Therefore work planners at first tended to set average work cycles at around five minutes, thus remaining within the range of their drifting experience, with a high variation between the different options, however. Planners in other companies, who designed dock assembly areas in later years, became bolder. The dock system for wiring harness assembly introduced at Mercedes Benz Sindelfingen plant in 1985 was set up for an average of 100 to 120 minutes. It was found out by management that workers divided tasks again and, in actuality, worked cycles of 20 to 30 minutes. This informal re-division of work was seen as a sign by management that, despite their high skill, the workers did not appreciate work cycles exceeding the 30 minutes. The company had assigned mostly skilled tradesmen with certified automotive related skills to these areas and it was expected that work cycles between one and two hours would be appropriate for their work motivation.

As stated earlier, many of the dock assembly systems were scrapped only a few years after being installed. This is true for all cases mentioned above, even for the most recent systems: the flexible assembly line at Volkswagen's Wolfsburg plant and the (actually brand-new) system at Rastatt. The new work structures are being based on the reconfirmed assembly principles with serial flow and shorter work cycles around the one-minute "ideal".

The economic justification of the modular work approach with stationary assembly systems - with expensive investments and space requirements for the automated guided vehicle system - was that it saved the line balance losses of assembly line production. Thus, the saved balancing losses would pay for the investments in technology and space. Further advantages were seen in a reduction of repair work and in the possibility of producing a broader range of variants under these conditions. Disadvantages lay in the parts supply system. Commissioning the parts sets for the dock system, besides being a complex task prone to worker errors, often led to a "parts tourism", as commissioning areas were usually far removed from the assembly areas. Problems with the control systems and the synchronous delivery of the modules to the station at the main line were further negative experience.

The reversal of the compromise approaches towards assembly work did not only hold true for stationary assembly systems but also for the drift work approach. In the case of one German company, drift work has been abandoned altogether and the principle of strictly working within the work station was reinstated. In other companies the range and the conditions of drift work are being specified and regulated.

One explanation for this U-turn back to classic assembly structures is related to benchmarking. The best practice plants at the beginning of the 1990s were all plants with serial flow and short work cycles, and most of them with just-in-time regime and team work following the model of the Toyota production system. The benchmarking explanation is not totally satisfactory, however, as different local conditions should allow for or even require different solutions. In addition the question arises as to the experiences which the manufacturers had made with regard to the effects of different process layouts on performance and humanization goals. Unfortunately, little efforts seem to have been made by the companies to collect data and evaluate the experiences with an alternative process layout.

Interviews carried out by the author among industrial engineers and process planners of German car manufacturers in 1995/1996 show that the U-turn was partially due to disappointments with the results of the compromise strategy and partially to the adoption of the lean production paradigm. The two major disappointments regarded:

(1) the expected attitudinal change which had served as justification for alternative work structures such as parallel stationary assembly. The level of absenteeism seemed unaffected by differences in the job characteristics of long and short cycle work. Workers in the modular production areas at Opel even showed a higher level of absenteeism [30: 28]. Employee turnover, which was the driving force for much of the job enrichment debate in the 1960s and 1970s, had ceased to be a problem anyway, as labor market conditions had drastically changed. Actually, a higher turn-over would solve more problems now than it would create. Finally, the expectation that alternative work structures would increase work satisfaction, particularly among workers with a skill certificate, was not fulfilled. These workers regarded the degree of job

enrichment as minor and they saw no principle change as to the paced character of work.

(2) Learning costs for breaking in new staff in parallel assembly systems were higher than expected. Problems resulted particularly from absenteeism. It was difficult to find multi-skilled persons able to replace absentees. Thus, qualification was difficult to replace and this became at times a serious bottleneck in meeting the production goals. In addition, long spells of planned absenteeism due to vacation or accumulated free days resulting from the reduction of the weekly working hours created problems when changes had been made in the meantime

Reasons related to the adoption of lean production as the new paradigm for process planning:

(3) With the introduction of just-in-time and total quality principles, short job cycles were expected to help better monitor faulty deliveries from upstream production stages and suppliers. Whereas long job cycles provided the latitude for the individual to solve problems and correct faulty parts, the stricter time regime would reveal such problems and force management to correct their root causes.

(4) With the introduction of group work, the task of minimizing cycle loss could be given into the responsibility of the group. This required, as benchmark plants such as NUMMI had demonstrated, that workers be multi-skilled for the tasks in their group and that work tasks be highly standardized. In order to keep the overall task of the group manageable and transparent to every group member the overall time of the group task and, accordingly, the group itself, should be small. As a consequence of this consideration one German company has recently revised its group work system and cut the size of groups in half.

In any case, the dominant view among process planners now is that a return to the classic one minute cycle will have a positive influence on major performance criteria. Table 4.3.2. shows the subjective assessments of a group of senior industrial engineers of a German car manufacturer, which was about to change its assembly process and shorten the job cycles in 1996. These industrial engineers bring in their experience with work cycles beyond 1,5 minutes including long cycle work in modular assembly areas. The factor "capacity" stands out among the factors which speak in favor of a shorter cycle time in their view, i.e. the expectation that daily production targets will be effectively met with the process design. Another area in which a major improvement is expected is the reduction of worker mistakes and of training efforts for breaking in new workers, whereas the shortening of cycle time is expected to have almost no effect on absenteeism due to illness and employee turnover. Work satisfaction is expected to suffer considerably.

Table 4.3.2. Factors in favor pro (+) or con (-) longer (> 1,5 minutes) and shorter (< 1,5 min.) work cycles (max. = +/- 80)*

	longer	shorter
absenteeism due to sickness	+ 5	- 5
employee turn over	+ 5	- 5
./. of skilled (apprenticed) workers on the line	+ 15	- 5
work satisfaction	+ 45	- 35
training effort for break in	- 30	+ 30
worker mistakes	- 40	+ 40
module mix	+ 15	- 25
JIT supply	+ 45	- 35 to + 10
capacity	- 80	+ 80
work time reduction	+ 5	+ 5
integrated/total quality control	+ 25	- 25
continuous improvement	+ 5	- 5
group work	+ 25	- 20

* As assessed by four senior industrial engineers of a German car manufacturer (Aug. 1996). The table shows the aggregated values in cases where the assessments pointed in the same direction and the aggregated positive and negative assessments where they were contradictory.

4.3.4
A Sustainable Solution?

The 1990s seem to mark the defeat of the alternatives to the classic assembly line. The closure of the work reform plants in Sweden as described in other chapters of the book are only the tip of the iceberg. Parallel assembly systems with stationary long cycle production are being abandoned on and reverted into serial processes, and even conventional assembly lines where cycle times had been extended are being redesigned towards the 1 minute ideal. On the whole, a modernization trajectory suggested and advocated by many ergonomic analysts and industrial sociologists has been fundamentally challenged. It relied on the assumption of the mutual complementarity and parallelism of the movements in the direction of job enlargement and job enrichment, i.e. from short cycle low discretion jobs to long cycle high discretion jobs. This assumption was challenged by Womack et. al. [37] and more recently by Adler [1]. Adler emphasized the job enrichment aspect related to the improvement activities and enhanced responsibilities for quality by the production groups. At the same time, these groups work in a highly repetitive structure with pre-determined times and motions, probably more strict than their counterparts studied by Walker and Guest fifty years before. From the background of the labor market situation and their globalized production structures, manufacturers in the West may not have to fear that problems with employee turnover will return. However, the decades of controversies about assembly line work provide many reasons to doubt that under paced repetitive working conditions, labor relations will not suffer in the long run if compensatory elements cannot be developed.

A way out of the dilemma could be to seek systematically for such possibilities of job enrichment, which do not require an enlargement of the job let alone the time sovereignty of work in stationary assembly cells. Thereby repetitive work with short cycle time may become sustainable in the eyes of the workers as compensatory mechanisms exist. The importance of involving workers in problem solving activities for motivation has been emphasized by Adler as an explanation of an NUMMI success [1]. Kishida, who investigated the problem of monotony at work in Japanese plants, stressed the role of personnel development and individual promotion perspectives in this regard [22]. Further possibilities of enriching jobs by giving workers off-line responsibilities and involving them in activities have been particularly developed by American companies in recent times. These include the involvement of work groups in the product and process development for new cars as well as taking up responsibilities in relation with suppliers and dealers. A synthetic approach, which brings together the different experiences from Japanese companies, American companies, such as Saturn, and European companies in order to further develop this trajectory seems necessary.

4.3.5 References

1 Adler P S (1992) The Learning Bureaucracy: New United Motors Manufacturing Inc. in: Staw B M, Cummings L L (eds) Research in Organizational Behavior. JAI Press, Greenwich, Conn.
2 Baldamus W (1951) Type of Work and Motivation. British Journal of Sociology Vol 2 No 1: 44-58
3 Caplan R et al (1975) Job Demands and Worker Health. National Institute for Occupational Safety and Health. HEW Publication No 75-160, Washington DC
4 Chinoy E (1992) Automobile Workers and the American Dream, edited and with an introduction by R. Milkman. University of Illinois Press, Urbana and Chicago
5 Coetsier P (1966) An Approach to the Study of the Attitudes of Workers on Conveyor Belt Assembly Lines. The International Journal of Production Research Vol 5 No 2: 113-135
6 Conant E H, Kilbridge M B (1965) An Interdisciplinary Analysis of Job Enlargement, Technology, Costs and Behavioral Implications. Industrial Labor Relations Review Vol 18: 377-395
7 Dudley N A (1968) Work Measurement. Some Research Studies. Macmillan, St. Martin's Press, New York, London etc.
8 Ellegård K et al (1992) Reflective Produktion. Industriell Verksamhet i förändering. AB Volvo Media, Göteborg
9 Engström T, Medbo P (1995) Data Collection and Analysis of Manual Work using Video Recording and Personal Computer Techniques, Manuscript submitted to International Journal of Industrial Ergonomics
10 Euler H P (1977) Das Konfliktpotential industrieller Arbeitsstrukturen.Analyse der technischen und sozialen Ursachen. Westdeutscher Verlag, Opladen
11 Friedmann G (1950) Où va le Travail humain? Librairie Gallimard, Paris

12 Friedmann G (1964) Industrial Society - The Emergence of the Human Problems of Automation, edited and with an introduction by H L Sheppard, The Free Press of Glencoe, New York
13 Globerson S, Crossman E R F W (1976) Minimization of Worker Induction and Training Cost through Job Enrichment. The International Journal of Production Research 1976 Vol 14 No 3: 345-355
14 Goldthorpe J H (1966) Attitudes and Behaviour of Car Assembly Workers: A Deviant Case and a Theoretical Critique. British Journal of Sociology Vol 17 No 3: 227-244
15 Johanson G (1991) Job Demands and Stress Reactions in Repetitive and Uneventful Monotony at Work. in: Johnson J V, Johanson G (eds) The Psycho-Social Work Environment: Work Organization, Democratization and Health. Essays in Memory of Bertil Gardell. Baywood Publishing Company, Amityville New York: 61-72
16 Jürgens U et al (1993) Breaking from Taylorism. Changing Forms of Work in the Automobile Industry. Cambridge University Press, Cambridge
17 Jürgens U (1995) Group Work and the Reception of Uddevalla in the German Car Industry. in: Sandberg A (ed): Enriching Production. Perspectives on Volvo's Uddevalla Plant as an Alternative to Lean Production. Avebury etc, Aldershot: 199-213
18 Kamata S (1973) Jidosha zetsubo kojo: Aru kisetsu-ko no nikki (*The Automobile Factory of Despair: Diary of a Seasonal Worker*). Tokio: Gendaishi Shuppan Kai. (English Edition: Kamata S (1983) Japan in the Passing Lane. George Allen & Unwin, Boston)
19 Karasek R (1991) The Political Implications of Psycho-Social Work Re-Design: A Model of the Psycho-Social Class Structure.. in: Johnson J V, Johanson G (eds) The Psycho-Social Work Environment: Work Organization, Democratization and Health. Essays in Memory of Bertil Gardell. Baywood Publishing Company, Amityville New York: 163-190
20 Karasek R, Theorell T (1989) Heavy Work: Job Stress, Productivity, and the Reconstruction of Working Life. Basic Books, New York
21 Kilbridge M (1962) A Model For Industrial Learning Costs. Management Science Vol 8: 516-527
22 Kishida K (1977) A Study on Subsidiary Behaviour in Monotonous Work. International Journal of Production Research Vol 15 No 6: 608-621
23 Kornhauser A (1965) The Mental Health of the Industrial Worker. Wiley, New York
24 Muramatsu R, Miyazaki H (1976) A New Approach to Production Systems through developing Human Factors in Japan. The International Journal of Production Research 1976 Vol 14 No 2: 311-326
25 Muramatsu R, Miyazaki H, Tanaka Y (1982) Effective Production Systems which harmonized Workers' Desires with Company Needs. The International Journal of Production Research 1982 Vol 20 No 3: 297-309
26 Rabinbach A (1992) The Human Motor. Energy, Fatigue, and the Origins of Modernity. University of California, Press Berkeley and Los Angeles
27 Salvendy G, Smith M J (eds) (1981) Machine Pacing and Occupational Stress. Taylor & Co. Francis, London
28 Schauer H et al (1984) Tarifvertrag zur Verbesserung industrieller Arbeitsbedingungen. Arbeitspolitik am Beispiel des Lohnrahmentarifvertrags II. Schriftenreihe "Humanisierung des Arbeitslebens" Vol. 52, Campus Verlag, Frankfurt/New York
29 Slack N D, Wild R (1995) Production Flow Line and 'Collective' Working: A Comparison. International Journal of Production Research Vol 13 No 4: 411-418
30 Spies S, Beigel H (1996) Einer fehlt und jeder braucht ihn. Wie Opel die Abwesenheit deutet. Ueberreuter, Wien
31 Touraine A (1955) L'Evolution du Travail ouvrier aux Usines Renault. CNRS, Paris
32 Ulich E (1972) Arbeitswechsel und Aufgabenbereicherung. in: REFA - Nachrichten Vol 25 No 4: 265-275
33 Upjohn Institute for Employment Research (1973) Work in America. Cambridge MIT Press, Cambridge Mass.

34 Walker C R, Guest R H (1952) The Man on the Assembly Line. Harvard University Press, Cambridge/Mass
35 Wild R (1975) On the Selection of Mass Production Systems. The International Journal of Production Research Vol 13 No 5: 443-461
36 Wyatt S, Marriott R (1956) A Study of Attitudes to Factory Work. Medical Research Council Special Report Series 292, HMSO, London
37 Womack J P et al (1990) The Machine that changed the World. Rawson Associates, New York

4.4
Rationalization also Involves Workers – Teamwork in the Mercedes-Benz Lean Concept

R. Springer

4.4.1
Teamwork – A Participative Approach to Rationalization

At the beginning of the 1990s, at a time when Mercedes-Benz was faced with the problem of how to catch up with Japanese competitors in terms of costs and productivity, the company compared the Japanese working and production model with its own organizational processes and structures. This comparison showed, among other things, that there was a considerable need to reform the way in which work was organized. In a number of respects and contrary to the original intention, the division of labour in production at Mercedes-Benz over the last few decades turned out to be an obstacle to working efficiently and flexibly. Operational sequences were mainly characterized by duplication of effort (e.g. in quality assurance), downtimes which were too long (e.g. machine downtimes), too many delays due to hold-ups in the cycle (resulting from a division of labour which was divided into work cycles which were too rigid) and a lack of flexibility on the part of the workers in general. What is more, due to the division of labour, many workers did not make maximum use of their skills and training.

When incentive wages are not used optimally, it is possible to increase the specified target performance (number of pieces) by way of standard time tests. However, unused manpower and reserve capacity can also be activated by redistributing tasks and integrating additional (e.g. indirect) activities into the production worker's range of tasks. This, of course, will only result in an economic benefit to the company when these activities are axed from other (indirect) areas, and when the manpower capacity related to these reduced activities in the other areas commensurates with the activities taken on by the production workers. When this happens, standard times for directly productive activities must also be adapted to the new conditions, i.e. they must be optimized accordingly. Increased output by individuals, however, will not be achieved simply by a higher number of pieces per day, but by extending the range of activities carried out by an individual worker.

In practice, these two approaches are nowadays often used side by side. However, a distinction must be drawn between the two: in terms of methodology, and particularly in terms of organizational policy. On the one hand we are dealing with an approach in the tradition of classical Taylorism (individual increase in the number of pieces), an approach which is geared towards a division of labour. On the other hand we are dealing with an approach towards rationalization (extending the range of tasks). This approach achieves productivity improvements by reversing the division of labour rather than increasing it. In this respect, it can be classified as post-Taylorist.

Contrary to a considerable body of scientific evidence and evidence from practice in industry, the main benefits for industrial companies using organizational measures such as job enrichment, job enlargement or even job rotation lie not just in the fact that the intrinsic potential to motivate workers is activated by such measures; what is more important is the fact that the redistribution of work improves the individual worker's capacity and overcapacity in the workforce is reduced. This of course presupposes that the workers concerned are prepared to take on additional tasks without their existing performance targets (i.e. in particular the relevant number of pieces per day) being reduced accordingly. In this respect, workers' motivation and willingness to improve productivity is crucial.

Nowadays, 68% of workers on average have vocational training at Mercedes-Benz production plants. Due to their experience, these workers have high expectations where the job content of their work is concerned and are not happy carrying out simple, monotonous (repetitive) tasks, tasks which have long characterized car production, even at Mercedes-Benz. Furthermore, the way these workers see themselves professionally frequently goes hand in hand with a well-developed business sense. A rationalization strategy which is strongly oriented towards a sharp division of labour and higher individual performance, as has been the case in the past, is totally incompatible with the professional expectations and interests of these workers. When they are at work, they do not simply want to use their bodies, they also want to use their minds. Or, more specifically, they are not prepared to give their best physically if they do not, at the same time, have the opportunity to use their knowledge and experience to help solve the numerous production and organizational problems.

This is why standard time tests and increases in the number of pieces meet with considerable resistance from these workers. Although this is not so obvious in times of economic recession, workers' dissatisfaction with their working conditions and the performance required of them becomes much more of an obstacle to improving productivity when the economy picks up. The 1980s proved this.

If, however, organizational rationalization now means not asking the individual first and foremost to produce more pieces but to carry out additional tasks, the question now is whether the professional expectations and interests of many workers can be better fulfilled than they were before. Higher productivity in return for more interesting and demanding work - isn't that an offer which the majority of workers would actively want to take up, given the pressing need to rationalize and the very high average level of training among workers these days? These were the considerations the company had in mind when it set about drawing up a new rationalization strategy for the company organization about five years ago. By eliminating such a division of labour, manpower reserves and reserve capacity which were not being utilized were able to be used again. Many people were of the opinion that the workers would prefer to help in this sort of approach towards rationalization rather than have strategies for rationalizing organization foisted upon them.

4.4.2 New Time Savings, Teamwork and a Continuous Improvement Process are Elements of a new Labour Policy

A company strategy which is geared towards extending workers' tasks and giving workers greater scope for action and decision-making is only acceptable when the workers use this in the interest of the company. Therefore, the new mode of work also requires a change in attitude and behaviour on the part of workers. This takes us away from the old entrenched positions inherent in (Taylorist) rationalization and permits not just a new type of flexibility, but also leads to rationalization becoming an objective, cooperative process in which workers actively participate. This can only be achieved when serious consideration is given to the explosive issue of performance with the cooperation of the workers. Both indirect tasks related to the work environment (for example, quality control, maintenance, materials transport, regulations concerning breaks and extra shifts) and the optimization of the relevant work system itself have become an integral part of the new form of work as part of the extended range of tasks and decisions at Mercedes-Benz. The continuous improvement process is now part of every work team's job.

If the performance data of a work system can be considerably influenced by the workers, they must also be able to influence the relevant performance standards for the work system in question and its workers. These standards (e.g. the target staffing levels of a system or the daily workload of a group) can nowadays be agreed on by the group and the supervisor, based on time-management data. Linking them with wages, as was often the case in the past, no longer occurs. A fixed monthly wage is paid, independent of individual performance. Therefore, the agreed performance standards are no longer strict standards, but are guidelines and can be higher or lower, depending on the specific situation. Any tem-porary additional output which may be necessary can now - with the team's consent - be produced without adjusting the manpower capacity. When individual remuneration is fixed, payment varies around the standard agreed with the team. Special credits and silent manpower reserves can be reduced.

Nowadays, when people talk about improving the work situation through teamwork, it should be pointed out that making a job more interesting for workers and increasing the scope for action and decision-making will mean greater demands on them in terms of performance. This is precisely how the new mode of work differs from those approaches towards organizing work which prevailed in the 1970s and 1980s. These did not have the express aim of increasing workers' capacity while at the same time humanizing work. When, at that time, people talked about economic effects, these related mainly to lower absenteeism by way of improved motivation. More far-reaching rationalization objectives, however, were not pursued. People neither recognised the rationalization potential of the new mode of work nor did they particularly want it. The champions of Taylorism were not interested in an alternative economic model to the division of labour or increased automation, as this might have called into question their status as experts and their monopoly on rationalization. The trade unions and works councils opposed all forms of rationalization as a greater form of exploitation and a higher potential for job losses. Most workers even regarded rationalization as something which, if they were not up to scratch, could only affect them adversely and could not be of any benefit to them. Despite all assurances, improvements in economic efficiency and humanization were seen as two principles which were diametrically opposed to each other and which had very little in common. At least, according to the underlying business theory at the time, one could only be carried out at the expense of the other.

Nowadays, no-one can dispute the fact that there is a considerable amount of tension between the principle of using the workforce as efficiently as possible and the objective of keeping stress at work to a minimum, especially where productivity depends mainly on humans and not machines. However, employees' working conditions are not the sum total of the performance demands made on them and the resulting physical and mental stress. What matters in the long run is the work situation as a whole, this too depends on the job content and scope for action and decision-making.

If we are to believe the research and evidence about the so-called change in values, these aspects of the work situation have taken on greater significance for many workers over the last few years. Therefore, we cannot ignore the fact that higher demands on performance are nowadays regarded as acceptable if they are accompanied by a more interesting job content and a greater scope for action and decision-making. But where precisely to draw the line remains a difficult task. It depends on many different factors and can only be defined by the people concerned. As part of a three-year pilot study into the different forms of teamwork, and in agreement with the Board of Directors and the central works council, a work-sociological assessment of several teamwork projects was carried out at Mercedes-Benz.

Professor Michael Schumann from the Sociological Research Institute (SOFI) in Göttingen was entrusted with this task. The following topics were investigated:

1. How do different forms of teamwork affect the employees' work situation?
2. How will the new forms of work be perceived and judged by those employees taking part?
3. Will there be performance conditions as a result of enriching the job, delegating decision-making and extending individual and collective scope for action which will allow the performance objectives that are operationally desired to be improved by consensus?
4. To what extent is a new understanding of work and performance being developed in relation to the new definition of responsibilities and tasks in which the employees themselves take the initiative to improve their work and the production process/production results?
5. To what extent has the basic assessment of business rationalization changed as a result of strategic objectives set by teamwork, and under what production circumstances do workers become aware that they play an active role in rationalization?

Table 4.4.1. Pilot projects studied

1. DNC lathe shop (mechanical production)
2. Exhaust manifolds (mechanical production)
3. Assembly line 2 - Group 3 (line assembly)
4. Swivelling jack-up platform (line assembly)
5. Completion (line assembly)
6. Brake disk production (mechanical production)

Source: SOFI

4.4.3 Teamwork can be Organized Restrictively or Offensively

A total of six teamwork projects were included in the study (table 4.4.1.), three of which were in capital-intensive mechanical production and three in labour-intensive car assembly. In the car assembly area, the projects were carried out on the assembly line, not in stationary assembly systems which are not connected to the assembly line. One very interesting aspect of the assessment was the question of whether the outlined approach to rationalizing the organization of work proved to be feasible on the manual assembly line where the work was dependent on the work cycle and where the degree of work freedom was particularly limited and the workload problem particularly marked.

A comparison between the projects in mechanical production and those in vehicle assembly showed that the scope for work organization in capital-intensive systems is greater than in labour-intensive systems (fig. 4.4.1. and 4.4.2.). The number of functions, for example, which can be included in the indirect tasks in the work environment is clearly greater in (semi-)automated mechanical production than in manual assembly. At the same time, because many activities are less dependent on the production cycle, the opportunity for reorganizing times for (semi-)automated work is greater than for manual work. Nevertheless, there is also scope for changing the organization of work on the assembly line.

In contrast to mechanical production work, the studies showed that in assembly work, however, people generally have a much more cautious approach to job enlargement, self-organization and empowerment. Even inside the production departments concerned, the way the teams are made up varies considerably. Both in mechanical production and in assembly, there are projects in which the objectively established scope for changing the work organization has been fully exploited, while at the same time, there are other projects in which this scope is only used very hesitantly.

On the whole, we can say that the scope tends to be used in those areas where it is relatively wide rather than in areas where it is more limited. This does not just apply to all the indirect activities which are included, the activities which are laid down or optimization tasks: the opportunities available to each team for self-organization, division of labour within the team, performance regulation or even further training vary considerably at times from project to project.

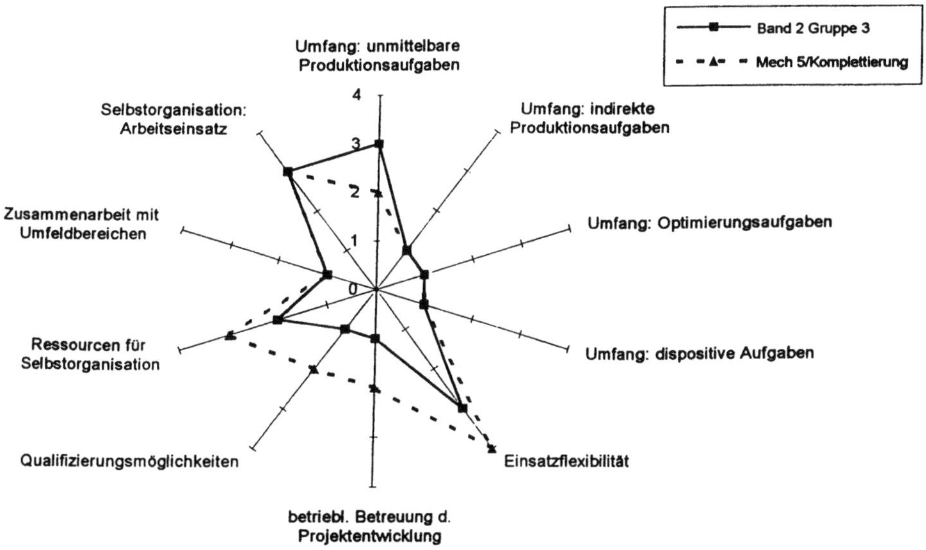

Fig. 4.4.1. Project profile, assembly design concept (Source: SOFI). Key to diagram, clockwise, from the top:

Scope: direct production tasks;	——	= assembly line 2 group 3
Scope: indirect production tasks;	---	= mechanical completion
Scope: optimization tasks;	5	= very high
Scope: tasks which have been stipulated;	0	= very low

Flexibility in labor deployment;
Support from project development staff at plant level;
Training opportunities;
Resources for self-organization;
Cooperation with adjacent work areas;
Self-organization in job assignment.

In this respect, restrictively organized teamwork projects must be differentiated from those which are offensively organized. This has a direct effect on the work situation of the employees. They very quickly recognise, for example, whether only the output of individual workers is increased through the grouping of several activities or whether this also makes the job content more interesting. They also recognise whether the group is really given greater scope for action and decision-making or whether the supervisor or planners are still trying to precisely define the details of the job. The workers pay particular attention to whether their ability to optimize their performance is really taken seriously and whether the improvements they suggest are put into practice. What management claims will happen when teamwork is introduced and what actually happens are closely compared, and if the reality of teamwork turns out to be restrictive and does not deliver what the workers were promised, based on claims of offensive teamwork, the workers will voice their complaints.

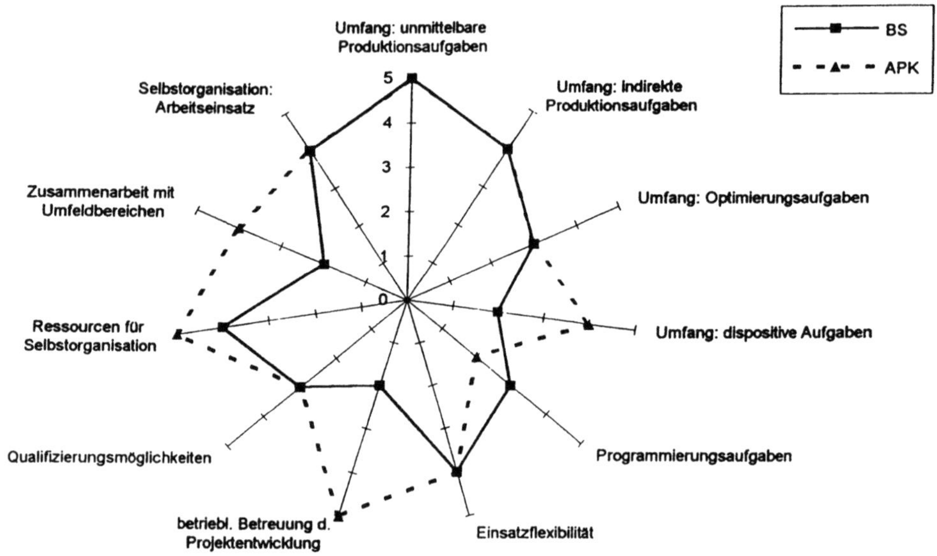

Fig. 4.4.2. Project profile, mechanical production design concept (Source: SOFI). Key to diagram, clockwise, from the top:

Scope: direct production tasks
Scope: indirect production tasks
Scope: optimization tasks
Scope: tasks which have been stipulated
Programming tasks
Flexibility in labor deployment
Support from project development staff at plant level
Training opportunities
Resources for self-organization
Cooperation with adjacent work areas
Self-organization in job assignment.

—— = Brake disk
- - - = Exhaust manifold
5 = very high
0 = very low

4.4.4 Higher Demands on Performance are Accepted if the Job is Enriched and the Group is Genuinely Allowed to Organize its own Workload

In the surveys on workers carried out by SOFI, the first question that was asked was how the work situation had changed overall in the view of the workers in the individual teamwork projects. It must be clearly indicated that the response to this question depends to a great extent on whether an offensive or restrictive

Table 4.4.2. Question: How has your work situation changed overall as a result of teamwork?

	improved	remained the same	become worse
Assembly line 2/Group3	17	35	48
Completion	49	28	24
Brake disks	48	28	24
Exhaust manifolds	73	10	17

Source: SOFI

route was taken in organizing the teamwork. If one initially compares the production of exhaust manifolds in mechanical production, this being the most offensively organized project, with the clearly more restrictive brake disk production project, 73% of workers in the exhaust manifold project are of the opinion that their work situation has visibly improved as a whole, while only 48% see an improvement in the brake disk project.

In the restrictively organized assembly project on assembly line 2/group 3 (interior assembly), only 17% report an improvement and 48% report a deterioration (table 4.4.2.). However, the fact that a positive balance is also possible in assembly work is also shown by the fact that in the clearly more offensively organized project which dealt with completion in the major assembly setup area, 49% of workers report an improvement and only 24% a deterioration.

If we now take a closer view of what exactly has improved the work situation from the employees' point of view and what has made it worse, we once again see how strongly the respective organization concept affects the attitude of employees.

97% of workers in the exhaust manifold project consider the work they do now to be far more interesting than before (table 4.4.3.). In the brake disk project and in the completion project, although not so many workers reported an improvement, 61% and 58% respectively (both clear majorities) believe the work to be more interesting. Only in the project which was the most restrictively organized of all the projects studied, are 50% of workers of the opinion that their work has become neither more nor less interesting as a result of teamwork. Of the other half, 36% report a deterioration in this project and only 14% an improvement.

A similar picture emerges when workers are asked about cooperation with their colleagues (table 4.4.4.). Here too, in both projects in mechanical production, the majority report an improvement and only a small minority report a deterioration while in the assembly project on assembly line 2, the majority feel that there has been no change.

Only in the offensively organized projects (exhaust manifold and completion) do more worker teams in both production areas report noticeable improvements

Table 4.4.3. Question: Has your job become more interesting as a result of teamwork?

	more interesting	remained the same	less interesting
Assembly line 2/Group 3	14	50	36
Completion	58	38	3
Brake disks	61	18	21
Exhaust manifolds	0	97	3

Source: SOFI

Table 4.4.4. Question: How has cooperation with your colleagues changed as a result of teamwork?

	improved	remained the same	become worse
Assembly line 2/Group 3	26	49	25
Completion	62	21	17
Brake disks	48	45	7
Exhaust manifolds	89	7	3

Source: SOFI

Table 4.4.5. Question: How has cooperation with your supervisor changed as a result of teamwork?

	improved	remained the same	become worse
Assembly line 2/Group 3	16	67	17
Completion	57	40	3
Brake disks	17	59	24
Exhaust manifolds	45	49	6

Source: SOFI

in cooperation with supervisors (by 45% and 57%) (table 4.4.5.). The overall impression here among the workers is that cooperation with supervisors has not changed. Even here, however, the workers see a particular need for change in all the projects in terms of extending the team's organization of its own work. Above all, they want the supervisors to keep the teams better informed, to give them scope for action and to involve them at an early stage in decision-making processes which affect their work.

Table 4.4.6. Question: How have the workloads changed overall as a result of teamwork?

	improved	remained the same	become worse
Assembly line 2/Group 3	2	14	84
Completion	21	24	55
Brake disks	7	21	73
Exhaust manifolds	12	28	60

Source: SOFI

Table 4.4.7. Question: Would you be prepared to work in teams in the future?

	yes	don't know	no
Assembly line 2/Group 3	43	31	26
Completion	58	17	24
Brake disks	52	21	28
Exhaust manifolds	76	10	14

Source: SOFI

4.4.5 Willingness to Improve Performance Through Self-Organization

Despite criticism that teamwork is not carried out consistently enough, the results of the survey shown so far point to the fact that for a large proportion of workers in the projects which were studied, there have been clear improvements in work content, cooperation and self-organization. This is particularly true of the offensively organized projects. The opposite is true, however, if one looks at the workload (table 4.4.6.). In all the projects, a clear majority report a deterioration which is primarily expressed in greater time pressure and, at times, in the form of greater physical stress.

Despite criticism by the workers that greater demands are made on workers to perform better under teamwork, a clear majority in the three projects (exhaust manifold, brake disk and completion), i.e. 76%, 52% and 58% of the workers, want to work in teams in the future (table 4.4.7.). Even in the project on assembly line 2, 43% of the workers believe in teamwork, although 84% of workers on assembly line 2 report increased workloads as a result of the new form of work.

How can we explain this behaviour which, at first sight, appears to be contradictory? Why do many workers prefer a form of work which, in their opinion,

Table 4.4.8. Question: How have the opportunities for helping each other at work changed?

	improved	remained the same	become worse
Assembly line 2/Group 3	28	37	35
Completion	52	28	21
Brake disks	66	31	3
Exhaust manifolds	87	7	6

Source: SOFI

Table 4.4.9. Question: How have the opportunities for organizing your work yourself changed?

	improved	remained the same	become worse
Assembly line 2/Group 3	12	57	31
Completion	40	23	37
Brake disks	41	46	13
Exhaust manifolds	38	31	31

Source: SOFI

Table 4.4.10. Question: How have the opportunities for coping with workloads on a permanent basis changed as a result of teamwork?

	improved	remained the same	become worse
Assembly line 2/Group 3	3	31	65
Completion	23	42	34
Brake disks	13	42	45
Exhaust manifolds	52	31	17

Source: SOFI

results in a higher workload? The studies showed that in the view of these workers, the increased workload, for example, cannot just be offset by the work becoming more interesting. The crucial factor for most workers is that teamwork can improve the opportunities for helping each other at work with the exception of those working on the assembly project on assembly line 2 (table 4.4.8.).

The same applies when we ask how the opportunities for organizing their own work have changed (table 4.4.9.). Here too, in three out of four projects, the majority report an improvement rather than a deterioration, despite the fact that some individuals have higher workloads.

By contrast, only in the very offensively organized exhaust manifold production project do a majority of the workers talk about improved opportunities for coping with permanently higher workloads (table 4.4.10.). In the other projects, people are more sceptical about this, although even here, as we have already mentioned, most of the workers believe in teamwork despite this. The workers seem to realize that because they do not organize their own work sufficiently as yet, and because there is not enough scope for regulating performance within the team, all the possibilities for making higher performance demands on people acceptable have not, in the long run, yet been exhausted. The example of the exhaust manifold project certainly shows that when teamwork is organized consistently, this is precisely what can be achieved.

Regarding the new approach to rationalization adopted by Mercedes-Benz, and in connection with the key issues of performance and workload, the following statement can be made: the increased performance demanded of workers and the resulting workloads will be accepted by most workers when this not only makes their jobs more interesting, but when it gives them the possiblity to achieve higher training and performance requirements themselves as a result of being able to organize their own work better.

4.4.6
A Willingness to Cooperate in the Rationalization Process

The work reforms in production, introduced with teamwork hinges on the idea of whether it is possible to convince workers to continually optimize their work procedures. To obtain the active cooperation of the workers in increasing productivity and in the rationalization process, the existing rationalization specialists in the planning department will need to give up their claim to a sole monopoly on rationalizing skills, and they must be prepared to cooperate with workers in the rationalization process. In addition, workers must also be prepared to bring their knowledge and experience to bear in this process.

The surveys conducted by SOFI produced perhaps the most surprising results in relation to employees' willingness to cooperate in rationalization. Many workers in the projects under study were very keen to be actively involved in measures oriented towards improving productivity. This can be seen in the basic question on whether, in the workers' view, the goals of "improving working conditions" and "increasing economic efficiency" were compatible in teamwork.

In all the projects that were studied, relative or absolute majorities consider these two goals compatible. In the offensively organized projects, these majorities are certainly larger than in those which were organized restrictively. On the whole, however, only a minority consider the two goals to be incompatible (table 4.4.11.).

The fact that the fundamental interest in the rationalization process is, by no means, just a reflection of economic necessity, but also documents a willingness to actively cooperate is clear from the answers to the question of whether cost cutting and rationalization should be the sole province of the company and management or whether the workers should also be involved (table 4.4.12.). Again,

Table 4.4.11. Question: Are the objectives "improving working conditions" and "increasing economic efficiency" compatible with teamwork in the long term?

	yes	don't know	no
Assembly line 2/Group 3	35	32	33
Completion	40	40	20
Brake disks	61	18	21
Exhaust manifolds	59	24	17

Source: SOFI

Table 4.4.12. Question: What do you think of the view that cost cutting and rationalization have nothing to do with workers, but are a matter solely for the company?

	disagree	don't know	agree
Assembly line 2/Group 3	39	31	30
Completion	53	36	11
Brake disks	59	35	6
Exhaust manifolds	48	30	22

Source: SOFI

Table 4.4.13. New attitude to the company and rationalization - all types of behaviour

34	Participants:	Rationalization has become something that is also a matter for workers
54	Don't knows:	Not yet totally convinced that the workers can also get involved in rationalization
12	Traditionalists:	Rationalization remains the province of the company

Source: SOFI

relative or absolute majorities in all the projects studied believe that they too should be involved in rationalization.

Eight typical statements by workers on rationalization, to which a yes or no answer was required, were used to test the attitude towards rationalization in the study results of all the projects (table 4.4.13.), we ended up with the following result: 34% of workers see themselves as "part of the rationalization process", 54% show an interest in becoming actively involved, but are not yet totally convinced that workers can also become actively involved in cost cutting and rationalization, and only 12% of workers take the traditional view of not wanting to have anything to do with rationalization, feeling that they need to be protected against it.Several things can be seen in the surprisingly positive attitude towards

Table 4.4.14. Question: What do you think of the view that workers should be mistrustful of everything the company does?

	disagree	don't know	agree
Assembly line 2/Group 3	30	35	35
Completion	26	33	41
Brake disks	19	26	55
Exhaust manifolds	44	18	38

Source: SOFI

rationalization. Firstly, the economic crisis Mercedes-Benz is going through is certainly a major factor in this case. The surveys were all carried out during a period which was characterized by a drop in the number of pieces produced and staff cuts. It wasn't just management who put thoughts and actions about economic issues high on their lists at this time, it was the workers too.

Something else that could be observed was that the attitude towards rationalization can have a clear effect on teamwork organization. In the most successful study project of all (exhaust manifold production), for example, the otherwise normal air of mistrust towards everything that emanates from the company was clearly reduced among employees (table 4.4.14.).

In addition, there was also a general change in attitude towards rationalization among the workforce which manifested itself in the strong interest in active participation in rationalization. The basic tenets of free enterprise and competition are only rejected by a dwindling minority these days. Alternative models based on planned economies hold virtually no attraction any more. What could be more obvious for workers than to be actively involved in the day-to-day running of the company, using their knowledge, experience and skills so that they can help shape their own lives, not just outside work but at work too?

4.4.7 Having Twin Objectives has Stood the Test of Time – Teamwork is now Being Extended and will be used Widely in Other Areas

The pilot phase of teamwork at Mercedes-Benz has now been completed. It was used to test a teamwork concept that was based on the idea that procedures could not only be designed more efficiently through the redistribution of tasks, but that the workers' work situation could also be improved so that they would abandon their previously rather distant attitude towards work, performance and rationalization. In this respect, teamwork has twin objectives.

This approach will have proved itself when workers' views are genuinely taken into account, when workers are allowed to organize themselves and when workers notice the changes in the daily routine. Therefore, teamwork will be extended to all production areas in future.

The basis for extending teamwork and using it widely forms a general agreement which replaces the existing pilot agreement. 40 % of all production workers at Mercedes-Benz already work in teams. In a year or two, the figure will be over 50%. Therefore, in the second half of the 1990s, teamwork will become the main method of operation in production.

4.5
Patterns of Work Organization in the German Automobile Industry

M. Kuhlmann · M. Schumann

While following the current discussion about lean production and group work, it is easy to forget that in the German automobile industry, there has been a trend towards alternative concepts of work organization since the early 80's. The basis of these early reorganization attempts, termed by Kern/Schumann in 1984 as the *new concepts of production*, was the assumption that the Taylorist principles of division of labor by task simplification and the separation of conception and execution were becoming obstacles to further increases in productivity [5]. In the face of marketing strategies, emphasizing diversified quality production [13], and due to the requirements of technically advanced production processes[1], many firms gradually realized that more integrated organizational structures were an efficient alternative. The typical German element in this discussion of the 80's was, that these attempts led to higher qualifications in the work-force, achieved by both additional training and the increasing employment of skilled workers (*Facharbeiter*) in the production departments. The goals of these new strategies were task integration and the transfer of responsibilities from specialized support departments into the production areas. A reorganization policy evolved which, in terms of work organization, ventured to expand and intensely utilize the qualification potential of the employees in the direct production areas; this signified a break with the organizational principles of Taylorism. An additional factor in this developmental phase in the German automobile industry was that, based on the goal of raising the status and qualification level of production work, a structure of compromise between management, the workers' councils and the unions evolved and in some cases, led to commonly supported reorganization policies for production and organizational structures.

Above we mentioned a few basic points in order to illustrate changes in the situation since the beginning of the 80's. In this text, we want to show in what form and to what degree new organizational concepts have been instituted in the German automobile industry. We will use results from our study *Trendreport Rationalisierung* [6,10,11,12] which was based on empirical research, carried out between 1988 and 1992, in several industries, including the German auto indus-

[1] Other reasons for the gradual change in work organization and labor management were the large increase in the level of automation in the 80's, based on new developments in robotics technology, and the related organizational efforts to secure and manage these process innovations. Both the strategy of product improvement and the application of new technologies in the face of comparative disadvantanges in wages and working-hours were reactions to the success of the Japanese auto industry.

try.[2] Our research was designed to determine to which extent change had already taken place in work and organizational structures, the dynamics and nature of the reorganization and how the employees had been affected. In the following section, we will give a short description of the different models of work organization, and discuss the range of the previous restructuring process. In a second section, reorganization trends on the shop floor, which are significant for the situation in the middle of the 90's, will be considered. Here, we will concentrate - based once again on our research [1,2] - on an assessment of current attempts to introduce group work and team concepts.

4.5.1 Reorganization of Work Until the Beginning of the 90's

One important finding of our study of reorganization concepts and their effects on employees is that the developments have to be differentiated along the lines of manufacturing departments or types of work activities, rather than from one firm to another. If one compares the technically advanced areas to the departments where manual work dominates[3], the focal points of reorganization show obvious differences and there are large variations in terms of the scope and speed of change. Consequently, large differences in the work situation and experience of the employees can also be observed.

4.5.1.1 Changes in Work Organization in the Technically Advanced Production Areas

The high-tech areas became both more numerous and qualitatively more important due to the rapid increase in automation during the 80's, and the application of new technology led to a transformation of manual production work. Firms realized that a rigid division of labor and the minimization of qualifications and scope of action of the production workers were not suitable for controlling and efficiently operating complex production technology. In the German automobile industry, this resulted in a phase of experimentation with various models of work organization, a

---

[2] In our research, we combined case studies on rationalization projects in the different areas of production with overall work structure analyses, where we collected data about the different types of jobs and the required qualifications. In the car industry, we covered the ten main plants of the three German car makers and we carried out 29 case studies on reorganization (58 workplace observations, 193 interviews with workers and 264 discussions with experts). Our analyses of the overall work structures in the auto industry are based on 79,200 manufacturing jobs.

[3] Technically advanced manufacturing areas include the mechanical parts production (engines, gear boxes, axles), large portions of the stamping plants, most of the welding areas, as well as parts of the assembly in the body shop and a large part of the paint work. In assembly, there are only islands of automation, although in the assembly of engines and gear boxes automation now plays a much more important role. The manual areas include most of the assembly work, the finish sections of the body shop and a number of prep and fine-finish jobs in the paint shops.

gradual transformation of the organizational structure towards task integration and an up-grading of production work (fig. 4.5.1.). In labor management, a process of reorientation began, from tayloristic towards more human-resource oriented policies. New concepts of production were established and, though their influence and the speed of their application varied widely, a new type of production work developed around the automated production processes. We term this work *system regulation*, and the production workers who are "on the spot" in technically advanced work environments and carry out a wide spectrum of functions are called *system regulators*. Process control, computer programming and troubleshooting are the most important functions, but simple tasks, such as quality control, cleaning or machine feeding are also typ-ical. In a different manner than the traditional production or maintenance worker, the system regulator is responsible for ensuring a continuous production process in its entirety. System regulation work can be organized at various levels of qualification, depending on the range of tasks and the process interventions allowed or demanded. In practice, however, there is a growing tendency to utilize skilled workers in production work (*Produktionsfacharbeiter*). One could explain this development as a result of institutional factors (labor markets, vocational training system) or the industrial relations system; from the point of view of enterprises, the higher qualified forms of system regulation seemed the best way of attaining high equipment performance, flexibility and developmental skills. In the end, these various factors led many firms to upgrade production work and to use production workers for more qualified tasks.

In a large majority of the plants and production areas we investigated, the traditional organizational structure (OT 1, fig. 4.5.1.) with its strong polarization and strict separation of functions, both within the departments (set-up man vs. operator) and between the departments (production vs. maintenance vs. quality control), had been restructured and the division of labor was reduced (OT 2-4). However, this process proceeded gradually, at times hesitantly and in each case, with a unique developmental dynamic.

1. A cautious reform of previous organization principles characterizes the organization type *limited integration* (OT 2). The traditional organization of the plant is left intact and integration is limited to taking over simple support department tasks (maintenance, quality control). Internal differentiation within the production team is reduced and a uniform profile of the system regulator evolves. In terms of personnel policy, this type relies on semi-skilled workers, since the decisive competence for production technology and process remains in the support or engineering departments.
2. With *extended integration* (OT 3), the traditional organizational boundaries become fluent. This type of organization is based on skilled workers. A qualified system regulator with extensive process and technology competencies evolves. The regulators are able to intervene in order to correct or optimize the computer program and production technology. This necessitates a systematic knowledge of the process and product technology. The system regu-lators have the possibility of using their knowledge, both independently and in conjunction with specialists (maintenance, logistics, quality, etc.) and engineers (industrial, process, product development, etc.) to insure and optimize production.

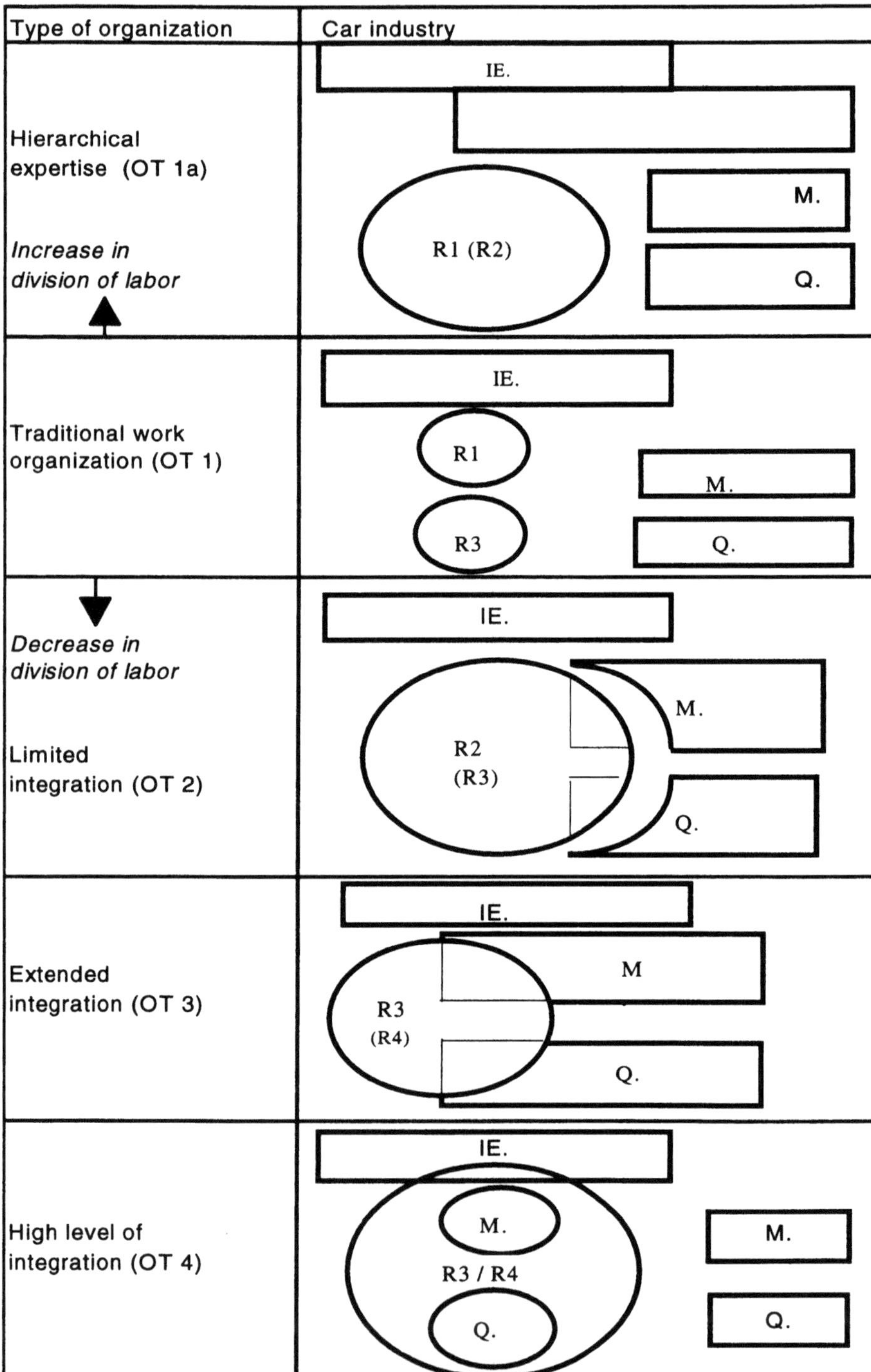
Type of organization
Car industry
IE.
Hierarchical expertise (OT 1a)
M.
R1 (R2)
Q.
Increase in division of labor
IE.
Traditional work organization (OT 1)
R1
M.
R3
Q.
IE.
Decrease in division of labor
M.
Limited integration (OT 2)
R2 (R3)
Q.
IE.
Extended integration (OT 3)
M
R3 (R4)
Q.
IE.
M.
High level of integration (OT 4)
M.
R3 / R4
Q.
Q.

3. With *high level of integration* (OT 4), the most important functions of process and equipment control are united under the aegis of multi-functional production teams. Only extensive repair, systematic quality control and complex system-wide planning remain unintegrated. This model is centered around highly qualified homogeneous work groups. The differences between skilled production workers, mechanics and electricians become smaller and are determined by specialized skills rather than hierarchical status. This concept aims at creating workers, highly-skilled both in mechanics and electronics, who perform not only far-reaching maintenance tasks, but have planning and optimizing responsibilities as well as other functions. In this case, the new type of qualification is achieved by building cross-functional teams.
4. In scattered cases, we observed forms of organization which led to an increase in the division of labor (OT 1a). In this type of organization, qualified task and process competence was not a component of the production workers' jobs, but the responsibility of on-site experts from the maintenance departments or technicians and engineers. One reason why this type of organization was not often utilized was that qualified production workers proved to be capable of handling complex process interventions. Another reason was that work on the production floor is usually not acceptable for technicians or engineers in the auto industry due to the working conditions (shift work, noise levels etc.).

The increase in task integration and the stabilization in the use of skilled and highly skilled work in the automated sectors of auto manufacturing represent an important step towards a more humane or holistic concept of labor management. At the same time, the combination of direct involvement in the work-process, practical experience and systematic, professional knowledge allows skilled production workers to make an independent contribution to solving the problems that arise in automated production processes, and offers a good foundation for improved cooperation with various specialists and engineers.

The task and performance situation of the system regulator is a fundamentally new one, since in automated production work, activities are detached from the production process. Their work is no longer direct work in the sense of hands-on manual manufacturing, but serves solely to ensure the functioning of a complex, seldom perfectly working production technology. The interventions required to solve arising problems cannot be fully anticipated, so this type of work is context-specific and situationally flexible - i.e. self-organized. Detailed job descriptions or instructions for "correctly" handling specific occurrences make little sense. The

◀ **Fig. 4.5.1.** Types of work organization and labor deployment in automated manufacturing
keys:
M. = Maintenance; Q. = Quality; IE. = Industrial engineering, planning.
Levels of system regulation:
R1 = system regulation with very restricted task integration;
R2 = system regulation with limited task integration/process competence;
R3 = qualified system regulation;
R4 = high-level system regulation

Table 4.5.1. Assessment of overall working situation by system regulators (in %)[4]

	Type R2	Type R3	Type R4
positiv	40	64	75
ambivalent	36	30	25
negativ	24	6	-

characteristics of this work situation - the unpredictablility and the wide range of competencies - suggests a new conception of work for the qualified system regulator. On the basis of his daily work activities, through his participation in the search for more efficient production procedures, he becomes an active participant in the process of optimization.[5]

The results of our worker interviews show that significant steps towards increasing qualifications and competence also have a positive impact on how the work is perceived. In comparison to traditionally organized manufacturing work, system regulators generally prefer their current work. The system regulators most common in organization types OT3 and OT4, those at the regulation levels R3 and R4 have an especially positive view of their work (table 4.5.1.). However, it would be wrong to assume that system regulation work harmonizes interests completely. We observed a realignment from typical issues of controversy and conflict, in that traditional conflicts, related to work intensification, lost relev-ance while a new terrain of conflict became more important. Current conflicts address issues like reducing the division of labor, increasing the scope of independent action and self-organization or the availability of further training possibilities and career development perspectives.

The criticisms voiced are often related to deficiencies that still characterized the reorganization in the automated areas at the beginning of the 90's. The shortcomings of the organizational reform were usually the result of one or more of the following problems:

- The use of skilled production workers often remained isolated or limited to special production areas, so the employees had little chance for further career development. Also, the difference in status of the production positions in comparison to the indirect departments was only haltingly reduced.
- The transformation of the corporate organizational structures also proceeded very slowly. The firms' efforts to decentralize ensued in small cautious steps. Changing the cooperation and communication structures between the various departments proved even more difficult, due to the still quite pronounced status differences. An elimination of hierarchical social relations between production and the indirect or engineering departments never got passed the

[4] Tables 1 and 2 are derived from in-depth interviews with workers (1,5 to 2,5 hours) concerning their work situation and their perceptions of work.

[5] An American description of the very specific situation of this new type of work is provided by Hirschhorn [3]. A good overview on the German situation in English can be found in Jürgens et al. [4].

starting point at any of the enterprises. Consequently, the integration of the production workers into the planning and control tasks remained precarious.

- When work groups with increased competence developed, very often this was a result of process necessities related to qualified system regulation work, rather than of corporate policies or a fundamentally changed view of social relations in production that might have led to group autonomy or empowerment. Usually, the additional competence was granted informally, due to the lack of technical competence and process knowledge on the part of the supervisors, as well as the inability of the engineering departments to ensure a smooth and efficient production process. The restructuring efforts undertaken in the 80's were far from conscious social innovations, such as the institutionalized self-organization in the new group work concepts (below). Task integration and the upgrading of qualifications had not yet, even in the highly automated areas, been combined with a deconstruction of rigid hierarchical structures. The most profound Taylorist drag on productivity, the separation of planning and execution, had been weakened but not eradicated.

4.5.1.2
The Dominance of Traditional Structures in Manual Work

The restructuring efforts in the manual labor areas have been less dynamic than in the automated areas. Although new ideas for changes in labor deployment related to new concepts of production have surfaced, the chosen applications have been more conventional. The shortcomings of the Taylorist work structures also became clear in assembly, but fundamentally new concepts of work organization were not developed. The changes actually instituted were at best, small isolated alterations, for example: a partial elimination of the assembly lines by using new transport technology (AGV), a cautious expansion of work cycles, ergonomic improvements, work groups with internal rotation (though short-cycle work remained), integration of materials handling and quality control. However, these separate ideas were seldom combined into concepts aimed at far-reaching work reorganization. In the still dominant assembly line areas, organizational change was limited to the integration of simple quality control tasks and a more flexible labor deployment through job rotation. Until the 90's, solutions that diverged from the traditional structures of short-cycle work and a high level of vertical and horizontal division of labor were tested only outside line assembly systems. They can be bundled into four basic types (fig. 4.5.2.):

(Type 1) In the model we term *modified line organization* the conveyor belt is replaced by detached carriers (AGV). Here, both stationary assembly and a more flexible assembly process is possible. From the point of view of the firms, this assembly layout allows them to achieve the flexibility they require and to more effectively plan and time the work processes. However, the principle of a centrally controlled assembly line process remains. Buffers are almost nonexistent and time sovereignty in general is very low. The rhythm of work is similiar to line work. Job descriptions in this type of system remain traditional: the work

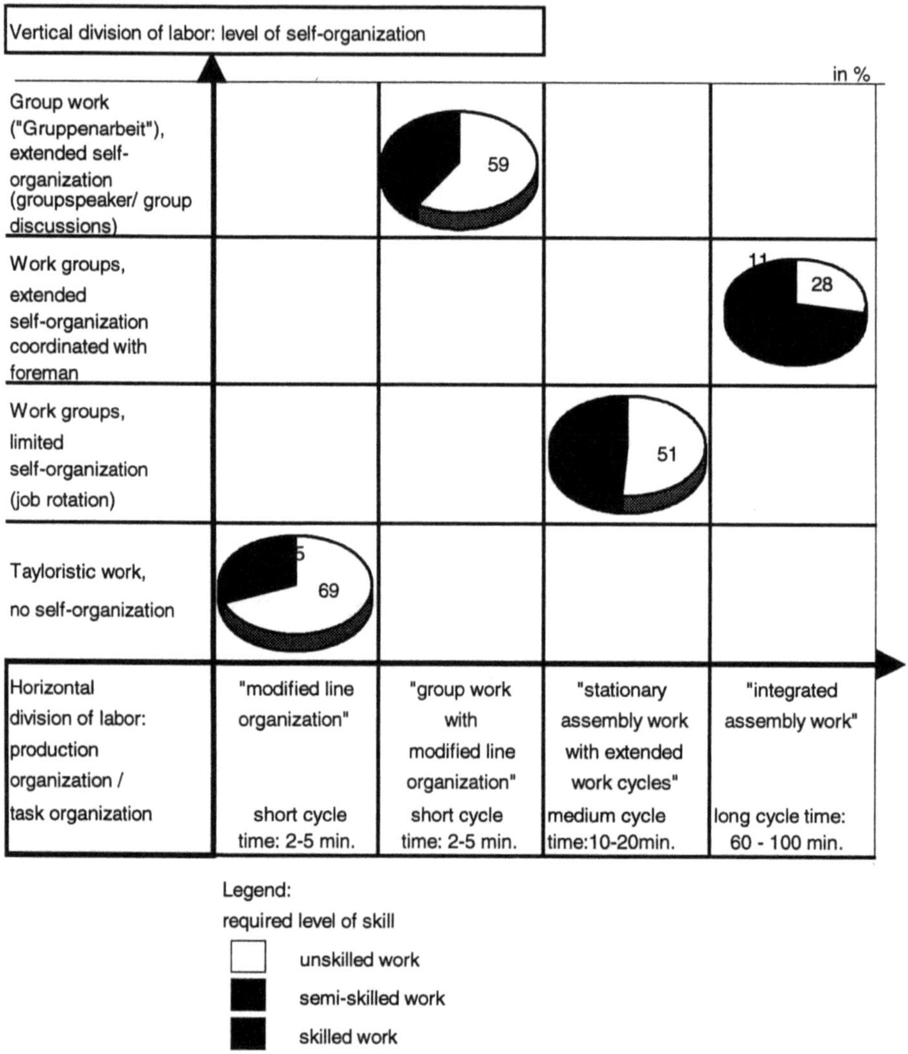

Fig. 4.5.2. Types of non-line assembly work

cycles are short (2 to 5 minutes), indirect activities are rarely assigned to production workers and the organizational tasks are handled by the supervisors. In the case of parallel production flows, the supervisors have to take on additional performance monitoring tasks, since the disciplining function of the line is missing. The employees are allowed little opportunity for self-organization. All in all, labor deployment is still based on the organizational model of the Taylorist assembly line.

(Type 2) Group work with modified line organization goes much further in reducing the horizontal and vertical division of labor, and begins to break with the mono-functional logic of repetitive work. The technical basis is, again, transport technology which allows for stationary work. In the cases we studied, the work cycles remained short, however, groups were formed to perform pre-assembly, rework and simple indirect functions (materials, quality). The groups were self-organized on the basis of elected group speakers and guaranteed time out for group meetings, and the workers assumed organizational competence (daily distribution of work, absentee planning). Due to self-organized rotation, expanded freedom of action as well as jointly negotiated performance norms, this type of assembly organization results in a much better work situation, although an appreciable upgrade in qualification did not take place.

(Type 3) In the *stationary assembly work with extended work cycles* line detached work is also possible, based on assembly docks. But here, the work cycles are much longer (10 to 20 minutes) than in type 1 and 2. In the cases we observed, the work groups' (3 or 4 employees) tasks included simple indirect functions, in addition to assembly work. However, group work in the sense of clearly defined self-organization was not institutionalized - the vertical division of labor was not reduced.

(Type 4) Finally, we observed one type of assembly organization that went much further towards increasing the scope of work and the level of task integration. This case of *integrated assembly work* involved a large (60 to 100 minutes) functionally connected work process, as well as the integration of indirect functions. In the case we observed, the work groups had no official self-organizing competence, informally, however, stable groups were formed and the workers were granted some freedom to independently organize work, performance and qualifications. The departure from the constraints of a rigid line organization was made possible not only through technical features like buffers, but also by the length of the work cycle itself. While even the type 3 organization required somewhat higher qualifications than traditional assembly work, only type 4 means a significant upgrading of qualifications to the level of "semi-skilled" work. However, the still existing characteristics of large-series assembly (high level of specialized planning, little trouble-shooting, much repetition) prevents the jobs, even in this case, from constituting fully skilled production work.

Our work analyses and interviews show that there are clear differences in the work situation and the perception of work between the types of assembly work we specified (table 4.5.2.). The employee assessments of the various assembly types are more positive, and improvements in the work situation are more pronounced when there is a significant reduction in the horizontal and vertical division of labor and when the departure from the rigid line organization is more complete.

Unlike the system regulators, for manual workers, the traditional problems related to work and performance are of great importance. For them, the most important issue is still the negotiation of an achievable performance compromise and the creation of tolerable performance conditions. The system regulators concentrate on more personal dimensions, such as the development of occupational and personal skills and competencies, or the desire for more responsibility and

Table 4.5.2. Assessment of overall work situation by assembly workers (in %)[6]

	Type 1	Type 2	Type 3	Type 4 (figure 4.5.2.)
positiv	9	38	43	64
ambivalent	36	38	24	14
negativ	55	24	33	21

the chance to act independently; in the manual areas, quantitative norms (a certain number of pieces of a specified quality) remain important. For manual workers, the assessment of the work situation is usually related to the conditions under which the performance norm must be fulfilled. Discretionary leeway is judged from the point of view of the opportunities to develop one's own work rhythm and better cope with the workload and other stress factors. The question of workpace, the problems of fatigue and of becoming ill or injured are in the foreground.

4.5.1.3 Outline of Work Structures

A look at our data on work structures in the German auto industry at the beginning of the 90's makes clear that the industry is still organized along the lines of a traditional mass-production industry (table 4.5.3.). Of the blue-collar workers, about 3/4 of the total employees, more than 1/3 are assigned to indirect departments (maintenance, quality control, logistics, development departments). In the production departments, manual work at a relatively low level of qualification still dominates. Inspite of the massive increase in automation during the 80's, system regulators are only a small portion of the total employees.[7] This is for the most part a result of the relative weights of the various manufacturing departments. The most personnel intensive, and numerically by far the most important, areas are the assembly departments and they have grown in proportion to other departments since the 70's.

Even though system regulators are an important group in the technically advanced areas, and in some sectors they are even numerically significant, they still play no more than a secondary role when one considers the entire work structure. When comparing various firms and production sectors, it becomes apparent that the proportion of system regulators and their level of qualification rises as more integrated organization types are instituted. We estimate that the proportion of system regulators will continue to increase, since simple manual work will be

6 See footnote 2 and 5.

7 According to our data on overall work structures, in the beginning of the nineties 5% of the system regulators performed unskilled ("watchman") and 41% still performed semi-skilled work. 41% of the system regulation work was skilled and 13% already highly skilled.

Table 4.5.3. Overall work structures in production (in %)[8]

Final assembly	Total	Press shop	Machine Work	Body shop	Paint shop	Subassembly (Engines etc.)	Final assembly
unskilled	59	49	43	58	64	49	67
semi-skilled	29	32	30	33	29	38	25
skilled	10	16	17	8	7	10	8
highly skilled	2	3	10	1	-	3	1
manual/ product	70	3	6	71	87	80	93
manual/ machines	21	73	64	23	6	11	6
machine operating	1	-	3	-	1	-	-
system regulation	8	24	27	6	6	9	1

further automated in some production sectors or contracted out to suppliers. Changes in the work structures are not only the result of factors like increasing automation, new types of work organization or the reduction of the vertical integration of firms, changes in the product itself also play an important role. Interchangeable components, platform strategies and design changes, aimed at reducing the number of parts and assembly operations, also lead to reductions in simple manual work. Product and process innovations will lead to an increase in system regulation, especially in the paint and body shops, where the proportion of system regulation work was quite low at the beginning of the 90's.

4.5.2 The Situation in the Mid 90's: Two Types of Group Work

The economic downturn in 1992/93, the publicity surrounding the book by Womack et al. [17] which asserted that the German auto industry was not nearly as

[8] We distinguished between four types of production work according to the main focus of work: "manual work on the product" (in pre-mechanized manufacturing: assembly, manual welding, manual rework, etc.), "manual work on machines" (depending on work organization in mechanized or partially automated manufacturing: feeding, set-up, etc.), "machine operating" (in mechanized manufacturing without computerized process control) and "system regulation" (partially or fully automated manufacturing with computerized process control). We defined the four levels of required qualifications as "unskilled" (short introduction, on the job learning up to one year), "semi-skilled" (systematic on the job learning up to three years), "skilled" (apprenticeship or several years systematic job training) and "highly skilled" (apprenticeship plus additional specialized training).

productive as the Japanese competitors and the resulting discussion about lean production have sped up developments in the auto industry during the last few years. The reorganization of work and corporate structures has been more radical and faster in the automated areas, and even in the manual sectors, work reform has become more far-reaching. The current attempts to introduce group work concepts serves as an obvious example. These attempts represent not only a continuation and intensification of the restructuring processes of the 80's, they also highlight some of the still unanswered questions about the basic principles and necessary prerequisites of post-Taylorist organizations.

The shortcomings of the 80's, related to creating functionally integrated structures and the difficulties of corporate decentralization, also influence the implementation of group work concepts. However, the current discussion is more advanced and explicitly aimed at finding solutions for the still unsolved problems of the partial modernization of the 80's. Today, the main issue is whether the German auto industry, under the pressure of increasing world-wide competition, will manage to continue making progress towards post-Taylorist work and structures. In order to develop new sources of productivity, firms will also have to transform their *social* organization [7]. Along with functional integration, the current discussion on group work in the German industry[9] concentrates on the importance and scope of institutionalized instruments for group self-organization, and on how far the policy of eliminating corporate hierarchies should be pursued. Although the priorities and decisiveness of the programs vary, almost all of the German auto manufacturers are now testing or instituting group or team work. The chosen solutions differ but the firms seem to have similiar goals.

- The formation of profit-centers and the decentralization of production related services are focal points of the restructuring processes. The new center-organization means more than just a formal regrouping of existing departments and it includes, for the first time, substantial planning functions.
- Functional integration is an important principle of the new work organization. In the automated sectors, the changes towards the organization types OT 3 and OT 4, begun in the 80's, are becoming more widespread and in the manual sectors, functional integration is, for the first time, being tested on a larger scale. The work roles are less specialized than in Tayloristic forms of labor deployment and indirect support tasks are becoming the responsibility of regular production workers.
- With the introduction of group work, firms are trying to activate and utilize the qualification potential of the employees. Simultaneously, the extended self-organization of the production groups strengthens cooperation and raises

[9] It is important to note that the German discussion about group work ("Gruppenarbeit") is not the same as the debate about team concepts in the North American industry. In the German context, group work is understood as a work organization that differs from the team concept as instituted, for example, in the NUMMI case. Contrary to team work, group work is more strongly oriented on socio-technical concepts of the 70's and ideas from the unions about the quality of working life (QWL) and autonomous work groups [8,16]. To understand the German discussion, it is necessary to know, that in the beginning of the 90's, a constellation developed, whereby group work for the management as well as for the unions and the workers' councils represented an efficient and more humane form of work organization.

the self-managing abilities of the decentralized productions units. The traditional separation of planning and execution is becoming more fluid. The objective of this reintegration of planning functions is to raise employee involvement in problem-solving and create better circumstances for process optimization.

Although the agenda of *organizational decentralization, functional integration of the work organization* and *raising worker involvement in problem-solving and process optimization* is widely accepted, there is a growing polarization within the discussion about group work concepts. Contrary to just a few years ago, the disparity of group work concepts being applied in the German auto industry is growing. In practice, group work is often a mixture of different elements, but recent developments make it useful to describe two antagonistic approaches: a structurally innovative and a structurally conservative one.

The concept we describe as *structurally conservative* or *Taylorized group work* is an attempt to implement team concepts used at the Japanese transplants in North America and Great Britain. Although there are important organizational changes even in these cases, the solutions remain structurally conservative due to the lack of an institutionalization of self-organization and a less thorough integration of planning and execution. If one focuses on the workplace and the work situation of production workers, several core principles of Taylorism are preserved.

1. The rigid line organization remains, along with repetitive short-cycle tasks. The basic principle of Taylorist production engineering - according to which high performance can only be achieved through standardization and routine - is preserved. Task and functional integration or rotation are used in some cases, but only in a limited way due to the principles of standardization and predictability. A broader use of the worker's qualifications occurs only on a very small scale. Task integration and more involvement affect only a few key workers (team leaders).
2. The institutionalized opportunities for participation and the degree of self-organization of production groups are quite limited by the temporal and material constraints of the tightly linked production process. Resources that make self-organization possible and encourage a more democratic and cooperative work culture are in short supply: Group meetings are limited to the presentation of corporate information and to solving specific production problems; team leaders are appointed by the firm and they, not the group as a whole, are responsible for handling organizational, indirect and support tasks or improvements.
3. Production and performance control remains the responsibility of the hierarchy. Labor deployment, as well as the supervision and evaluation of job performance, are done by supervisors. Because of the foreman-like position of the team leader, who is not an ordinary member of the work group, the hierarchy is not only strengthened but also expands into the work place.
4. Overall, the organizational structures remain expert-oriented. Planning and execution are still allocated to different groups of employees. The limited upgrading of qualifications of the production workers, along with the principle of tight manning levels, does not eradicate the insurmountable boundary be-

tween the shop floor and staff activities - inspite of all endeavors to change corporate culture. With Taylorist group work concepts a departure from functionally separated organizational structures takes place only at hierarchical levels above the shop-floor. Here, the departure from traditional structures is focused on a reduction of the distance between the technical specialists and the production process, as well as more team work between engineers and managers. Normally, the production groups are not included.

With *structurally innovative* or *self-organized group work* the traditional structures change fundamentally: the explicit objective is to upgrade and activate the production group as a whole. This concept concentrates on empowerment and an increase of responsibilities for production workers. It aims to develop and utilize the already high level of qualification of German auto workers, by instituting much more self-organization, reducing hierarchies and expanding cooperative relationships, even in areas like assembly, where the possibilities of work reorganization through functional integration are limited..

1. Tasks as well as the range of discretion and decision-making by the groups are expanded. Indirect and planning functions are explicitly included, as they are particularly apt to increase the degree of freedom on the job and to launch an upgrading of qualifications. A combination of tasks both line-bound and line-detached is generally pursued, although the scope of task integration differs depending on the specific production technologies.
2. The formation and development of the groups center around the goal of self-organization and empowerment. They are given extended responsibility for the organization of their work area (distribution of work, job rotation, planning absentees) and more influence on what happens on the shop-floor. Self-organization and a more democratic work culture are encouraged and supported by the institutionalization of necessary resources. That means providing both time and money, but also a lasting reform of the social structure on the shop-floor, which guarantees that groups are allowed to organize a substantial part of their daily work according to their own judgement. Group speakers are elected and they remain full members of the group while acting as its speaker and moderator. Self-organized group meetings serve to foster mutual coordination, strengthen communication and to solve operational and social problems.
3. In the Taylorist model, lower management is characterized by authority based on their hierarchical position and their control functions. In the concept of self-organized group work, first-level managers are not only supervisors but have supportive and motivational responsibilities as well. They have to support the self-organization of the production group and, along with the group speaker, represent the group in dealing with other departments. This concept also strengthens the position of lower management, but here, the hierarchy is not expanded into the work group as in the structurally conservative model.
4. Another element of this concept is a more complete decentralization of administrative and technical functions. Although here, as opposed to the Taylorized group work, this coincides with the formation and extension of cooperative and communicative structures which ensure that the groups themselves take on additional responsibilities. The change of corporate hierarchies

is one important element of self-organized group work, and is illustrated by the groups' right to invite specialists to their discussions and their increased possibilities to force these specialists to take action.

Our studies on the effects of various group work concepts show that the concept of self-organized group work is backed by the employees and that they view it as a fundamental improvement of their work situation.[10] In comparison to the structurally conservative model, it also became clear that group work, based on self-organization and the expansion of qualifications and responsibilities, raises employee engagement in problem-solving and provides a basis for participation in production optimization. Greater worker involvement demands such a more radical approach to worker self-organization and empowerment. In the end, this could lead to both an expansion of genuine human resource management and an improvement of the incomplete modernization of the 80's

While reliable empirical results about the effects of various group work concepts are now available, the data on the actual industry-wide application of group work is less precise. A survey of workers' council representatives carried out by the metalworkers union shows that 22.2% (auto industry mid 1994) of the blue-collar work is in groups, and that the rate is increasing [9]. Unfortunately, this data base does not reveal which forms of group work are being implemented. In our studies, we observed that in cases where firms quickly instituted group work on a large scale, structurally conservative solutions were applied or in the end emerged. The far-reaching changes necessary to institute a structurally innovative reorganization were not accomplished.

Contrary to work organization changes in the 80's, in the introduction of group work, corporate policies play a much more important role. Now, firms are trying to form corporate-wide strategies by negotiating agreements with the workers' councils, although within the plants or production sectors, even in the future, leeway will exist. It does appear, however, that in some cases, firm-specific management strategies are being developed: while the GM-owned Adam Opel AG is trying to institute a team concept like NUMMI or CAMI, which they first tested in their greenfield plant at Eisenach, Mercedes-Benz seems to be making the most obvious attempt to develop a structurally innovative group work concept [14,15]. There are a number of factors which indicate that the different developments between high-tech and low-tech sectors which we already specified for the 80's will also remain relevant in the case of group work. The specific conditions associated with system regulation or manual work will also affect the future of group work in these areas. In the manual labor sectors, the conditions of production (limited possibilities for functional integration or upgrading of qualification; importance of the conflict about performance norms) will slow the development of self-organized forms of group work. The current tendency in production engineering to return to line production will probably have structurally conservative effects, although our studies prove that self-organized group work is also possible on the line. In sum, our results suggest that the future development of work organization and the question of whether structurally innovative forms of group work will be

[10] Gerst et al. 1995, forthcoming; results are also discussed by Springer in chapter 4.4 in this volume.

instituted, depend on the ability of the firms to transform their social organization: will firms seriously pursue decentralization and make significant steps towards less hierarchical structures? Will management dare to truly empower workers? The social forms of work shaped by Taylorism are becoming more open to question. Whether management and worker representatives are willing and able to initiate a process of social transformation towards a post-Taylorist social organization remains unclear.

4.5.3 References

1 Gerst D et al (1994) Gruppenarbeit in der betrieblichen Erprobung. Angewandte Arbeitswissenschaft Vol 142: 5-30
2 Gerst D et al (1995) Gruppenarbeit in den 90ern. SOFI-Mitteilungen Vol 22: 39-65
3 Hirschhorn L (1984) Beyond Mechanization. MIT Press, Cambridge/Mass
4 Jürgens U et al (1993) Breaking from Taylorism - Changing Forms of Work in the Automobile Industry. Cambridge University Press, Cambridge
5 Kern H, Schumann M (1984) Das Ende der Arbeitsteilung? Beck, München
6 Kuhlmann M, Kurz C (1994) New Work Structures an Industrial Relations in the German Automobile Industry. In: ISVET, SOFI (Eds) New Work Organisation and Industrial Relations in the German Industry. Franco Angeli, Milano
7 Kuhlmann M, Kurz C (1995) Strukturwandel der Arbeit? Betriebliche Reorganisation und die Bedeutung sozialer Strukturen. SOFI-Mitteilungen 22: 31-38
8 Muster M (1990) Team oder Gruppe? In: Muster M, Richter U (Eds) Mit Vollgas in den Stau. VSA-Verlag, Hamburg
9 Roth S (1995) Wiederentdeckung der eigenen Stärke? IG Metall, Frankfurt am Main
10 Schumann M et al (1991) The Spread of the New Model of Production - A Halting Transformation of the Structures of Work. Intern. Journ. of Political Economy Vol 20: 14-41
11 Schumann M et al (1994) Trendreport Rationalisierung. Edition Sigma, Berlin
12 Schumann M et al (1995) New Production Concepts and the Restructuring of Work. In: Littek W, Charles T (Eds) The New Division of Labour. De Gruyter, Berlin New York
13 Sorge A, Streeck W (1987) Industrial Relations and Technical Change: The Case for an Extended Perspective. WZB-discussion paper IIM/LMP 87-1, Berlin
14 Springer R (1993) Neue Formen der Arbeitsorganisation - Ursachen, Ziele und aktueller Stand in der Mercedes-Benz AG. Angewandte Arbeitswissenschaft Vol 137: 19-37
15 Tropitzsch H (1994) Effizienzsteigerung durch mehr Partizipation. Angewandte Arbeitswissenschaft Vol 142 :1-4
16 Turner L (1991) Democracy at Work. Cornell University Press, Ithaca
17 Womack J et al (1990) The Machine that Changed the World. Rawson Associates, New York

4.6
The Current Social Form of Automation and a Conceivable Alternative: Experience in France

M. Freyssenet

A tool has always been the materialization of the intelligence of producers to attain more efficiently their goal. However, the end pursued, the social conditions to attain it and the social modalities of the materialization of the intelligence have not remained unchanged throughout history and neither are they the same in different societies[1].

Aims, conditions and modalities have varied and do vary, depending upon the type of social relationship which links participants to the activities concerned. This would explain why the material form of the means of work not only bears the mark, but also symbolically represents and delimits the practical use that can be made of these means within the social framework, at the heart of which and for which they were conceived. And in our case here, the social relationship is the wage relationship.

We hence tried to identify and question the objectives, principles, presuppositions and social images which determined the technical choices characterizing some automated installations (especially in automotive industry: robotized welding lines, mechanical assembly lines, automatic testing equipment and expert control and maintenance systems) reconstituting or following their design process and utilization. It appears that these choices are subject to a specific manufacturing philosophy and explain some social and productive problems.

This does not suffice, however, to demonstrate that other technical forms are possible. It is still necessary to verify that - by pursuing different social objectives and by changing presuppositions - one is in fact defining other processes and other social forms of automation. One French automaker agreed to explore what changes in work organization principles might mean through modifications of technical specifications of machines, materials and automated installations and consequently of their use.

4.6.1
The Economic and Social Presuppositions Behind the Current Processes and Social Forms of Automation

We identified three major economic and social presuppositions: the return on investment depends on workforce reduction and the rate of this reduction; prompt

---

[1] This chapter is a modified and abridged version of an article published in *Sociologie du Travail* (4/1992: 469-496). Translator's notes marked with asterisks.

repair is the key to the availability of automated lines and the greatest uncertainties in production are human and social factors.

4.6.1.1
First Presupposition: the Profitability of Investment Depends on Workforce Reduction and the Rate of Reduction

Reducing the number of workers in manufacturing and maintenance is considered to be the favored means of achieving an immediate increase in financial performance. Other criteria to justify automation have appeared in reports on investment since the mid-1980s. However, the reduction in labor *costs* remains the determining variable in the formulae for calculating profitability. As a result, planners are preoccupied with fully *utilizing* the workforce in order to limit their numbers.

Among all productive automated activities, surveillance by humans is quite often perceived to be highly non-productive time. It can easily and advantageously be replaced by automatic signals and stops in the event of faults or problems. The operator can therefore be assigned either to associated tasks of quality control, recording information, preparing tools and performing preventive maintenance or to the supervision of multiple automated lines from a central control room.

But the elimination of human supervision and consideration is subject to several conditions, in order not to be counter-productive. Firstly the greatest possible number of faults, problems and breakdowns must be taken into consideration during the design process and must be capable of being automatically identified at a reasonable cost. Then signals and stops must be neither too frequent nor should they occur simultaneously. And finally, the identification of the primary causes of these faults and problems must also be possible while machines are not working. Experience shows that these three conditions are difficult to fulfill. In fact, the medium-term performance of an automated installation depends more upon the ability to eliminate primary causes of stoppages and faults, than upon rapid repair or adjustment. What this capability really requires is the immediate availability of enough operators and maintenance personnel in order to observe and analyze the actual working condition of the machines.

The calculation of the productivity of work and of the return on investment by taking the size of the manufacturing workforce into account, multiplied by a constant coefficient in order to take the indirect workforce into account, has become meaningless. It is therefore being challenged today and we are witnessing attempts, albeit unsuccessful, to substitute other modes of calculation.

4.6.1.2
Second Presupposition: Rapid Repair is the Key to the Availability of Automated Lines

In the case of stoppages due to faults, problems or breakdowns, the output, quality and timing of automated production depend upon the speed with which operating and maintenance staff intervene. So the task is to act on all the time fragments of an intervention.

1. The amount of time needed for pinpointing an incident is thus reduced through the automatic and immediate display of the problem's location on a screen.
2. The amount of time needed to diagnose a problem (the longest and most uncertain parameter which varies most strongly from operator to operator) is reduced by *automatic testing equipment* or expert systems which indicate the underlying cause of the problem.
3. The amount of time needed for dismantling, repair and re-assembly is substantially reduced by improving accessibility to parts and mechanisms, but, above all, by standard exchange procedures which permit the off-site repair of a faulty part while production continues. If such a standard exchange routine is not possible, the time is reduced by limiting the scope of the repair work to the minimum necessary in order to resume production.

Treatment of breakdowns, which requires shutting down the means of production, is usually postponed until the night-shift or the weekend. Moreover, measures to increase reliability are only implemented after automatic recordings of the amount and nature of downtime have been analyzed. A specialized department can then determine which stoppages are the most critical in terms of length and frequency, so that priority can be given to the appropriate action.

Automatic instruments for identifying, diagnosing and recording problems, as well as the modularization of machines and the standardization of parts, all designed for prompt repair, permit identification and clear distinction between four levels of maintenance, as a function of the length and complexity of interventions and the allocation of each level to a particular category of staff.

The first level consists of short and simple intervention (two to three minutes maximum), i.e. removing blocked products, cleaning production cells and restarting production after automatic stoppage. These activities are performed by line operators; their proximity and continuous presence ensure that the interventions are kept as short as possible.

Second-level intervention must be performed as quickly as possible. This work consists of diagnosis of the immediate cause of the breakdown through automatic identification of the part, mechanism, electrical circuits or electronic boxes out of service, supported by automatic testing equipment or expert systems. This is followed by a standard exchange procedure or by limited repair. Maintenance workers, electricians, mechanics, fitters, are now assigned with this type of trouble-shooting work, excluding what was formerly their task: i.e. identifying causes of breakdown and in-depth repair.

Repair is carried out at the third stage, off site, in central workshops or on location, but not during production periods. The electronic boxes and circuit boards are automatically tested in central workshops in order to identify defective components. Parts and mechanisms are examined there in order to choose between repair or replacement with new parts, depending upon the cost of the specific option.

Searching for and resolving the primary causes of breakdown, the fourth level of maintenance, consists of activities which are increasingly deferred. They are initiated when a part or mechanism is changed or repaired too often, and when automatic recordings of stoppages reveal repetitive and costly breakdowns. These tasks are carried out by an engineering department and may or may not involve coordination with maintenance staff or workshop technicians.

At a first glance, the priority given to rapid repair as against analysis of causes and increasing reliability only seems to be profitable. It permits machine utilization rates to be increased in the short term, but a limit is soon reached because not enough is done in order to increase reliability. The curve then tends to regress because of the premature material wear and tear and the increasing number of problems and breakdowns.

The *long circuit* used to make machines reliable which is implied by this maintenance philosophy is, in the final analysis, costly, discouraging and rather inefficient. It is costly because breakdowns continue as long as their primary causes are not eliminated. Down-times, the time for dismantling and re-assembling, even when very short, altogether represent a considerable period of inactivity. The stock of mechanisms, modules and parts in rotation increases. Repeated stoppage leads to other problems, faults and breakdowns. It is discouraging for the staff because they have to live with constant and repeated problems. They become discouraged seeing that the problems are never fully resolved, and they tire of correctly *documenting* them for engineering departments. Their tasks are often reduced to merely drawing up a summary or "blind" description, without actually recognizing what is significant.

Finally, it is not very efficient, because in the final analysis neither one single group nor one single person possesses all the practical knowledge of the actual functions of the equipment. The solutions envisaged (away from the real conditions of production) to eliminate the causes of breakdowns turn out to be insufficiently adapted and often needlessly complicated. Dialogue and the breaking down of barriers between factory, maintenance functions and engineering departments, advocated and introduced by certain companies, are often just palli-ative measures for the consequences of an inadequate technical design process which is rarely questioned.

This maintenance philosophy is not specifically *taylorist* in nature. It does not imply analysis of time and action in order to establish and lay down the best way to work, the method which is at the heart of Taylor's doctrine and which constitutes its original contribution. On the other hand, it forms part of the two hundred years old development of the separation of knowledge from work.

4.6.1.3
Third Presupposition: The Greatest Uncertainties in Production are the Human and Social Factors

This presupposition is probably the most important one. Technicist in formulation, it is linked to a fundamental preoccupation of the company. Companies need to reduce uncertainty in all areas and make the production process transparent in order to control it. The efficiency of the technical system is said to be constantly threatened by the major elements of uncertainty, i.e. the productive worker himself who, as a human being, is subject to errors and, as an employee, motivated by his own interests and the social life within the factory, characterized by tolerance, arrangement and compromize which question the rational nature of the system.

Hence the tendency, during the design process, to limit the field of possibilities and to pre-determine the operations to be performed. The specific framework for work enables the operators to understand only that part of the automated plant which the designers consider necessary and sufficient for them. It also limits their intervention scope according to the modalities considered *a priori* to be logical and coherent with the theoretical principles of the operation of the system. The above-mentioned economic and social presuppositions thus lead to a process and to a social form of automation which, for the operating and maintenance staff, seems to be rigid, externalizing, excluding and substituting. Should some of these presuppositions seem likely to disappear, others appear more lasting, however.

This social process and form of automation are contradictory - in principle and in practice - whilst some tentative approaches to implement work organizations which really promote skills are currently being implemented.

4.6.2 Compatibilities and Incompatibilities Between the Current Form of Automation and new Forms of Work Organization

The process and current social form of automation are compatible with certain new forms of work organization and are contradictory to others. In fact, not all new forms of work organization are - despite appearances and discourse - really skill-promoting forms of organization.

4.6.2.1 Organizations which "Enrich"

These are the forms of organization, particularly in final assembly, where operators of automated installations are in charge of primary maintenance, quality control, tooling and production supervision and where they are sometimes asked to organize themselves autonomously in order to fulfill these functions. The socio-technicians project launched in the 1960s in order to *reorganize* work and create autonomous groups hence emerged at the same time when automation was introduced. What happened in fact? The new activities assigned to the operators were first simplified by automation designed for this purpose. We have already seen how current forms of automation - presupposing that human supervision is nothing more than waiting for a problem to occur and hence non-productive "free" the operator from having to wait, thanks to automatic signals and stops (automation has in fact often rendered human supervision non-productive, permitting only such knowledge of the installation which designers *a priori* consider to be necessary and sufficient for operators).

The operator is now free to carry out other tasks. Now, the priority given to prompt intervention in order to resume production as soon as possible leads to a division between the tasks of maintenance, repair, quality control, setting up, tooling and production supervision. Part of this knowledge (notably recording and diagnosing) is naturally incorporated in the automation system, and the above-mentioned division allocates the remaining operations to different categories of staff, according to how long operations take and how complex they are. We already saw this in the case of machine repair. As far as quality control is concerned, the automatic detection of faults reduces the operator's activities to extracting the product or to marking it for downstream rework or to immediate touch-up, if the latter can be done easily and within the time of the work cycle. Moreover, tooling operations are limited today. Tooling is usually performed during an automatic stop after *X* cycles, by positioning a jig with the tool which has already been set up at the factory or elsewhere by means of set-up tools.

These new tasks, frequently regarded as a contribution to *reskilling* work, do in fact often lead to *training* and classification at a higher level and even to the title of skilled worker. However, they represent the juxtaposition of operations which have become partial, and their implementation does not allow an understanding of the operation as a whole nor does their implementation permit the acquisition of practical knowledge - a precondition for real and lasting development of skills.

This type of task transfer to operators, which has eliminated the jobs of setters, inspectors and touch-up workers, continues. The automation of breakdown diagnosis (as part of maintenance work which requires formidable skills) by means of automatic testing equipment, as well as the generalization of standard exchange procedures, in effect allow the future shifting of an increasing part of second-level repair activities to system operators to be envisaged, without this transfer requiring a true development of skills for the operators.

On the one hand, a deepening division of labor is in gestation with the training of a blurred category of polyvalent *operator-repairers* and, on the other hand, with the development of a smaller group of specialists to deal with rare or new types of breakdowns which cannot be automatically diagnosed. Rigid and externalizing automation, linked to new forms of work organization which are limited to enriching the operators' work, leads to a deeper division of labor. It is this linkage which is the most widespread today.

This entire process is taking place as if the scenario observed when specialized mechanization and Taylorism began were being repeated. At that time we saw, on the one hand, the *reskilling* of workers by their reassignment to the *operating* of specialized machine tools and, on the other hand, the creation of the category of maintenance workers, replacing the skilled workers who operated, set up and maintained the universal machine tools on which they worked[2]. This

[2] F.W. Taylor points this out in *Scientific Management* (Westport, Conn.: Greenwood Press, 1947 edn.). On page 146, he notes: "It is true, for instance, that the planning room and functional foremanship, render it possible for an intelligent laborer or helper in time to do much of the work now done by a machinist. Is not this a good thing for the laborer or helper? He is given a higher class of work, which tends to develop him and gives him better wages. In the sympathy for the machinist the case of the laborer is overlooked." It is interesting to note that

meant a *real reskilling* for workers in the sense that, at the beginning, decomposition of tasks and specialization of machines were still far from what they would later become[3]. This reskilling was accompanied, moreover, by salary increases for those concerned and subsequently by the creation and allocation of a higher classification, i.e. specialized worker.[4] But this reskilling was relative and temporary, as will be that of the operators of automated installations if the above-described techno-organizational model, which is the most widespread one today, should prevail.

4.6.2.2 Organizations that "Skill"

Forms of work organization which are really *skilling* are characterized by the formation of teams for operation and maintenance; these teams organize themselves autonomously and have effective responsibility not only for achieving production program goals but, above all, for improving the performance of the plant section for which they are responsible in terms of output, quality and timing. These forms of work organization -in principle and practice- are contradictory to automation in its current form. We studied this specifically in welding shops.

Improving the performance of an automated line by using a basic team implies that the team has an understanding of the actual operation of the line beyond its physical form. Now automated lines and their machines are often designed in such a way that anybody - skilled or unskilled - would find it practically impossible to monitor, during production, prospective problem areas, mechan-isms which may

---

he adds "This sympathy for the machinist is, however, wasted, since the machinist, with the aid of the new system, will rize to a higher class of work which he was unable to do in the past and in addition, divided or functional foremanship will call for a larger number of men in this class, so that men, who must otherwize have remained machinists all their lives, will have the opportunity of rising to a foremanship." We know since that this was the case for only a small proportion. Today we are beginning to hear the same argument regarding maintenance workers and technicians who are replaced by polyvalent maintenance staff or by operators of automated equipment, formerly unskilled workers, when automated systems are put into service, equiped with automatic error or breakdown diagnosis.

3 F.W. Taylor, in 1902, did not imagine the machine tool operator would be reduced to his later role. In *Scientific Management* p. 101-102, he writes: "The repair boss sees that each workman keeps his machine clean, free from rust and scratches and that he oils and treats it properly and that all of the standards established for the care and maintenance of the machines and their accessories are rigidly maintained, such as care of belts and shifters, cleanliness of floor around machines and orderly piling and disposition of work." In short, F.W. Taylor recommends that the operator should do what is today called primary maintenance, which for the person performing it should be classified as skilled work, whereas the worker whose activity he describes is trapped in the labourer category and will keep that label until the period between the wars! Regarding this classic phenomenon of the opposite development of the classification of individuals and the real skill required by the work they do, see Michel Freyssenet, "Peut-on parvenir a une definition unique de la qualification?" in *La division du travail,* Paris: Galilee, 1978, pp.67-79.

4 [Translator's note: *ouvrier specialize*: normally today translated as *unskilled worker*, for reasons the text is explaining at this point.]

fail, tools which may become misaligned and movements which may lead to desynchronization. The working parts, i.e. those parts which drive the machines and the transfer of the product within the machines, are no more visible than the overall kinematics (mechanical movements) are readable[5]. Electronic and electromechanical screens at control and signalling stations permit the presentation of just that part of the plant for which the operator is responsible to intervene. Now the transparency, the intelligibility, the analyzability of working machines are preconditions for the team in charge of improving performance. The paradox is that it is just at this point that machines become even more compact and opaque. Their design even discourages or dissuades efforts to understand their weaknesses and their deviations while they are being used and hence to anticipate breakdowns and problems by preventive action. This is true even for operating and maintenance teams made up exclusively or mostly of skilled maintenance workers.

The discourse on skilling organizational forms and on calls for workers' initiative and autonomous organization lose their credibility in the eyes of those who are supposed to benefit from them and take part in them. The conviction that production techniques are socially neutral is widely accepted by the promoters of these forms of work organization and prevents them from perceiving the contradictory situation in which the operating and maintenance staff are placed[6]. They interpret the reticence of the latter and the merely average or short-lived results of these new forms of organization to result from the deep-rootedness of *Taylorist mentalities* and from insufficient financial compenzation offered for the effort expected, without perceiving the need to reconcile the principles behind technical design and organizational form if a skilling process is to begin.

It may occur, however, that despite the material obstacles encountered, staff in charge of a given technical system implement modifications which are more or less authorized and exceed their orders, with the result that they acquire a good knowledge of their line and improve its results. But the pursuit of automation in its current social form and notably the introduction of computers for operating and forecasting breakdowns, which substitutes operators' abilities, undermines their motivation to participate.

The very design of automation, as it is currently taking place, is hence clearly in doubt. Are alternative processes and forms of automation conceivable and achievable? That is what we wanted to find out by participating in the design and the implementation of several automated systems, especially the automatic placement of mechanical components under the carbody.

---

[5] [Translator's note: French *lisible:* literally, *readable*. Here and later the metaphor *read* is used as in *reading machines* (i.e. being able to understand how machines work by looking at them).]

[6] The critique of the "technological determinism", without distinguishing between the thesis of determinism by the technique itself as an autonomous force and the thesis of determinism of productive techniques because these themselves are socially determined, has contributed greatly to the notion that changing work organization is a sufficient measure in order to reverse the division of labour.

4.6.3 A Process and a Social Form of Automation Aimed at Financial Performance and Real Skilling of Work are Conceivable and Achievable in a Localized Way, but can they Be Generalized?

We tried to recognize the consequences of a skilling form of organization for the automation process and for the technical characteristics of automated systems. Due to certain difficulties, the application of these new design principles has only been partial. Some of these difficulties can be overcome, but others which probably originate in the wage relationship as we know it today still represent major obstacles. Process and social form of automation described below are probably achievable in some places and for a certain time - the question, however, is their durability in the current wage relationship.

4.6.3.1 Giving Priority to Increasing Reliability Promptly is a Strategy for Financial Performance and Real Skilling Work, but is Only Achievable Under Certain Social Conditions

Now the performance in terms of output, quality and timeliness depends on the rate of utilization and stability of the setting of automated and integrated machines in production lines. It is more efficient and more profitable to foresee or add a worker, if his work contributes to increasing the actual utilization rate of the automated system than to try to eliminate a job in order to increase the theoretical ratio of output per worker. An additional skilled worker is profitable if the output rate of an automated system is increased by one or two percentage points during the year as a result of his work.

The prompt analysis and elimination of the primary causes of faults, problems and breakdowns by an operating and maintenance team may be a method to continuously and permanently increase the utilization rate of automated lines. The *short circuit* of increasing machine reliability that characterizes this scenario implies, in the first place, that workers are available for monitoring, and that they are placed at key positions where possible problems may arize. In contrast to the current philosophy of production and maintenance, it can be said that the availability of the system is proportional to the availability of workers to ensure permanent monitoring, thus permitting the identification of the primary causes of misalignment, anomaly and problems .

The establishment of this short circuit to increase reliability may also be a starting point for a process to really reverse the division of labor, under certain social conditions, as we shall see later.

The operating and maintenance team would then become an indispensable and respected partner for the engineering department, because they possess new knowledge which is absolutely necessary for the design of systems which are properly adapted to their conditions of use.

4.6.3.2
This Production and Maintenance Philosophy Allows a Progressive and Non-Excluding Automation Process

The reduction of the workforce can be accomplished progressively as the team manages to increase the reliability of the production line for which they are responsible.

Moreover, the complete substitution of one type of workforce by another, which often takes place today, is not only unnecessary, but would even be counter-productive. Increasing machine reliability in fact requires a sound knowledge of how products and machines perform, as well as of the given conditions of production. Operating and maintenance workers who have worked on previous systems usually have this kind of know-how. Regarding such a process, one can even imagine that the automation of a function or of an operation would only be introduced if the team is able to identify, with the help of the engineering department, the relevant parameters and the events which may occur while they are performing this function or operation.

We then have a process of automation which is much more efficient and socially "smooth", where the basic team becomes an active and indispensable partner. Above all, this process leads to an alternative social form of automation.

Workers are no longer considered to be the unreliable element in a technical system but, on the contrary, as an element to increase reliability. But in order for them to acquire practical knowledge of the actual operation of the production line and its theoretical problems, the technical design of the system must permit this: it must be readable and intelligible, capable of being tested and analyzed, adapted and modified.

4.6.3.3
Machines Must Be Readable and Intelligible...

If the essential function of members of the basic team is to prevent faults, problems and breakdowns and to eliminate their causes, then the primary attribute of the machines and lines must be visibility of their actual operation during use. They must be readable, understandable and intelligible. This is an essential pre-condition for the workers, individually or collectively, to acquire knowledge of the operation of the machines which no prior classroom training can replace.

As long as a high level of reliability has yet to be reached, there are no economic reasons to make machines compact and opaque. On the contrary: in order to reach this point, it is necessary to proceed via *extroverted* and *transparent* machines.

..Testable and Analyzable

It is impossible to predict *a priori* all the places, the mechanisms, the movements which should be recorded in order to pinpoint the primary causes of problems. They must be accessible and capable of being fitted with analytical and testing tools. The overall structure of the system must also be *open* in order to make its operation readable and intelligible, and it must make possible to physically *listen to* the component parts and to *analyze* the product flow.

Making the search for primary causes of breakdowns a priority also eliminates the distinction between known types of breakdowns (which can be diagnosed by automatic testing equipment or by expert systems, the instructions of which have to be followed by the staff) and new or rare types of breakdowns, the analysis of which is left to specialist technicians. All breakdowns, simple or complex, frequent or rare, receive the same treatment: elimination of their causes. The function and the design of automatic testing equipment and expert systems is hence modified. By definition, these instruments cannot generate instructions, because their function is to help find causes which are still unknown.

...Adaptable and Modifiable

Increasing reliability is often achieved by adapting to the particular conditions in which a system is used, i.e. through modifications which are only complex and costly as they increasingly require a process of dismantling, rebuilding, rewiring and re-writing which is long and complex because of the mechanical or electronic structure of the production line. It follows that complete modularization of machines, as well as systematic standardization of parts, cannot be implemented, especially when the aim is to simply permit rapid repair. An alternative type of modularization can be designed which meets the requirement of adaptation and modification necessary to achieve the desired increase in reliability, rather than the requirement of prompt and limited repair.

Giving priority to increasing reliability promptly, i.e. the pre-condition of and first stage towards a real reverse of the intellectual division of labor, has significant implications, both for the automation process and for the design of automated systems. The resultant technical options are limited neither to the ergonomic adaptation of work stations to reduce physical burden and "enrich" the work content, nor to arrangements to promote a more collective and autonomous approach to work organization. They affect the structure and arrangement of machines and production lines and the very functions and purpose of automation.

How far has it been possible to go in the projects in which we have participated towards the application of the principles for design of automation just described?

4.6.4 The Difficulties of Implementation and the Social Preconditions for a Generalization of the Process and Social Form of Automation Already Discussed

Some implementation problems were sufficient in the experiment to prevent certain technical recommendations from being adopted. However important, these difficulties are logically surmountable in the course of time.

When companies purchase standard machines and equipment from suppliers, all modifications mean high, excess costs. The scenario proposed now implies more than just simple modifications. It requires nothing less than a complete rethinking of equipment. It then requires convincing the suppliers that it is in their interest to do this. The principles of technical design, as explained, are hence more likely to materialize in the short term in individual production plants. The cost of creating a new technical family of automated machines is not the only difficulty. Even when convinced, engineers are still reluctant to proceed beyond improving the man-machine interface or increasing the social acceptability of automation.

The above-described social process and form of automation imply to be lasting under two social conditions: to be able to guarantee employment and job promotion, and to accept the social dynamics of reversing the division of knowledge from work.

First condition: employees will not participate in increasing reliability unless they are guaranteed not only employment, but - even more - employment in which the new abilities which they have acquired through their work of increasing plant reliability can be used and further developed. With an anticipatory type of employment, management must consider not only age and seniority of workers, but also all their different skills. To do this, companies will have to plan their future, not only in terms of development of their markets and ways to ensure return on capital, but also in terms of development of staff abilities - a development likely to lead companies far from their original activity.

Second condition: operating and maintenance teams who permanently analyze the real operation of their systems and increase reliability create genuine knowledge which nobody else really has. This "on-site" knowledge becomes indispensable for the design of the next generation of machines and may enter into a complementary or contradictory relationship with the more theoretical knowledge of the engineer. Workers are no longer just asked by designers for their comments or suggestions, as it is - at best - the case today. They are in a position to participate effectively. A real reverse of the division of labor can then and only then be set in motion. It is possible to envisage a transition from know-ledge which is socially divided to a socially shared knowledge, founded on co-operation without subordination. Are the companies in a position to accept this and to bear the weight of its social dynamics and consequences?

4.6.5 Conclusions

It has hence become possible to describe a process and social form of automation which can lead to a real and lasting reverse of the division of knowledge from work. But the type of company required for this process leads to doubts about its generalization, especially in the absence of a thorough transformation of the wage relationship itself, because merely abandoning Taylorism is in itself not sufficient.

From a scientific perspective, the exercize has the advantage of confirming that production techniques are not only sociologically, economically and culturally conditioned in their development and diffusion, but that they are also socially constructed and constituted by a set of objectives, principles, images, economic and social presuppositions, which are themselves rooted in the wage relationship and the division of knowledge from work, linked to it for two centuries.

The division of knowledge from work has two sides: one material, the other organizational. It is now transmitted more efficiently via production techniques, because most of the necessary knowledge has been incorporated into these techniques, rather than via work organization in the factory, which only distributes what remains of the knowledge. Production techniques are not simply marked by the social conditions of their design. They are also, in the context for which they were designed, an active instrument in the type of division of labor which is at work there.

Technology is obviously *malleable*, if considered in general. However, the specific techniques which are implemented and in particular the production techniques discussed in social science research on work, are materially constraining, rigid and substitutive, because this is how their presuppositions are today. They determine the content of work, not because techniques are determining in themselves, but because they themselves are socially "constructed". They only possess the *hardness* or the *malleability* of the social framework in which they materialize. The opposed theses of *technological determinism* and of the *social neutrality of techniques* here have in common that they grant techniques a status of social extra-territoriality. Productive techniques belong to the realm of sociological analysis, with nothing special to mark them out, like any other social product.

It is, of course, necessary to see the social component not as a separate area of analysis (alongside the economic, the technical, the political components) but as the specification of a limited number of social relationships (each with its own economics, techniques, symbolism) in which we are historically called to act.

4.7
Worker-Generated Production Improvements in a Reflective Production System – or *Kaizen* in a Reflective Production System

K. Ellegård

4.7.1
Introduction

Workers are often well aware of the problems that give rise to waste in production. In Japan, the workers' efforts to tackle such problems are called *Kaizen* activities. In Sweden, workers' involvement in everyday production improvement is not as systematical or organized. Workers' efforts are concentrated on technically oriented suggestions.

Many Swedish workers have found out that the strongest arguments for change are those which point to definite and objective phenomena. In the same way as a material product provides a stronger argument than a service rendered, a piece of paper with the work process clearly illustrated on it provides a stronger argument than the sound of the words saying exactly the same thing. The issue of this paper is to present a method which was developed to make worker-generated improvements easier by visualizing the work process itself, whereby strong arguments are developed.

In respect to worker-generated improvements in production, I argue that a reflective production system has a greater potential than any production system leaning on the assembly line technique.[1] To realize this potential in a reflective production system, however, certain means and methods have to be developed. My argument is based upon the different potential results yielded from the following situations:

1. A multiskilled worker, who works in the early stage of the assembly line, finds a suggestion for an improvement.
2. A worker in a team assembling complete products, working in a plant according to the reflective production system principles, finds a suggestion for a certain improvement.

Both workers may, skillfully and well informed, argue for an improvement with production engineers, but the reflective producton worker (2) has advantages because he/she controls the totality of the production and therefore, the cost of evaluating the suggestion from the case of the first worker (1) will be higher than for evaluating the second case (2).

[1] A reflective production system is described in more detail in chapter 3.10 by Ellegård and is shortly summerized in the following section.

The reason for this is that the worker (2) in the reflective production system team can perform a preliminary evaluation of the consequences of his suggested improvement for the rest of the production process. Further, he or she can discuss the effects for the entire completion of the product with his or her workmates in the team. The multiskilled worker (1) on the assembly line, however, does not have a clear conception of the work performed in the rest of the assembly line process, and he has no natural daily communication with the workers at the end of the line. The distance in the assembly line plant, measured in terms of organization and geography, raises many obstacles for such communication, and no preliminary evaluation concerning the complete production process made by workers can take place.

Therefore, the worker in the beginning of an assembly line has to leave even the most simple and preliminary stages of the evaluation process to others, for example, production engineers or supervisors. As specialist engineers are required for all evaluation efforts, including the most trivial pre-assessment of consequences, the cost in an assembly line plant exceeds that of a reflective production one.

However, even in a reflective production system, many tools have to be developed for examining work performance systematically in order to improve the production performance. Techniques for workers' own evaluation are not immediatly available, and a number of tools for the workers' and the teams' evaluative efforts must be developed. In this paper, one tool for detecting production-related imperfections in reflective production systems will be presented. The tool was developed during 1992, in an iterative process by a team in which I was involved at the Volvo Uddevalla automobile assembly plant.[2] This tool can be utilized by workers in their teams to suggest improvements in production and further, to pre-evaluate the suggestions within the team. The tool is, in addition, a good ground for discussions on production related problems, as the arguments can be supported by illustrations of the *work process*.

4.7.2 Organization and Initiatives for Improvements in Different Production Systems

4.7.2.1 Potential for Improvement

Traditionally, in hierarchically organized production, initiatives for change normally come from above. Management and production leaders form the strategy

[2] Volvo closed down the Uddevalla plant for reasons discussed by Ellegård in chapter 3.10. However, the potential of the reflective production system was recognized by Volvo, and in 1995, a desicion was taken to form a joint venture with the British Tom Walkinshaw Racing, and to re-open the plant in Uddevalla. The name of the joint venture is Autonova AB. This time, a complete plant will start operations in late 1996, with a body shop, a paint shop and an assembly shop. The reflective production system principles will be implemented.

and decide what to do when changes are needed. Production engineers are engaged in the planning process as well as in the implementation process. In the last few decades, "soft ware" specialists, for example, the personnel department, have also been involved in the process of change, in order to help the production department to form a new work organization. The idea of *integrated work groups* which is widespread today, means that work groups on the shop floor handle not only production, but also, among other things, quality, preventive maintenance, some materials handling and short-term planning for production, training and attendance. When this kind of organization was introduced for the first time in the 1970's, there were many protests from middle management and specialist departments, who felt their power base was being threatened.[3] Volvo was the forerunner in the introduction of integrated work groups and most other automobile producers by that time were sceptic of the organizational innovation.[4]

However, changes in work organization combined with raising the workers' level of competence, individual multiskilling and integrated work groups, have almost always proved to be profitable when introduced. The same result is apparent in many companies all over the world. When workers on the shop floor receive better tools (mental, organizational and technical) in order to do things correctly from the beginning, the result improves: for example, better quality and more efficient production. The need for adjustment afterwards is reduced, and the maintenance department can concentrate on other things than routine maintenance which henceforth, is carried out on the spot by the teams. Such improvements occur in all kinds of production systems, whether the work stations are serially linked, on the assembly line or are parallel, as in the reflective production system. One explanation for the favourable results obtained is that the organization gives the workers a better understanding of the production conditions, and also enables them to exert some control over their own work situation. The question now is: Can we achieve even better results?

Before this question is answered, it is necessary to explain what a reflective production system is, because the answer differs depending on whether we are dealing with a reflective production system or with another kind of production system.[5, 6]

---

3 Many people in industry are fostered into the old hierarchical tradition. One result is that thinking and acting are influenced by the old ways of thinking, even in modern production units where one would perhaps not expect to find it.

4 Autonomy is a fundamental prerequisite for successful results of integrated work groups. However, even with limited autonomy, there are often some positive results, but the potential is much greater where teams are autonomous.

5 The Reflective Production System is described in more detail in the chapter 3.10 by Ellegård in this book.

6 We have, as researchers, introduced the concept and the term Reflective Production to label the innovation. It is described in more detail in English in [3], in Swedish in [1] and [2]. A presentation in French is given in [3].

4.7.2.2
Reflective Production - A Brief Presentation

A reflective production system differs quite radically from other production systems: When it was created, inherent human needs, abilities and ways of learning formed the starting point. Therfore, the technical equipment in a reflective production system is built up for a much enlarged work content, so that each worker may in fact carry out all the work needed to produce the whole product within one and the same team. Autonomy is thereby created. As a consequence, the product flow is highly parallelized. These points are vital for understanding the potential of reflective production.[7] When focussing on the production process in a reflective production system, therefore, the following characteristics can be seen:

- *A parallel flow of products*; this calls for a new materials handling system that offers *individual materials kits* for every individual product. In the materials kit, the components are placed on a rack, grouped in an order that helps the assembly worker to put the components of the product in its functional context and located in the right place. The kit is taken to the work station, where the product is then completed.
- *Long cycle work tasks*; this means that the work consists of enlarged and qualified tasks. Consequently, each worker does not repeat the same work task more than a couple of times per day. The theoretical basis for long work cycles is called *holistic learning*.[8]
- *The work task for a team comprises the total work needed to make a complete product* from the start to its completion. This means that each worker experiences the birth of a complete product in an autonomous team.

Workers in a reflective production system thus have to learn a lot about the product, materials and about production methods. They also need information channels to pre-production and marketing. Increased competence in these fields is an asset that yields positive results in ordinary production, as well as when performance does not follow the production plan. The extended competence of each team in a reflective production system enables the team members, if they have adequate tools, to detect problems in their work environment by themselves.[9] They learn to introduce improvements in production methods, in materials handling, in tools or other equipment, or organization in the plant.

---

[7] A reflective production system focusses on the entire production chain; from design and pre-production, production process to marketing and customers.

[8] See Chapter 3 by Nilsson in [3]

[9] In a plant with a serial flow, where the work is segmented functionally as well as geographically along the production line, it may be hard to find a good solution, since nobody has the necessary overview. Accordingly, there is danger of suboptimization.

4.7.3
When what there is and what there Ought to be do not Correspond — Strategies to Solve a Disparity Problem

One of the main problems in industry frequently facing workers in every-day work situations is that what *is,* is seldom what *ought to be.* The difference between the norm and the reality can be called a disparity problem. This problem can be tackled by employees in different ways:

1. not bothering about the differences, i.e. accepting it
2. ignoring it, to „explain away“ the problem
3. leave it to someone else
4. to take responsibility oneself, to close the gap between what there is and what there ought to be.

In industry, as traditionally organized, workers are trained not to deal with anything else other than their own direct production work tasks, and they are accustomed to acting only upon orders. To wait for an order before a new step is taken, is the accepted and expected behavior on the part of workers.[10] Therefore, the workers strategy in plants without integrated work teams is usually 1., 2. or 3. above. They do not have the formal rights nor the necessary competence for analyzing and solving disparity problems. Someone else, a specialist, is expected to do that. In most production systems, workers do not have enough overview and, accordingly, not enough competence to identify the consequenses of the changes they propose.

In most production systems with a serial product flow - with or without integrated work teams -, workers who have ideas about how to improve production are asked to turn to a department specially set up for assessing ideas from the shop floor. This is a logical way to take care of ideas generated by those who cannot assess the full consequences of the idea by themselves, i.e. identify its consequences for the whole production process. As we suggested in the introductory section of this chapter, a worker (1) who is performing his or her short cycle work task in the beginning of a serial flow (on an assembly line), and who has an idea about how to achieve the goal set for his own part of the production flow more easily, may not be able to judge the consequences of his/her idea for the workers at the end of the same line. The worker may have found something which would perhaps revolutionize production on his part of the line only, but which would at the same time, cause severe problems at other places on the line. The risk of suboptimization must be eliminated, and this is done by production engineers in a specific department set up for this purpose.

As suggested above, a worker in a reflective production system (2), on the other hand, can assess the consequences of an idea for the entire production process. The reason is that he/she and his/her team are in control of the entire pro-

[10] This does not at all imply that they do not recognize problems. On the contrary, being quiet about problems is the same as letting somebody else detect them, sombody who has the responsibility to give orders for change. For instance, this can give some extra time off when a machine break-down has to be mended.

duction process because this takes place on one and the same spot, and immediately affects the members of the same team. Suggestions for improvements, generated by teams in a reflective production system, are "pre-assessed" in the minds of highly skilled workers and the risk of suboptimization is negligible.

4.7.4
The Development of a Worker Controlled Holistic Method to Follow up Work Performance in Order to Improve Production Conditions

4.7.4.1
An Example of a Disparity Problem

This is an example of how a disparity problem was tackled by a team working in a reflective production system, the Volvo automobile assembly plant in Uddevalla. The team's contribution was to participate in the development of a new method to follow up work performance. They solved what initially seemed to be a problem of their own, but they did, in addition, put the finger on a much deeper organizational problem. Their contribution was greater than expected from the beginning, and consequences for the whole plant were discovered. At the same time, it led to increased co-operation and understanding between workers and between the various departments in the factory.

The goal for the production volume, as set up by this team, was to complete 12 cars a week (= what ought to be), but they produced a total of 11 cars (= what is). Thus, there was a genuine disparity problem. It was underlined by the fact that the other teams in the same shop were able reach their corresponding goals.

Theoretically, and in terms of the time-measurement system utilized in the Uddevalla plant (SAM), it should be possible for the team to produce 12 cars, given that they possessed the normal competence.[11] The workers in this team were of the opinion that they met all these needs. Why did they fail, and not the others? There were many different explanations, either related to the team and its performance (competence, working pace) or to the prerequisites for work (preparation for the work process, equipment, tools, material feeding, organization, responsibility).

The shop manager, for example, attributed the failure of the team to lack of "working spirit". One argument for this conclusion was that the other teams in his shop did achieve their goals. Then, the team members in our team asked for a production engineer to measure their time use.[12] This was done during one week and the result showed that the team held the working pace required to produce 12 cars a week. But still, they produced only 11 cars per week. This result surprised the shop manager. But it did not surprise the team: they had always believed that they fully met the expected level of work performance.

---

[11] SAM is a system for measuring time in production, much less detailed than MTM.

[12] Time is not necessarily a controversial factor in a reflective production system.

Thus, the conclusion of the time measurement analysis was that the disparity problem did not emanate from a lack of working spirit or from bad work performance. The disparity problem had to be explained by other factors. A comparative analysis, based on some of the teams which fullfilled their production goal, showed that they in fact worked faster than the norm set up for the individual tasks by the SAM-system.[13] They achieved their goal at the cost of working faster than they were expected to do. There was something wrong here.

4.7.4.2
Different Methodologies to Measure Time Use in Production

The underlying question put by the production engineer, who made the time measurement studies, was: Did our team work above or below the standard time set for each operation? Therefore, the production engineer measured time use only when work actually was being done. Accordingly, all kinds of interruptions had been excluded from the analysis. In other words, *the analysis had been partial and not valid for the total working time.*

My methodology differs in some fundamental ways from that used by most production engineers: I take into consideration continuity, sequential order and thus, the whole context of the work tasks. In such a way, all the time used is measured in sequential order, time used for assembly as well as time used for interruptions (various events and incidents such as material problems, equipment or tool problems, meetings, someone from within the team, or from the outside, comes along to get some information etc, etc). Thus both ongoing work and the time when work is interrupted are taken into consideration for analysis. The approach can be called holistic and its focus is on the *assembly work exactly as it is actually experienced by each individual worker.*

I talked with the team and the members asked me to suggest a way to analyse the disparity problem. We discussed their situation, and decided to find out what was really happening when they assembled cars. They thought that such an analysis might help, not only themselves but also other teams in the plant, as the same underlying problems probably existed in all teams, but were not detected in the other teams, since they met their production plan by working harder than intended. The plant management supported our joint effort to solve the disparity problem.

4.7.4.3
The Interactive Development of a Holistic Method

The team and I decided to handle the analysis in a simple and worker-oriented way. I asked the team members to write a „work diary“ every day for one week. In the diary they were asked to write:

---

[13] The norm was set by production engineers by means of the SAM-system.

Table 4.7.1. Contents in a diary for one work day[1]

time	code	car no
06.54	1B	10
08.08	4	‘’
08.13	1B	20
13.06	2	‘’
13.09	1B	‘’
15.05	1B	30
15.15	2B	‘’
15.21	1B	‘’

[1] 1 = assembly work, 2 = material problem, 4 = giving information to others; B = on the car body (not pre-assembly work); 10, 20, 30 = different cars.

- the time when they started to do a work task (an activity), which kind of work they were doing and the identification number of the car worked on,
- the time when they changed the work task (changed activity) and what the new task was and the identification number of the car worked on,
- the time when an interruption occurred (that is also a kind of activity!) and which kind of disturbance they were interrupted by,
- the time when they started to work on another car, which work task they performed and the identification number of the new car,
- and so on during the week.

We discussed the easiest way to write the diary, as we did not want it to take much time. Otherwise, the method itself would have been an interruption of the work. The team members developed a code-scheme for the activities (based on work tasks and the interruptions) that could occur during the work day. Activities could be related to assembly work, material problems, equipment problems, meetings, clearing up the work shop issues and giving information to others. The assembly activities could also be directed to different components, namely, assembly on the body of the car or pre-assembly work on the doors, the engine, the instrument panel and the sun-roof. Activities could also be directed to the pre-paration of the body for assembly.

The diaries were small booklets, easy to put into a pocket. An example of the contents in one diary day for one of the team members is illustrated in table 4.7.1.

What exactly can be seen from the application of the holistic diary method? The five team members wrote diaries for one week. The number of notes in the diaries differ a little, but not very much, there were about 10 notes per day. The smallest number was 5 notes in one diary for one day, and the highest was 14 notes for one day in another diary. There were more differences between the members of the team than between the days for each individual member. This is explained by the team's internal work distribution rule: One person did most of the pre-assembly work, and she had the largest number of notes in her diary as she shifted work objects the most. Two other team members made some pre-assembly, but most of their work was performed directly on the car and the two remaining team members worked directly on the car all the time.

I collected the diaries and made a survey of their contents. During this week, 11 cars were completed, 2 of which were started up the week before. Another 2 cars were started up on Friday, to be completed the following week. Thus, altogether 13 cars were worked upon during the week, and nine of them were both started up and completed.

I completed a detailed analysis of the diaries as quickly as possible. I drew a diagram showing the entire work week. The diagram had two axes: time was shown on the vertical axis, and the different work tasks, work-related activities and types of interruptions were shown on the horizontal axis. In the diagram, each individual team member and his/her activity pattern was represented by a continuous line; this formed a *trajectory* for each individual.[14] The line went right up when the person was working on a particular task and it changed direction in order to switch position when a new task started, or when an interruption occurred. I used several different colours for the diagram, each individual's trajectory line has a different colour, so that it was easy to follow the changes in activities for each person in the team. In figure 4.7.1., the principle is shown for one car only.

When I showed the diagram to the team and explained it, their immediate reaction was that it clearly illustrated when and in which context interruptions appeared in their work. It also clearly showed that one of the main reasons why they fell short of the required production volume target set up, was the time spent on solving material problems and the time spent on meetings. Another relatively time-consuming activity was that of giving information to a small project group, set up for designing an ergonomically better work-place. In comparison, I also made up some tables of a more traditional kind. The results shown in the tables confirmed the impressions from the diagrams, see table 4.7.2.

▶

Fig. 4.7.1. A team's work on one car. Note that the work is performed over two days, break of day is indicated by a broken horizontal line. The rectangle is an illustration of the car and when the lines (individual paths=symbols for the individual workers performing their work tasks) are inside the rectangular symbol, work tasks are performed as expected. When the individual paths are outside the symbol of the car, this indicates either a material problem, a meeting or break for coffee or lunch. The time axis is vertical and the activity axis is horizontal. See bottom of figure for legend. When the individual path moves vertically, parallel to the time axis, an activity is performed, and when the individual path moves horizontally parallel to the activity axis, a change of activity is indicated

[14] The method, and its principle for illustration, is based on the assumptions underlying the time-geographic approach, which has developed within Human and Economic Geography by professor Torsten Hägerstrand and his research group in Sweden. I was a member of this group for several years.

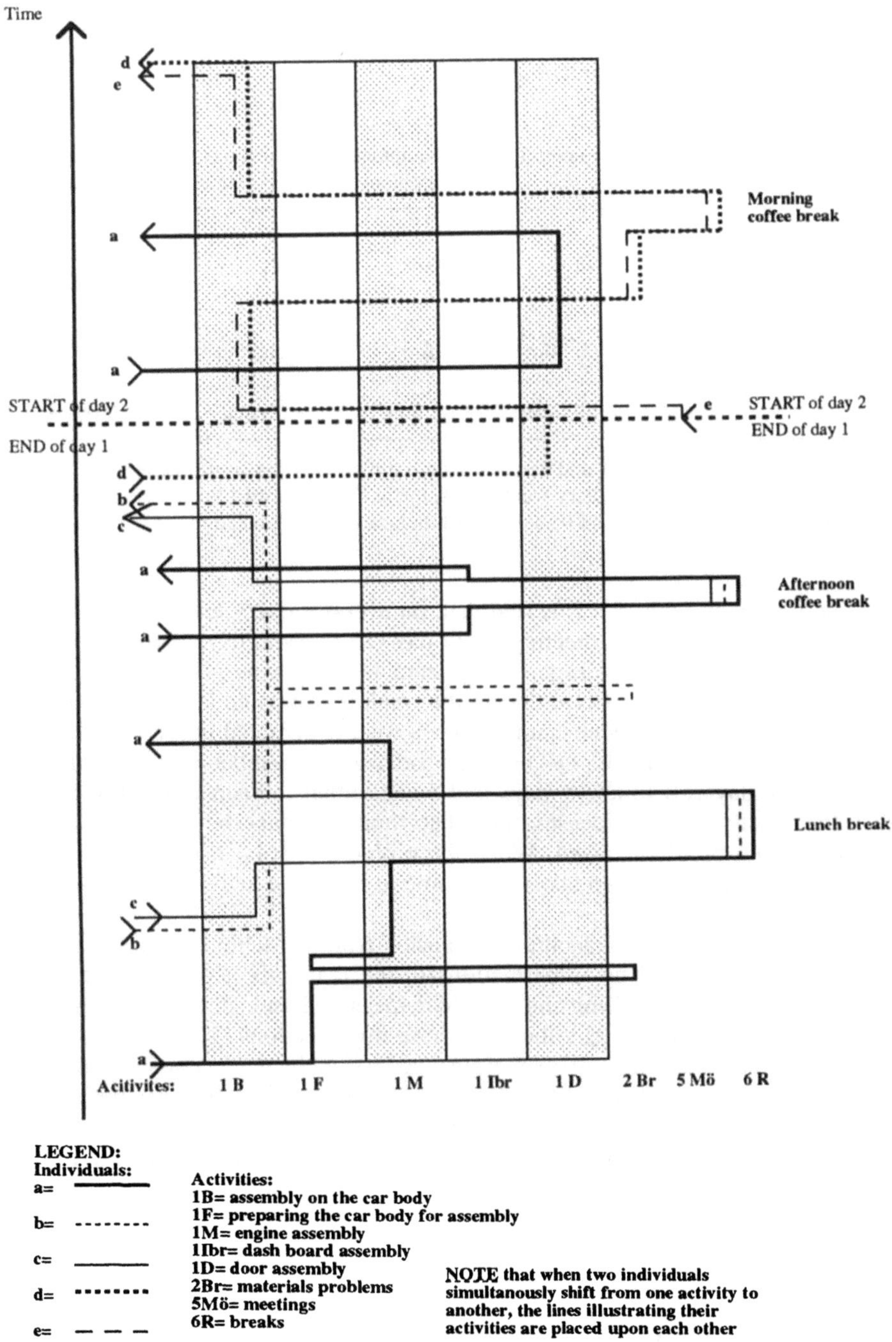

Time
d
e
a
a
START of day 2
END of day 1
d
b
c
a
a
a
c
b
a
e
Morning coffee break
START of day 2
END of day 1
Afternoon coffee break
Lunch break
Acitivites: 1 B 1 F 1 M 1 Ibr 1 D 2 Br 5 Mö 6 R
LEGEND:
Individuals:
a=
b=
c=
d=
e=
Activities:
1B= assembly on the car body
1F= preparing the car body for assembly
1M= engine assembly
1Ibr= dash board assembly
1D= door assembly
2Br= materials problems
5Mö= meetings
6R= breaks
NOTE that when two individuals simultanously shift from one activity to another, the lines illustrating their activities are placed upon each other

Table 4.7.2. Time use specified for different activities during one week for the team.[1]

	Assembly work	*Interruptions*			
	assembly work	material	equipment problems	meetings problems	giving information to others
week, all cars	153h	4h	0h	4h	2h
per car, mean value	17h	0h 30m	0h	0h 30m	0h 10m
no of times		25 times	1 time	25 times	10 times

[1] The figures are rounded off and are based on the nine cars that were started up, as well as those completed during the diary week. Thus, the two cars finished from the week before and the two cars only started up on Friday are not included in the calculation

The figures in table 4.7.2. reveal that, on average, every team member had one material problem and one meeting every day (25 times, 5 days and 5 individuals). It also shows that, on average, every team member was giving information to others, outside the team, twice during the week investigated (10 times and 5 individuals). Altogether, the five team members were interrupted in their work routines about 60 times that week. On average, each team member was interrupted 2,4 times every day (60 times, 5 days and 5 individuals). Each interruption implies that the intended work ceases and that thinking must be concentrated upon the immediate problem raised. When the problem is solved, one has to start the ordinary work again. It also takes some time to recall the activity performed before the interruption.

Therefore, a more far-reaching conclusion from the table is this: there are two types of losses caused by imperfections in the materials handling. There is one direct type of loss, equalling the time loss measured and, in addition, there is an indirect type of loss, brought about by those imperfections. The time lost because of indirect losses is difficult to measure, as assembly work starts again after an interuption and goes on as intended, *but at a slower pace,* as it takes time to speed up the work pace to the same level as before the interruption. This indicates that if there had been no interruptions, not only the time used for solving the problems would be eliminated (four hours), but also less time would have been spent on the assembly work itself.

4.7.4.4
Conclusions on the Team's Performance

Looking back upon the work performance of the team over a whole week, some obvious features show up:

1. The work tempo in the team appeared to be rythmic and it looked very much the same from day to day.
2. The division of labour within the team was stable. There was one pair of workers building one half of the car and a second pair of workers was building the other half. The fifth worker made most of the pre-assembly work tasks (for example, engine, instrument panel and doors). This division of the work tasks was the same throughout the week.
3. There was no "working up" tactics in the team in order to get a time-buffer. In general, speeding up work can result in a time-buffer which can be used for eliminating time-losses caused by various interruptions (this was what the other teams in the shop did), or for a Friday afternoon rest. Our team preferred to work at a stable pace, and to work comfortably throughout the work. This strategy, in general, probably yields better quality results than the "working up" strategy.
4. The work was interrupted by material problems more or less on every car. This means that there were frequent, but not very long-lasting, stops in assembly work. When such stops are added to the indirect losses due to material problems, however, a substantial problem may arise.

The organization in the Volvo Uddevalla factory was imported from the mother plant in Torslanda. Therefore, the product flow was, organizationally, cut off in the middle, forming one materials handlings department and one production (assembly) department.[15] Two deeply embedded inconveniences, due to this division into two departments, were revealed by the analysis made from the diaries of the team. All of a sudden, it became evident to a great many people that the traditional administrative partitioning counteracted the smooth functioning of the production process.

a. Who is responsible for the information flow between the two parts of the product flow? When the core activity (the entire product flow) of an enterprise is divided into parts, a major question about communication between the parts arises. For example, communication is needed when there is something missing in the materials kit or when a component is destroyed during the assembly. Now it is relevant to ask, on the one hand, what the materials handlers know about the final product. They know it not as a whole, but as its parts, and not as the parts related to each other but as the parts' identification numbers and where they are located in the store. The assemblers, on the other hand, know the product as the process of putting the components together and thus, see the relation between the parts. But for them, in their assembler role, the identification number of the parts is usually irrelevant. A genuine communication problem occurs when the assemblers are in a hurry, and when they use another language than the materials handlers. For this reason, it is of decisive importance to make the organizational borderlines in a reflective production system permeable for the workers in each of the departments concerned. This can be done by letting workers from each department learn from

[15] In the Autonova plant (see chapter 3.10 of this book), there will be a production department responsible for materials handling as well as for assembly. The team members will get their own materials.

the other's work. One good way to achieve this is to create an overlapping competence, so that they can communicate easily without the risk of misunderstanding.

b. The materials handling department and the assembly department respectively, in their every day assessments, did not use the same measure of internal quality.[16] In the assembly department, one internal (and indirect) measure of performance quality was time use for assembly work. In the materials handling department, quality was measured by the number of incorrect details and components delivered to the assembly department. This measure may indeed indicate a very good quality performance internally, in the materials department itself; but even one single critical component wrongly delivered can cause severe problems (read time-consuming problems) in the assembly department. Thus, the consequences for the assembly department should be taken into consideration when the materials department is measuring its own quality.[17] The number of wrongly delivered components is not a good enough measure for this purpose.

The team members were convinced that their suggestions for improvements would be met with interest and respect if they could base their arguments for improvement on the method we had worked out together, and which has been briefly presented in this paper.

▶

Fig. 4.7.2. Monday 24 Febr. 1992, the team at work. The five team members performing their work tasks are illustrated by lines (differently shaped for the different team members, see legend down to the right). Lunch and coffee breaks are indicated by shadowed areas. Work tasks performed by the workers are: *Body* = assembly on the car, *Pre ass* = pre-assembly on the body, *Engine* = pre-assembly engine, *Instr panel* = pre-assembly instrument panel, *Doors* = pre-assembly doors. Interruptions indicated by the workers are: *Matr probl* = material losses, *Mat stands* = moving and fetching material stands, *Proj* = project group getting information from the team, *Meeting* = meeting, *Break* = lunch and coffee break

---

[16] The official quality measuring was performed when the car was completed at the quality audit. Of course, the quality measures used for materials handling and for assembly in this audit were the same.

[17] The effect is indirectly supposed to be the same whether the wrong gearbox or an incorrect screw is delivered. Of course, the gearbox causes more problems for the assemblers: they will have to wait for the correct gearbox to be delivered from the materials handling department (which means that they lose as much time as the number of team members multiplied by the time they have to wait); or they will borrow a gearbox from another kit of components meant for another car. Both strategies to overcome the wrongly delivered components yield the risk of generating new difficulties and quality problems. If the wrong screw is put in a kit for one car, it is much easier to go on without far-reaching consequences for the team members: reserve quantities of frequently used screws are often put in boxes near the assembly place, as screws are often sometimes lost or destroyed during the assembly.

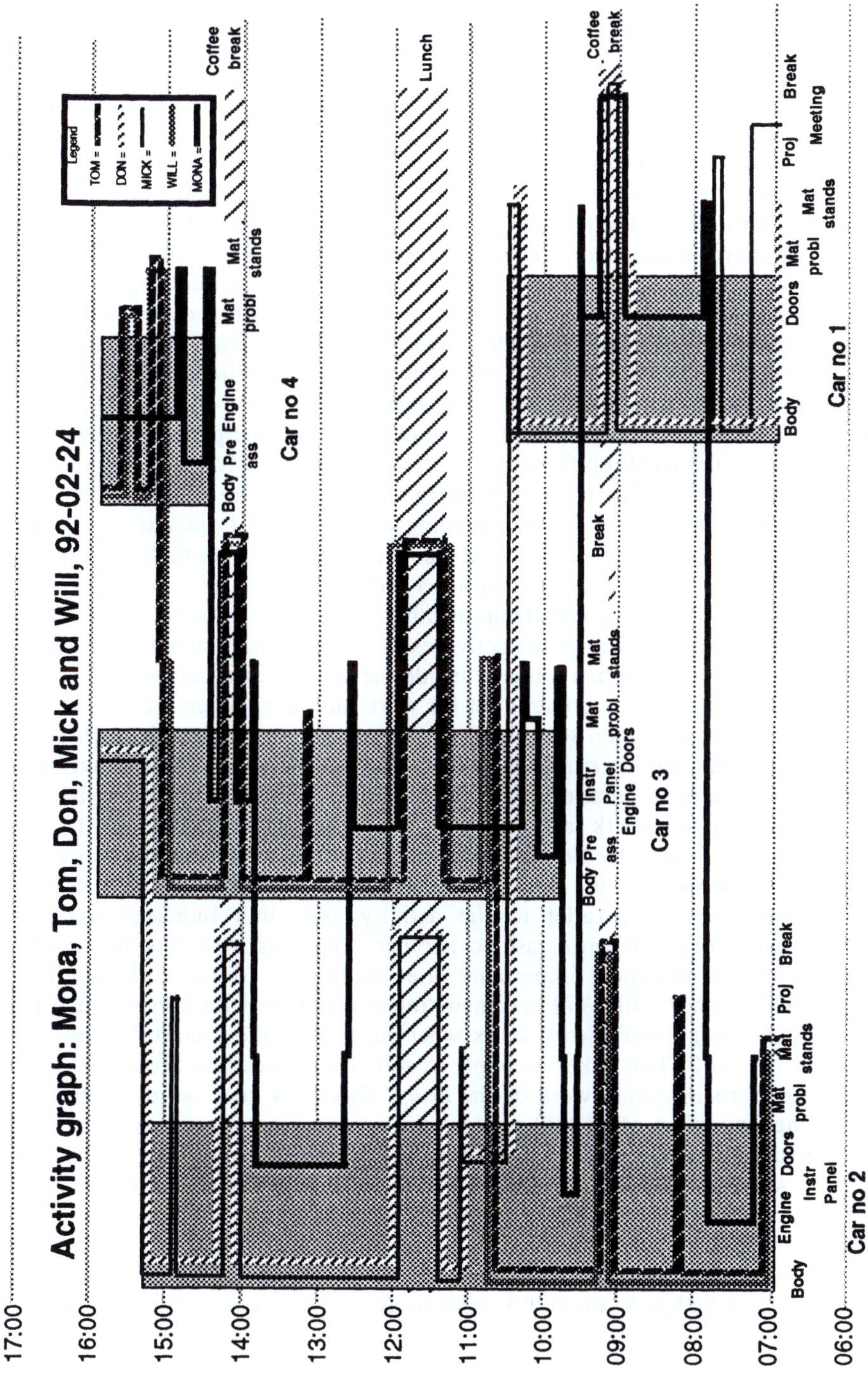
Activity graph: Mona, Tom, Don, Mick and Will, 92-02-24
Legend
TOM =
DON =
MICK =
WILL =
MONA =
Body
Engine
Instr
Panel
Doors
Mat
probl
Mat
stands
Proj
Break
Car no 2
Body Pre
ass
Engine
Instr
Panel
Doors
Mat
probl
Mat
stands
Break
Car no 3
Body
Doors
Mat
probl
Mat
stands
Proj
Meeting
Break
Car no 1
Body Pre
ass
Engine
Mat
probl
Mat
stands
Car no 4
Coffee
break
Lunch
Coffee
break
17:00
16:00
15:00
14:00
13:00
12:00
11:00
10:00
09:00
08:00
07:00
06:00

4.7.4.5
Further Development - A General Tool for Worker-Generated Improvements

Both the team and the management of the factory wanted a computer program to be developed, basically on the principles of the diagram and table just mentioned. In that way, the members of the team would be able to use the method themselves: they could make diary based, context-related time-use studies on their own work in the team. Such descriptions could then be used for analytical purposes, resulting in better control over their own work and work time. Thus, the results could become a common platform for the team's reflections about improvements in production-related topics. The computer model was developed between 1992 and 1993. Unfortunately, the Uddevalla factory was closed down in spring 1993, so the team could never use the computer model, though they did get help from the actual process during which the method was developed.

In figure 4.7.2., a time-use diagram, drawn with the help of the computer model, illustrates the work performance in the team for the Monday in the diary-week in February 1992. The cars which were worked upon are illustrated by rectangles: two cars were started up the preceding week (car no. 1 and car no. 2), and they are completed on the Monday, and 2 cars were started up on the Monday and left to be completed on the Tuesday (car no. 3 and car no. 4).

The work performance by each team member, as explained above, is illustrated by a single line, which illustrates the activities of one person by means of a trajectory (see also fig 4.7.2.). When a line is inside the symbol of a car, work is being performed according to the standards set for work on a car. When the line is outside the symbol of a car, then some kind of interruption has occurred. Coffee and lunch breaks are planned interruptions, and are marked separately in the diagram.

When a trajectory is parallel to the time axis, the worker, illustrated by this trajectory, performs the work task indicated below the symbol of the car, for example, work tasks related to the *Body*. The length of the section of the trajectory which indicates the work on the *Body* tells us the time necessary for this work task. When the trajectory is parallel to the activity axis, this indicates that the worker changes from one work task to another. The length of this movement, however, is of no analytical importance. Let us follow Tom's workday (Tom usually works together with Will and their trajectories hence look very much the same). This morning Tom's work starts when he fetches the material stands for his work on car no 2. This takes a couple of minutes, before and after 7.00 a.m.. After this, he starts assembly work on the *Body*. This work goes on, uninterrupted for about one hour. Just after 8.00 a.m. he starts talking to someone in a project group, preparing a new work place. Then his trajectory leaves the symbol of the car and stops just above the task *Proj*, where it stays for about 10 minutes. After finishing this discussion, he continues working on the *Body* again until 9.00 a.m., time for his coffee break. The coffee break is indicated by the changed direction of his trajectory, it leaves the symbol of the car and stops just above the *Body* of car no. 2 and goes on with it until he is ready; this occurs at 10.40 a.m.

Now, Tom leaves car no. 2 and turns to car no 3. Before he can start assembly work, he has to bring the material stands into the work station. This is indicated in the figure by his trajectory going to *Mat stands* for some minutes and after this, the trajectory goes into the symbol of the car and the assembly activities on the *Body* begin. Tom carries out assembly work on the *Body* until his lunch break at 11.20 a.m.. After lunch, he continues assembling the *Body*, but at 1.05 p.m. he is disturbed by a material problem. This is illustrated by the movement of the trajectory from *Body* to *Mat probl.* Tom solves his material problem and goes back to the *Body* assembly at 1.15 p.m. He carries on with assembly work on the *Body* until his afternoon coffee break, when he leaves for coffee for about 15 minutes. After the coffee break, he carries out final assembly work on car no. 3, he is finished at 3 p.m., after which he leaves the empty material stands outside the work station. He goes from car no. 3 to car no. 4, taking the material stands for car no. 4 and starts his assembly tasks on the *Body*. but just fifteen minutes later, he has a material problem, and he has to start solving it (his trajectory leaves the inside of the car symbol and enters the space where *Mat probl* is indicated). this material problem lasts for about 15 minutes, and after this, Tom continues his assembly tasks on the *Body*, until the work day is over, at about 3.50 p.m.

Besides drawing diagrams, the computer program can calculate the exact time used for the work tasks and the interruptions. Thus, it can also show the total time used for work tasks, and the total time needed to cope with interruptions during the work on each car (compare with table 4.7.2.).

I presented the computer program to the team many times during the development process. The team found the computer-drawn illustrations of their work useful for describing the work week and for reflections about it. They considered it as a manageable tool for formulating their own arguments for changes in production. For them, the program was a device which helped them visualize the work process itself. Now, they had a clear and definite picture of the work itself, and could point at the diagrams when arguing for changes. Previously, the strongest argument was to count the number of cars produced (this was the reason for discovering the disparity problem when the goal was fixed at 12 cars a week, whereas the result was 11).

4.7.5
General Conclusions

In most work places, workers are interested in improving their work and work performance. But there are only a few plants where workers themselves have adequate tools to argue effectively for the improvements they suggest. In this paper, I have presented such a tool, a method for detecting problems related to the work process (disparity problems) and for analysing the circumstances under which these problems can be solved. My main objective has been to develop a method that makes it easier for workers to argue for improvements, a method suited for describing the work process. That is why I have called it a holistic method, concentrating on the sequential order of the work process. This kind of analysis points out when and where an interruption occurs in the process, and which immediate consequences each disturbance generates.

In the paper, it has been clearly shown that a substantial number of problems were detected directly by the illustrations, and the analysis also made it possible to detect systematic underlying problems.

One example of an underlying problem is embedded in firms' traditions: there is a flow of materials and products moving through the plant. The flow is constantly changing (every day the workers perceive a stream of new products, new sets/kits of materials) and therefore, the teams constantly perceive the time dimension. The upper levels of the organizational hierarchy, above the team level, are at a greater distance from the product flow. Therefore, they may not perceive the flow in the same way as the workers do. The organizational scheme in a company is comparatively static. Tradition - or inertia - is one reason for this. Administrative partitions set up on the shop floor level in a company also tend to persist - they live a life of their own. Ususally, the same partitions are set up in new plants belonging to the same company. One reason for this organizational cementation is that the top management of the company wants to be able to compare the different production units. The plant management's view on the partitions to be set up has, of course, been conditioned by the existing organizational set-up in the old plants.

The general conclusion from this paper is that when people take a keen interest in their work and its development and when they get tools to handle their disparity problems, the result from their efforts can lead to unexpected and far-reaching, positive final results.

4.7.6 References

1 Ellegård K (1989) Akrobatik i tidens väv. Choros 1989:2, Dept of Human and Economic Geography, Gothenburg University

2 Ellegård K, Engström T, Johansson B, Medbo L, Nilsson L (1992) Reflektiv Produktion, industriell verksamhet i förändring. AB Volvo

3 Ellegård K, Engström T, Nilsson L (1994) Actes du Gerpisa 9, March 1994, in the section: La Réforme du travail industriel - principes et réalités de la planification de l'usine de montage d'automobiles Volvo à Uddevalla (in English in Ellegård K, Engström T, Nilsson L (1990) Reforming Industrial Work, Principles and realities in the planning of Volvo's car assembly plant in Uddevalla. The Swedish Work Life Environment Fund, Stockholm)

4 Hägerstrand T (1985) Time-geography: focus on the corporeality of man, society and environment. Reprint from: The Science and Praxis of Complexity. The United Nations University: 193-216

5 Sloan Management Review (1994) Winter and Spring issues. Massachusetts Institute of Technology

6 Wild K (1975) On the Selection of Mass Production Systems. The International Journal Production Research Vol 5 No 5: 443-461

4.8 Advanced Automation or Alternative Production Design? A Reflection on the new Japanese Assembly Plants and the Alternative Approach of Volvo Uddevalla

C. Berggren

4.8.1 Introduction

For decades, increased levels of automation have been regarded as the logical and necessary solution to the dual problem of monotonous work and stagnating productivity in car assembly. Whereas the automation drive among Western auto producers in the 1980s largely failed, a new generation of plants in the Japanese industry in the early 1990s promised to be more successful. In this chapter, it is argued that, even in Japan, automated assembly has proven to be highly dependent on full capacity utilization and hence an excessively costly way of improving human working conditions. The Japanese high-tech approach is confronted with the different approach of the Volvo Uddevalla plant, and its deliberately "low-tech" manual assembly system. The author demonstrates that this model was a particularly flexible way of assembling and delivering customized vehicles to individual customers. It is interesting to note that, during the past few years, there has been an "automation backlash" in most of the car industry and a return to more simple manual assembly systems. The radical parallelization and customer orientation of the Uddevalla system still constitute highly innovative features, however.

4.8.2 The Automation Drive of the 1980s and the new Japanese Wave

From an engineer's perspective, all human work is residual, waiting for automation in the next technological wave. In the car industry, the machining sections were mechanized in the 1950s and body shops and stamping lines were automated during the 1970s. The next logical step was to automate the final assembly process. In the early 1980s, GM and Volkswagen took a great leap forward in the trim and final departments. FIAT soon followed suit. Volkswagen's automated assembly lines in Halle 54 at Wolfsburg became world-famous, and figured prominently as the new trend-setter among industrial sociologists [5]. America's GM soon experienced "a rude awakening", however, and the advanced automation plans for its early Saturn project were abandoned [6]. In terms of profitability,

European carmakers did not fare well either. In order to be economically viable, automated plants had to operate very close to full capacity. In the crises of the early 1990s, this turned out to be extremely difficult.

Why did the Western producers not succeed in automating final assembly in an economically sound way? Was it because of flaws in the basic, deterministic engineering perspective, or because of weaknesses in preparation and execution? Jürgens, Malsch and Dohse identified the risks and problems of the German automation drive to be, for instance, insufficient preventive maintenance, the "exploding variety of parts and specifications" and lack of coordination between product engineering and manufacturing [4: 353-362]. The authors of the well-known MIT study *The Machine that Changed the World* [7] maintained that Japanese lean production, with its emphasis on standardized operations and an integrated problem-solving approach, would be more successful and hence open the doors for large-scale automation: "...once lean production principles are fully instituted, companies will be able to move rapidly in the 1990s to automate most of the remaining repetitive tasks in auto assembly - and more. Thus by the end of the century we expect that lean-assembly plants will be populated almost entirely by highly skilled problem solvers..."

Lean production would do it differently, because the starting point was different - a thoroughly streamlined assembly process and great precision in parts manufacture. Add to that the inexpensive capital in Japan in the late 1980s, the incessant introduction of new products and the high investment in new plants and equipment. In Japan, at last, it seemed reasonable to expect a successful foray into the land of automated assembly. From this perspective, it is interesting to present its vintage of automated car plants of the early 1990s in some detail. Three manufacturers and their plants are discussed below: Mitsubishi Motors and its Mizushima plant, Nissan Kyushu and Toyota Kyushu. Economic justifications and issues of capacity utilization are emphasized.[1]

Mitsubishi Motors was one of the few automakers in Japan which managed to make a profit during the 1992-1995 recession. One factor is successful and prudent product policy, another is the company's heavy emphasis on design for manufacture. By contrast to other Japanese automakers, Mitsubishi did not add new plants during the Bubble Boom. As a result, its main Mizushima plant has been a very well utilized site. In 1993, the plant produced 700,000 fully assembled cars and 250,000 knocked-down sets. Without constructing any new facilities, Mitsubishi Motors did bring in new technology in final assembly, however. The most automated process is the high-volume Mirage/Lancer line, where the automation degree totaled 19%. In order to justify investment, downtime must be virtually eliminated. In 1993, the target for the maximum downtime per shift was *60 seconds*. The actual level was close to the target: 70 seconds. In reality, this implied the abolishment of the traditional andon-system (the chords used by operators to stop the line). In spite of the high capacity utilization at Mizushima, the automation of the Mirage-line was not very profitable. Many of the robots had

[1] The following section is primarily based on plant visits in Aishi and Kyushu in October and November 1993, supplemented by interviews in September 1995, with Japan Autoworkers and managers in several Toyota Group companies.

not been introduced in order to save costs, but rather to eliminate dirty and heavy operations.

In contrast to Mitsubishi Motors, both Nissan and Toyota inaugurated completely new plants in the early 1990s. *Nissan* started operations on the southwestern island of Kyushu in the 1970s. The second assembly plant, called *assembly pavilion*, came on stream in 1992 and has a capacity of 250,000 cars per year. The plant is officially designated as *Nissan's Human Land*. Its philosophy is based on the three goals of high technology, human orientation and protection of the environment. In 1993, the degree of automation in final assembly was 20%, and the goal was to increase automation to as much as 50%. Assembly automation is costly, however. The new body and paint shops were 20-30% more expensive to build than conventional operations with the same capacity. The assembly pavilion was 2.5 times more expensive than an equivalent conventional plant. These high fixed costs meant that operations were very susceptible to variations in capacity utilization. During the long post-Bubble recession, the plant operated at only half its capacity. Capital may be less expensive in Japan than in Europe, but the basic problem is the same: high automation means high fixed costs, and a heavy dependence on strong sales. In times of recession, robots cannot be sent to training classes or dealer outlets.

Toyota Motor Corporation has often been seen to be more conservative and cautious than other Japanese car makers. The legacy of Ohno Taiichi meant a strong focus on low-cost, incremental rationalization. In the late 1980s, Toyota also suffered from increasing labor problems, and its production engineers feared that the company was falling behind the technological development of the Western automakers. In Tahara, at a psychologically important distance from Toyota city, an ambitious modernization program was launched, resulting in the construction of very advanced body and assembly shops. Inaugurated in 1991, the Tahara No. 4 assembly plant combined high automation with a system of buffered line sections in order to improve on-line quality control as well as working conditions. In the 1993 recession year, only half the capacity was utilized and here, too, ambitious automation turned out to be excessively costly.

Toyota's new *Kyushu* plant builds on the experiences of Tahara, but has scaled back the automation level. (In contrast to Mitsubishi and Nissan, Toyota does not disclose any figures of automation degrees, so exact comparisons are not possible). The first shift at Kyushu commenced production in 1992, a second shift was added in 1993. The total annual capacity of the plant corresponds to 200,000 cars. This target has not been reached yet; instead Kyushu workers had to be sent to the Japanese mainland on *shukko* in 1994. Toyota Kyushu is separately incorporated and strongly committed to developing a new way of manufacturing cars, not just copying the mainland Toyota methods of the 1980s. Attractive working conditions figure prominently among its goals. A new management style is introduced, putting much more emphasis on voluntary worker involvement than the compulsory approach of Toyota Aichi. As a result of its reduced level of automation, construction costs of the Kyushu plant were lower than at Tahara - but certainly higher than for a conventional plant of the Motomachi type. Consequently, the Kyushu plant is dependent on high capacity utilization in order to break even. Sophisticated equipment in final assembly is just a minor part of the increased fixed costs, but it does not make it easier to achieve the profitability task.

Now there is another motive for assembly automation at Nissan, Mitsubishi Motors and Toyota. According to company spokesmen, robots were introduced in order to eliminate strenuous and repetitive operations, such as fitting front seats, mounting wheels or installing batteries. The Nissan union emphasized that substituting robots for men in these areas was an important union demand when the new plant was planned. However, if improving working conditions is the main motive for assembly automation, then it could be worthwhile to give a thought to other options, including the Swedish experience of radical alternatives to assembly line work. In the lower-volume segments of vehicle production, this approach seems to be a cost-effective way of humanizing assembly work and at the same time closing the gap between factory operations and individual customer requirements.

4.8.3 The Uddevalla Option: Skill-Based Manual Assembly and Increased Customization

The Swedish car industry of the mid-1980s faced the same difficulties in recruiting a stable workforce as their Japanese counterparts experienced a few years later in the same decade. When sales and production volumes increased and planning for new plant capacity started, Volvo engineers first tried to solve the personnel and productivity problems by rapid automation. Inspired by GM and VW, the central staff of manufacturing engineering at Volvo produced elaborate, long lists of items suitable for the introduction of robotized operations. Most of the experiments and try-outs turned out to be excessively costly, however. When planning for a new assembly plant at Uddevalla started in 1985, reports of the problems at GM's and Volkswagen's automated plants had started to circulate. After a year of controversy, planning and replanning, the members of the Uddevalla project team determined that they could not wait for the technologies of an unspecified future to solve the problems of the present. If the improvement of working conditions on the assembly line is a basic issue, there are two basic alternatives: to *reduce* this work as much as possible by the use of robots and assembly machines; or to *change* the nature of work by means of different techno-logy, design and work organization. In short, the alternatives could be phrased as liberation from work, or liberation of work. Uddevalla chose the second strategy.

The rejection of the traditional mechanization perspective made it mentally and materially possible to focus on the development of manual assembly procedures in their own terms. (For details about the implementation, see chapter 3.9 by Kajsa Ellegård). Taking advantage of its highly flexible assembly process and skilled teams, the Uddevalla plant also planned downstream integration. Most Volvo cars are sold with a lot of "extras", such as radios, telephones, tow hooks, etc., which are installed by dealers. This helps dealers keep their workshops busy and gives them an additional source of revenue. To customers, this division of labour between plants and dealers means high costs, less reliable quality and additional delay. It is, for instance, much more efficient to mount a tow hook, including the necessary cable connections, while the car is being assembled, than doing this at a later stage when several components first have to be dismantled. Uddevalla could install the tow hook in a third of the time which it took the dealers, and the reduction in lead time was equally impressive. Beginning in mid-1993, Uddevalla's market and delivery planners planned to integrate almost all of these traditional dealer operations into the factory process in order to produce cars that could be delivered directly to the customers.

In order for small-scale assembly to be economically feasible, the expensive, dedicated machines of traditional high-volume lines had to be replaced with simple and cost-effective technology. An important aspect at Uddevalla was to improve assembly ergonomics at the same time.

This was closely related to the fact that 40% of the assembly workers were women. In order to avoid an increased rate of occupational accidents, radical improvements in methods, tools and working postures were needed. A case in point was the new type of tilting equipment introduced in order to permit working in an upright position. The technical demands on these tilters were strict: rapid handling, vertical adjustability, no protruding parts, good stability. At the same time, the price should not be too high, because each of the 40 assembly teams had to be equipped with two tilters. In a spirit akin to the famous Ohno tradition, Uddevalla hence promoted a number of low-cost development projects. With regard to the assembly of windows, the idea at the start of planning was for each shop to have the same type of automated facility as the equipment installed at Volvo's main plant in Gothenburg. The final solution was to fit the rear window at a central station in the materials shop. A robot applied the adhesive tape to the window, which was subsequently fixed manually to the car body. The windshield was mounted by the assembly teams using simple mobile presses. The total cost for the equipment was just a tenth of that at the Gothenburg line plant.

In 1992, after just three years of operation, Volvo decided to close its two Swedish branch assembly plants, Kalmar and Uddevalla. All Swedish assembly operations were concentrated at Volvo's main assembly plant in Gothenburg. Does this mean that the new assembly concepts were not viable? No. At the time of its closure, the Kalmar plant was the best-performing of Volvo's Swedish assembly operations, the quality of its cars approaching the world leading Lexus level. Uddevalla had a slow start, but then displayed a most impressive learning curve. When Volvo decided to close the Kalmar and Uddevalla plants, it was after a total collapse of the Swedish automobile market. In just fours years, registrations plummeted from 340,000 to 150,000 in 1992. That year, Kalmar and Uddevalla produced less than 30,000 cars each and the Gothenburg plant was operating at less than 50% of its capacity. These sales problems were partly due to a deep recession in Sweden, but partly also the consequence of Volvo's highly inefficient process of product development. In the late 1980s, when sales of Volvo's mainstays, the 200 and 700 series, started to slide the company had no new model to offer to the market. Uddevalla was a brand-new plant, but had to build an outdated car. This situation sealed its fate. Nevertheless, the performance development at Uddevalla during the three years when it sought to catch up with most efficient European mass production plants is most interesting. This is indicated by the following four parameters.

1) *Quality:* When J D Power released statistics for the 1993 model year, almost all automakers had improved, but progress at Volvo was more rapid than the average. The 940 model improved from 132 to 87 complaints per 100 cars, making it the best European import. Within Volvo, the two small-scale plants, Uddevalla and Kalmar, improved most of all and had a clear lead in comparison to the Gothenburg plant.
2) *Productivity:* Already in mid-1991, Uddevalla equalled Volvo's main Gothenburg plant - which was not very efficient at this time, however. Inspired by Japanese methods, new managers in Gothenburg launched a comprehensive productivity program, but Uddevalla also put in a second gear. In the month before the shut-down decision, Uddevalla had a clear edge in worker hours per car. From 1990 to 1992, assembly time per car was reduced by one hour per month on average. According to adjusted figures calculated by Paul Adler and John Paul MacDuffie, Uddevalla needed 25.9 hours per car in 1992, compared to an average of 30.8 for European luxury producers as reported in 1989. A further reduction to 17.1 hours per car was projected in 1993 [1: 46]. For special vehicles, such as police cars, Uddevalla had a productivity advantage of 2:1 compared to Gothenburg, where such cars were first assembled on the main line, then partly disassembled and refitted in a special workshop.
3) *High flexibility* was a third import advantage of Uddevalla's advanced worker qualification. One indication is the low effort needed to break in model changes. The annual model changes in 1990, 1991 and 1992 were introduced with 25% lower tooling cost per car compared with the Gothenburg plant. In spite of its two hour-work cycles, the cost for training and information was 60% lower per car than in Gothenburg. Following the model change, the skilled teams at Uddevalla returned to normal productivity in half the time needed at the assembly line.
4) *Thorough customer orientation:* Traditionally, most Volvo cars are assembled according to a central scheduling system. If customers wanted a combination of options, which was not preplanned in this master schedule, they had to wait an average of two months for delivery. Uddevalla changed these rules. The plant started to assemble all cars for the European market on individual customer orders only. Uddevalla planners told dealers they would guarantee delivery within four weeks of any custom-ordered car. In one year, Uddevalla reduced the total lead time by half and planned a further 50% cut in 1993. The customer orientation at Uddevalla built on the strength of its flexible production system. In order to remain productive, the many parallel teams and materials handlers did not need any specific sequencing of cars with different option contents (for example, every second car a turbo, every third an automatic transmission, every fourth a 16-valve engine). Moreover, the introduction of customer order-planning provided an additional motivational advantage for the teams. Now the teams knew that the cars were not to be stored in a warehouse somewhere, but delivered directly to individual customers. The plant took a pride in taking on difficult requests.

In May 1993, Uddevalla was closed down. Only 18 months later Volvo announced a reopening of the plant. This time, the commercial context was entirely different, however. The reopened plant is operated by a joint venture where the British racing firm TWR Engineering controls 51%. It is devoted to the flexible production of exclusive niche vehicles in small series. The first to be launched in 1997 are a coupe and a convertible. The maximum volume will be lower than the previous 40,000 cars per year, but the factory will now be complete, including the body and paint shops. *Uddevalla I* challenged the parameters of mass production, whether American or Japanese. The challenge meant that Uddevalla had to match the productivity of mass production, which created enormous pressure on the plant to implement constant efficiency improvements. Had it been allowed to continue, the potential consequences might have been far-reaching. *Uddevalla II*, as well as the similar systems of low-volume assembly of special cars, is confined to a niche business, clearly separated from Volvo's high-volume units. Even if the new Uddevalla is a success, it will hardly influence Volvo's mass production system. There has been much more diffusion within Volvo Trucks, which is the world's second largest producer of heavy trucks. The Truck business is characterized by a much broader scope of functional variation and customer adaptation than the car division. Consequently, the *Uddevalla principles* were adopted as an integral part of the assembly system of Volvo Trucks in Gothenburg.

4.8.4
Europe in the 1990s – A Return to Manual Assembly

An important motivation for assembly automation in Japan in the late 1980s and early 1990s was the need to reduce and improve difficult and potentially damaging manual operations.

At Uddevalla, the method for eliminating such operations, for example installing front seats or batteries several hundred times a day, was to introduce a radical parallel design, where each team installed seats only four times a day, so that advanced automated equipment was not necessary. This solution involved more worker hours per car than the Nissan Kyushu approach, but less expensive equipment and less fixed costs. In Japan, where automakers, notwithstanding the current recession, still fear labor shortages as a long-term problem, a high degree of automation may still be the right way to go. In Europe, however, there has recently been a clear break with the previous automation perspective. All over Europe, automakers are not only putting a stop to further investments in assembly automation, in several cases, they are even reducing the automation level already reached in the late 1980s. At Poissy, Peugeot has replaced its automated glazer line with manual stations. A few years ago, dashboards were fitted by robots, but in 1994, the robots were taken away and replaced by men.[2] In the same way, Saab of Sweden will reduce its automation level in body assembly when the successor of the current 9000 model is introduced, following the trend set by GM Europe

[2] Visit at Peugeot Poissy, April 22, 1994.

(which effectively controls Saab). At its main plant, Opel in Rüsselsheim also recently reduced the automation level. In an interview with the journal of Swedish engineers, Opel managers explained the new credo: "We are back to basics. Now we try to keep the automation degree of the technical equipment at the lowest level possible".[3]

The same trend is operative in Japan. In the 1980s, the degree of factory automation increased for each new model launched - in Japan as well as in Europe. After the burst of the Bubble Boom, this trend reversed. When Toyota launched its new RAV4 recreational vehicle at the refurbished Motomachi plant in 1994, the level of automation was significantly lower than at Tahara. The automation degree in Motomachi's body shop, for example, is only 70%, compared to 98% at Tahara. One reason is that robots did not save as many workers as expected. Another is that it is easier to adapt to volume changes in plants with lower automation, and that less advanced equipment means a much lower break-even level.

The current trend by European and Japanese automakers vindicates the basic skepticism of Uddevalla engineers concerning automation as a solution to assembly problems. Unfortunately, increasing European distrust of technological solutions means a return to manual assembly lines. On-line problem-solving and quality inspection are emphasized, but at the same time assembly work is intensified. There are few attempts to develop alternative designs, and the high level of unemployment implies that there is no real labor market pressure to upgrade job content and working conditions. Uddevalla combined low-cost technology and innovation with a genuine reskilling of assembly work. Its flexible system for producing and delivering customized vehicles was adopted by Volvo Trucks. It deserves the interest of carmakers who are exploring the possibilities of building and delivering customized cars with minimum lead times and highly compressed learning loops.

Acknowledgement
The Japanese fieldwork for this chapter was conducted during a stay in 1993 as a visiting scholar at the Okayama University, financed by the Japan Society for Promotion of Science. I am deeply indebted to Professor Nomura for his extraordinary generosity in sharing his time and knowledge with me.

4.8.5 References

1 Adler P, Cole R (1994) Rejoinder. Sloan Management Review Spring: 45-49
2 Berggren C (1992) Alternatives to Lean Production. Work Organization in the Swedish Auto Industry. Cornell ILR Press, Ithaca
3 Berggren C (1993) Mästarprestationer eller mardrömsfabriker? (*Excellence or disaster?* An evaluation of Volvo's assembly plants in Uddevalla and Kalmar) Arkiv 56-57: 31-78

[3] Ny Teknik (1994): 36.

4 Jürgens U, Malsch T, Dohse K (1993) Breaking from Taylorism. Cambridge University Press, Cambridge
5 Kern H, Schumann M (1986) Das Ende der Arbeitsteilung? Verlag C. H. Beck, München
6 Keller M (1989) Rude Awakening. William Morrow, New York
7 Womack J P, Jones D T, Roos D (1990) The Machine that Changed the World. Rawson Associates, New York

4.9 A Misguided Trajectory? Automatically Guided Vehicles in Auto Assembly

K. Mishina[1]

4.9.1 Introduction

Automatically guided vehicles (AGVs) arrived in the auto industry at the beginning of the 1980s as a functional alternative to the assembly line. For good reasons, their arrival was met with high hopes from all concerned. When Henry Ford installed the assembly line, it was a blessing to a large number of unskilled workers who secured a high-paying job and saw their standard of living rise considerably.[2] The assembly line, however, soon alienated workers on the shop floor and encountered unusually persistent resistance. Workers first sought union representation successfully, and expressed their disapproval of the working conditions under the assembly line through formal grievances and strikes. Their deep-seated frustration also manifested itself in dismal product quality, absenteeism, turnovers, walkouts, alcoholism, drug abuse, and sabotage. These problems were alarming enough in their own right, but they began to disrupt production so seriously that auto companies were forced to seek a real solution in the 1970s. AGVs arrived against this backdrop with a promise to relieve workers from inherent properties of the assembly line, to which all evils were commonly ascribed, and, thereby, release auto companies from the intolerable burden of the labor problems.[3]

Did AGVs fulfill their promise? To what extent did they in fact improve the working conditions in auto assembly? These questions motivate the present contribution which revisits the AGVs introduced to Plant X, an auto assembly plant in Europe, a little more than a decade ago and examines their net effects through plant observations, interviews, and a large-scale survey questionnaire involving all workers. The findings are obviously of great importance to whomever is interested in AGVs and their future role, if any, in the auto industry. The data, however, speaks to a much broader audience. The investigation of AGVs required that new light be shed on the assembly line, to which AGVs are an alternative, and resulted in a serious doubt about the literature that does not distinguish what the assembly

[1] The field study underlying this manuscript was funded by the Division of Research, Graduate School of Business Administration, Harvard University. Its support is gratefully acknowledged, as is the hospitality of Plant X, the research site for the study.

[2] See [6] for historical details.

[3] There is a long list of books and articles that are dedicated to the subject of so-called "assembly line blues." Representative examples include [2,3,7,8].

line is and what it is made out to be by poor management. This study is therefore about management failures and false convictions as much as it is about AGVs.

The rest of the article is organized as follows. The next section begins with a general discussion of AGVs: what they are, how they work, and why they are used in the auto industry. It then reviews two attempts at defining a trajectory moving away from the assembly line and characterizes AGVs as an intermediate approach between the two. The third section introduces the research site, Plant X, and describes its AGV installations along with the design of the field research. Following these preparations, I will present major findings that are consistent with the expectations placed upon AGVs in section four, and findings that are not consistent with the expectations in section five. The sixth section returns to a basic question that was raised by the field research at Plant X: what on earth is really wrong with the assembly line? The answer presented here explains why AGVs worked effectively at Plant X in certain dimensions and, at the same time, refutes popular beliefs about the assembly line. Section 7 concludes the article.

4.9.2 AGVs as a Trajectory

Let me begin with an explanation of the defining features of AGVs. AGVs serve the same purpose as the conventional conveyor system in that both are carriers of the work pieces, such as engines and cars, being assembled. AGVs are unique, however, in that they permit fine control over individual work pieces. In the conveyor system, once work pieces are set in place, it is difficult to alter their sequence, interval or routing. Work pieces simply move through a set of predetermined stations at a predetermined point in time in the order in which they start at the beginning of the work area. Every work piece consequently visits an identical set of stations in the particular order in which they are arranged. With AGVs, any of these elements are variable. Work pieces can visit any stations, as many times as they need to, and in any desired order, while their routing can vary from work piece to work piece. The difference in the degree of freedom of control is rooted in the power source. The conventional conveyor system moves all work pieces with one motor - a centralized power source, whereas AGVs, to warrant the name "vehicle," are powered individually by their own dedicated battery and motor.

AGVs in practice are only as flexible as the process design calls for. They run on a system of non-obstructive rails, following the commands issued by a central computer with which they maintain two-way communication. AGVs, by design, cannot go where neither the rail system nor the computer program "guides" them to go. Once these process design elements are firmly fixed at the time of installation, AGVs are able to move flexibly so long as they stay inside this rigidly set boundary.

The process design in auto assembly does not call for much flexibility. AGVs are commonly used elsewhere to execute a jumbled process flow in which work pieces move back and forth freely amongst a limited number of general-purpose stations. This setup is ideal for custom production which requires a variety of process combinations. A jumbled process flow is also economical in low volume production because it avoids duplicative investment in expensive stations, such as flexible machining centers, by sending a work piece to the same station repeatedly at different stages of production.[4] In the world of mass-produced automobiles, a jumbled process flow is doubly unnecessary. Because of the highly standardized product design, demand for process variations is kept minimal. High production volume also justifies dedicated stations if they help simplify the process flow and speed it up. Auto assembly thus employs one-way process flows in which all work pieces pass through an essentially identical series of operations, severely limiting the room for the flexibility of sequencing and routing by design.

Why then are AGVs used at all in auto assembly? AGVs are expensive to install and maintain. Why pay for the freedom of control if it is not needed? The answer has to do with the flexibility of timing. In the sequential process flow, of which the assembly line is a classic example, interdependency governs all stations. Every station, performing a unique set of operations á la division of labor, depends on its upstream neighbor to feed a new work piece and its downstream neighbor to dislodge a finished work piece. It can be idled - either starved or blocked - when it is ready to do the work assigned to it unless its neighbors are ready as well. This interdependency creates a need for synchronization whereby all stations work to the same rhythm. The assembly line is extremely efficient, in particular, because it enforces synchronization to its logical extreme by building this feature into its unalterable mechanical setup. This very essence of the assembly line backfired on the labor front, however. It was singled out as "mechanical pacing", and was deemed the most undesirable feature of the work on the assembly line with such familiar expressions as "inexorably moving assembly line" and "the line never stops no matter what." AGVs demolish mechanical pacing in auto assembly because they are a stationary work platform and stay with a worker until he or she actively releases them. No doubt this is a trivial use of AGVs given what they are capable of doing. But one thing AGVs do well was sufficiently attractive to the auto industry given the labor problems it faced with the assembly line.

AGVs are employed in auto assembly in combination with a specific process flow. This point is best seen when they are compared with other trajectories leading away from the classic assembly line. One such trajectory is pursued by Toyota. It retains a sequential process flow and, as a matter of fact, most of the elements of the classic assembly line. Its presumption is that the assembly line works just fine most of the time and workers resent mechanical pacing only under abnormal circumstances, e.g. when they cannot fit the parts snugly, when they find something wrong with the car in front of them, and when they fall behind with too much work to do. All it takes then is to make the assembly line stoppable whenever these abnormal circumstances present themselves. Toyota brings this idea into reality with the Andon cord, a string that is always found within easy reach of workers and is connected to a toggle switch which, when turned on, stops the line at the boundary of two adjacent stations. This provision fundamentally converts

[4] Machining and semiconductor fabrication are two good examples in which AGVs implement a jumbled process flow effectively.

the nature of the assembly line even though it is not particularly visible. Another trajectory is pursued by Volvo.[5] It turns the assembly line upside down and adopts a parallel process flow whereby all stations are perfectly interchangeable. In each station, a small team of workers builds the complete car undertaking the whole set of necessary operations. This wholesale approach eliminates interdependency as well as the need to synchronize production altogether. It also solves another problem of the assembly line, repetitiveness, with long cycle times that are measured in hours as opposed to seconds. Interestingly, Toyota again retains a controversial feature of the assembly line: short cycle times. It instead addresses the issue of repetitiveness with enormous product variety, under which few workers repeat an identical task set on two consecutive cars, and, especially in overseas assembly plants, bi-hourly rotations.

AGVs combine Toyota's minimalist approach and Volvo's wholesale approach in a mixed process flow.[6] A mixed process flow comprises a sequence of islands, each of which contains a small number of parallel stations. AGVs are allowed to choose a station at each island opportunistically without committing themselves to a particular station in advance. Because of this feature, production is not seriously disrupted, as it would be in any sequential process flow, when a station is unexpectedly tied up with a troublesome work piece. AGVs simply avoid that station in favor of the other parallel stations with subsequent work pieces until the troubled station restores the normal condition. This aspect reduces the pressure applied on the workers who are given a new role of discharging every work piece with a press of a button when they are ready for the next piece. Workers therefore regain control over pacing. Yet, unlike in a parallel process flow, work content remains relatively simple so that workers do not have to remember an enormous number of operations. Table 4.9.1. summarizes the comparison of the three process flow types. The trajectory AGVs represent is an intermediate approach in between two other radically different approaches.[7]

4.9.3 Plant X and the Data

Plant X, the main research site for this article, is a full-scale auto assembly plant in Europe. It has a long history of prosperous operations and is considered a good plant by the management of the company. Although the plant is unionized, its labor relations have been cordial all along. It stands out in its pioneering use of

---

[5] See [1] for details.

[6] Note that AGVs are flexible enough to replicate any process flow in principle, but a more economical carrier, conveyor or dolly, just suffices for a sequential process flow and a parallel process flow.

[7] Jürgens, Malsch, and Dohse [5: 362-69] offer an alternative interpretation. They view AGVs as a dispensable tool for modular assembly which represents a trajectory leading to complete automation. They do acknowledge that modular assembly does not necessarily require AGVs, but do not discuss the role of AGVs in isolation from modular assembly. AGVs, taken alone, have no logical connection with the automation of assembly work.

Table 4.9.1. A comparison of three process flow types.

Type	*Sequential Process-Flow*	*Mixed Process-Flow*	*Parallel Process-Flow*
Schematic Representation		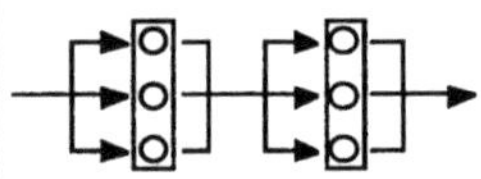	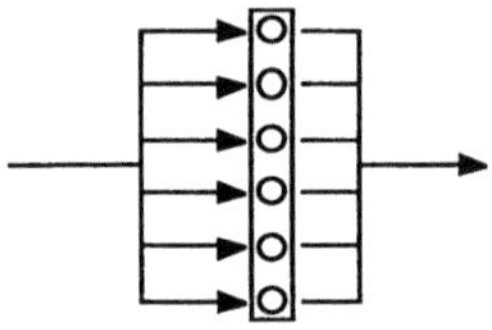
Example	Toyota Trajectory (Stoppable Assembly Line)	Plant X Trajectory (Call Button AGV System)	Volvo Trajectory (Team Build Cells)
Typical Carrier	Conveyor	AGV	Dolly
Pacing Mechanism	Keep moving until somebody pulls the Andon Cord	Stay stationary until the worker pushes the call button	Stay stationary until the team brings in the next work piece
Cycle of Repetition	Short cycle time, but many different cycles due to variety and rotation	Medium cycle time	Long cycle time

AGVs in certain operations, but it is otherwise indistinguishable from the majority of the auto plants that escape special media attention.

Plant X introduced AGVs in the early 1980s. The decision came naturally as part of a capital investment campaign called "modernization program" in which automation and new technology were emphasized above all else. The program was undoubtedly motivated by the arrival of new technologies, of which AGVs are one, and the relative absence of investment in the years that directly preceded it. Plant X, however, was confronted with another problem at the time the program was conceived. It experienced considerable difficulties in hiring direct workers in the 1970s to the point where it began to hire women for the first time in its history for the factory jobs. AGVs were thought to be a step in the right direction to reduce the dependence on direct workers in the long run, while making assembly jobs more attractive in the mean time.

AGVs were adopted in three areas of operations: engine dress-up, instrument panel (IP) subassembly, and door subassembly. The three AGV areas together account for 22% of all direct workers in the assembly shop. It is commonplace today to pre-assemble the IP, doors, and engine accessories more or less as an independent subassembly unit off the main assembly line with or without AGVs. In the early 1980s, however, these units were customarily assembled piece by piece directly onto the car on the main assembly line in virtually all auto plants. Plant X, among a few others, pioneered the shift toward modularization by altering the product design from the ground up. Having undergone one extensive overhaul, the AGV areas in Plant X continue to be one of the most advanced in the industry.

All three areas are organized as a mixed process flow with a few islands of several parallel stations. Complete with their own parts kiting operations, they occupy substantial floor space relative to their counterparts in other auto plants where modular subassembly is performed on a conventional assembly line – as much as ten times more in my own rough estimate. Consequently, the AGV areas resemble a miniature factory of their own, isolated from the main assembly line as well as from each other. They also provide workers with more work content than the main assembly line which, in Plant X, produces a car well under every minute. The cycle times – time allotted to a work piece in each station – are from four to eight times longer in the AGV areas, even though, several parallel stations in an island collectively make engine, IP, or door modules as fast as the main line builds cars. The difference, however, is somewhat overstated because many workers in the assembly shop are engaged in a practice in which they work on every other car, every three cars, etc. on the moving assembly line with correspondingly longer cycle times. With automation, at best, sporadic throughout the assembly shop, the contrast between the main assembly line and the AGV areas converges on the process flow.

The analysis that follows below relies heavily on data obtained by a questionnaire. This questionnaire was originally drafted by myself for the purpose of studying the determinants and the effects of worker morale in auto assembly. Although my main interest lay in discerning the extent to which plant performance depended, through worker morale, on detailed process design elements rather than more obvious managerial factors such as supervision, the result is equally instructive to the present inquiry concerning AGVs. What follows hence presents a relevant part of this much larger study, the full analysis of which goes well beyond the scope of this short article.

The survey was conducted with utmost care. It was preceded by a preliminary plant observation in the summer of 1994, and subsequently approved by the management and the union of Plant X. In November of the same year, the personnel department, on my behalf, distributed the questionnaire, with an optically readable answer sheet and a cover letter that explicitly spelled out its purpose, to all Plant X employees who were paid by the hour (non-salaried employees). The non-salaried employees were asked to answer the questionnaire in the privacy of their home, both anonymously and voluntarily. Their answer sheet was later collected in a box placed at an unattended plant gate for three weeks. The questionnaire had 50 short questions, and the pilot runs conducted by union representatives confirmed that it could be answered in 20 minutes. Of the several thousand non-salaried employees, 40.3% returned the answer sheet. These respondents represented all shops, departments, shifts, age groups, and genders roughly in proportion to Plant X's population. In addition to the technical check of sample biases, major findings were scrutinized with plant observations and interviews at all levels in the winter and the summer of 1995.

The questionnaire was designed for cross-sectional analysis. The data was therefore compiled by shift and by section (a section for direct workers corresponds to a segment of the assembly line or an AGV area), i.e., by the smallest meaningful organizational unit that contained an average of around 50 non-salaried employees.[8] This design allowed me to look at Plant X with any desired slice. It merely meant creating a larger group in which I was interested as a particular combination of the small units. In this article, two such groups are most pertinent: the AGV areas and the rest of the assembly shop. The ensuing analysis focuses on detecting statistically significant differences in the ways in which

---

[8] The questionnaire was distributed to all non-salaried employees, as opposed to relying on the sampling method, to ensure that this smallest unit contained a sufficient number of respondents.

these groups reacted to a relevant subset of the questions. Since analysis like this primarily draws inference from relative comparisons, it escapes usual criticism of questionnaire-based research, e.g., the influence of precise wording. In what follows, it simply does not matter how positively or negatively a question is answered as long as wording affects all groups symmetrically.

The individual questions were kept as simple as possible and as direct as possible. Of the 50 questions, 32 asked the respondents to express the degree to which they agreed or disagreed with a simple statement – my job is stressful, for example – on a five-point scale. Four asked a yes-or-no question with a third choice, not sure. Six presented a list of ten criteria – wages, for example – and asked the respondents to choose three top reasons why they chose to work for Plant X when they did and why they might quit Plant X if they ever would, respectively. Then, the ten criteria on the list were given a score based on a simple scheme – three points for the first choice, two points for the second choice, one point for the third choice, and zero for the rest – as if each respondent answered 20 questions. The remaining eight questions were used to define the profile of the respondent such as age and gender. To the best of my knowledge, confusion, if any, was confined to a few of these profile-related questions.

The world, nonetheless, is not perfect. The research suffered from a deficiency for the purpose of studying the effects of AGVs. Before I contacted Plant X for the first time, it had already decided to do away with AGVs in favor of a conventional conveyor system in the three subassembly areas in the near future.[9] When I arrived, the decision was public knowledge in Plant X. Therefore, I was asking the non-salaried employees in the AGV areas to evaluate the work they were doing with AGVs, knowing that they would be moving back to the assembly line. This knowledge potentially introduced a bias to the data, although it was difficult to predict in which direction the bias would distort the results of the analysis. Some people may have reasoned that there must be something wrong with AGVs if the plant was giving up on them. Others may have wanted to let it be known that they liked AGVs and resented the move back to the assembly line. It is unfortunate but the problem was beyond my control. I could only hope that careful interpretation would mitigate the problem wherever doubt prevailed.

4.9.4 Good News

The AGV areas demonstrated a statistically significant improvement over the rest of the assembly shop, most notably in the following arenas:

– higher overall satisfaction with Plant X (YN, 1%)[10]

---

[9] The decision, I was told, reflects a realignment of the company's manufacturing strategy rather than the plant's evaluation of AGVs. AGVs in Plant X were seen as neither a catastrophic disaster nor a spectacular success.

[10] The information provided in the parentheses indicates the question type (YN: yes-no questions, 5P: a five-point scale question, 10L: a 10-choice question) and the significance level (either 1% or 5%).

- higher job fulfillment (5P, 1%)
- less frustration with work content (10L, 1%)
- less frustration with work climate (10L, 5%)
- weaker conviction that production takes precedence over safety at Plant X (5P, 5%)
- higher satisfaction with the training provided (YN, 1%)
- better ergonomics (5P, 1%)
- better workability (5P, 1%)
- fewer safety hazards (5P, 1%)
- better defect reportability (5P, 1%)
- better feedback on reported defects (5P, 5%)
- better overall supervision (5P, 1%: all three related questions).

The response rate to the questionnaire was 38.9% in both groups.

Of these, the first point needs some explanation. Asked whether Plant X is a great place to work until retirement, 77% of the respondents said "yes" and 16% said "no" in the AGV areas. In the rest of the assembly shop, 68% said "yes" and 25% said "no." This gap turned out to be statistically significant – but only when seniority was not taken into account. Overall satisfaction increased with seniority in both groups, regardless of the process flow, and there were simply more senior workers in the AGV areas – not only in my sample but also in the plant population. Does this mean that AGVs, in themselves, had little to do with the satisfaction of workers? No, not at all. Workers with a service record of 10 years or longer at Plant X accounted for 85% of the respondents in the AGV areas and 48% in the rest of the assembly shop. One may well ask why the majority of the workers on the assembly line arrived in Plant X after the AGVs, whereas most workers in the AGV areas stayed on with the plant – actually all along with the AGV areas – ever since the AGVs arrived. High turnover rates are precisely one of the problems facing the assembly line, and the AGVs seemingly solved this successfully.

The list of improvement is lengthy and looks impressive. Is it biased in favor of the AGVs? Three reasons suggest otherwise. First, the listed items are, to a large extent, predicated on the defining feature of the AGVs: absence of mechanical pacing. This result logically makes sense. Second, the improvement in perception is indeed reflected in action. The data shows that the respondents in the AGV areas report the defects they find more faithfully to their supervisor than elsewhere in the assembly shop – a difference born out by the end-of-the-line quality data I examined at Plant X. Third, although the data used in this analysis is powerfully discriminating because of its large sample size, it did not find any statistically significant difference for the questions that have little to do with process flows or pacing. This fact reassuringly suggests that the improvement listed above is more than incidental. Specifically, the AGVs did not affect:

- pride in product (5P)
- criteria for job selection (10L: seven of the ten choices[11]).

[11] These seven choices are: wages, benefits, specific bosses, work climate, plant location, job security, and personal growth opportunities. Another choice, specific friends, was significant

In these questions, no statistically significant difference was detected at 10%. All in all, it is rather difficult to question the superiority of the AGV areas over the classic assembly line as far as working conditions are concerned.

4.9.5
Bad News

Are AGVs a panacea for all the problems of the assembly line? Not quite. For one, the improvement in the AGV areas came at a price. For another, the AGVs did not bring about noticeable improvement in certain arenas where improvement was surely expected. To say the least, the assessment of AGVs in auto assembly is not as straightforward as it appeared in the previous section.

On the downside, the AGV areas suffered a statistically significant deterioration in comparison with the rest of the assembly shop in the following arenas:

- lower perceived productivity of team meetings (5P, 1%)
- weaker teamwork in helping adjacent workers who are falling behind (5P, 1%)
- weaker perceived contribution of the team concept to work environment (5P, 5%)
- weaker kaizen attitude (5P, 5%).

AGVs were welcome in auto industry because they eliminated built-in interdependence among workers: an inherent trait of any sequential process flows. The other side of the same coin, however, proved to weaken the impetus for teamwork. Plant X introduced a team-based, shop-floor organization after it installed the AGVs, but the team concept did not take hold in the AGV areas as vigorously as it did in the rest of the assembly shop. On a similar note, as much as the workers liked the process design in the AGV areas, they passively took it as given and unchangeable. To be fair, they may be intimidated by the high engineering content in the AGV areas. But it would be unfortunate if this prevented them from taking control of their immediate work environment. Although it remains to be seen how high a price these problems represent, whatever the price, it is, to a large degree, embedded in the nature of the mixed process flow.

More important than the downside is disappointment. The AGV areas did not show a statistically significant improvement over the rest of the assembly shop, most notably in the following arenas:

- motivation to do a good job (5P)
- accuracy of performance evaluation (5P)
- stress on the job (5P)

---

at 10%, but not so at the 5% level. The other two exceptions are work hours and work content. The follow-up interviews found a reasonable explanation for them. In the assembly shop of Plant X, a non-negligible number of young workers take on a second job elsewhere and they come to Plant X in the first place, knowing that its work hours allow them to do so. Newcomers to this shop are also disillusioned about the assembly line when selecting a job.

- work pace (5P)
- boredom (5P).

These questions yielded no statistically significant difference at the 10% level, except the one concerning work pace that was 10% significant but not 5% significant. These deficiencies are central to the AGVs' raisons d'être in the auto industry and call for closer examination.

Let me begin with boredom. The problem here is not the AGVs. It proved to be the conventional characterization of the assembly line. It has been customarily claimed that the assembly line job is simple and repetitive – so much so that even a monkey could do it – and insultingly boring to human beings. Assembly line workers at Plant X could not disagree more. Asked whether they agreed with the statement "I get bored while performing my process," their reaction on a five-point scale (from 1="strongly disagree" to 5="strongly agree") averaged at 1.8. Almost everyone either "disagreed" or "strongly disagreed" with the statement because they are too busy to get bored on the job. The AGVs, which were supposed to make work holistic and dignified with the prolonged cycle time of repetition, did not fare any better than the assembly line where there was not a problem of boredom to begin with. In the next section, I will have more to say about the general misunderstanding of the assembly line

The other arenas listed above are a result of split responses amongst the three AGV areas. Such is the case with motivation and performance evaluation in the engine dress-up area. The AGV system in this area was somewhat dated and reported only the number of processed engines and the actual processing time by station. Because of this limitation, the supervisors deemed the productivity data of the workers useless – data unadjusted for the differential labor content of various engine types. Table 4.9.2. shows a consequence. This table reports a 90% confidence interval of the output, island by island, as a percentage around the mean based on the data I collected from all AGV stations on a typical shift. In plain English, this interval shows how wide a variation of output one needs to expect in order to accommodate nine out of ten workers in a given island. Such a variation is exceptionally large in three of the five islands in the engine dress-up area (Island 1, 4, and 5) and noticeably large in another (Island 3).[12] This is a serious problem where everybody does the same work and receives the same pay in parallel stations. Workers in the engine dress-up area were aware that somebody was abusing the deficiency of the system and keen to report others suspected of shirking. No wonder they had a problem with performance evaluation and their motivation plummeted as a result. Although the differences were not statist-ically significant, this area did not fare well in relation to the assembly shop in these two dimensions and bogged down the whole AGV group.

The IP area and the door area presented a stark contrast for two of the variables in question: stress and work pace. For these two variables, the IP area showed a statistically significant improvement and the door area a statistically significant deterioration relative to the non-AGV part of the assembly shop. Since cause-

[12] It is true that some stations receive more time-consuming work pieces than others by chance. According to the data in the IP area and the door area, this mix variation is normally in the order of ±2% and at most ±6%.

Table 4.9.2. The 90% confidence interval of output as percentage of average. The base data shows the number of modules assembled during a typical shift by each worker, from which the average and the variance were calculated by island to derive this table.

	Engine Dress-Up	I/P	Door
Island 1	± 18.1 %	± 7.3 %	± 6.1 %
Island 2	± 3.5 %	± 6.5 %	± 5.6 %
Island 3	± 9.1 %	± 11.4 %	± 13.0 %
Island 4	± 17.5 %	± 7.3 %	NA
Island 5	± 19.3 %	NA	NA

effect relationship apparently runs from work pace to stress, a key question of the follow-up phase became why pace was perceived so differently in the two AGV areas. It turned out that this was not just a matter of perception. The IP area was full of visible idle time with workers waiting for an AGV. Even after an AGV arrived, work pace seemed, at best, leisurely. From what I found, the generous staffing level in the IP area was dictated by the AGV system designed for a higher volume of production. No wonder the IP area workers were content. But one question remained unsolved. Why were workers in the door area stressed by the work pace much more than those on the notorious assembly line? Why did they feel stressed when they controlled pacing in the absence of a moving conveyor belt?[13]

The answer was not entirely apparent. Only a few contributing factors came to my attention. First, the door area alone staffed each station with two workers, one working on the right door and the other on the left door. This mirror setup created interdependency because an AGV could not leave a station until both workers were ready and may have introduced a degree of competitive pressure.[14] Second, the door area circulated more AGVs in the system per station. Consequently, an AGV almost always waited in front of every station, ready to move in as soon as the station released the AGV currently in position. This situation prevented the workers from banking their time altogether. The faster they built doors, the more AGVs kept coming at them. Third, it was possible that the door area was deliberately short staffed. When there were less than 10 stations in an island, adding or subtracting a station meant more than a 10% change in capacity. The production

[13] The AGV system in the door area was not designed to apply pressure on the workers. When they fell behind with a particular door, the AGV turned on a light to inform them that they had used up the allotted standard time for the door. When they failed to produce as many doors as they should have for the day, a supervisor informed them that they had fallen behind the time standard and they typically replied with a comment "I hope tomorrow is a better day." Nothing further happened. Productivity was noted in performance evaluation, but was not linked to compensation. Under this relatively pressure-free circumstance, workers held a button which had to be pressed before they received another AGV.

[14] In this setup, one worker is always idled until the partner is finished. The waiting time proved to be usually quite minor—in the order of a few seconds. The workers used longer waiting periods to stock up their parts bins most of the time.

volume may have been such that the current staffing level is a stretch but adding a station would result in excess capacity. If so, the door area is a testimony to the fact that the AGVs cannot be the solution when the problem is that of work load and productivity.

No matter what the case is, the sharp contrast between the IP area and the door area poses a question: to what extent is the relaxed pace in the IP area responsible for the good news for the AGVs? Not much is probably the right answer. The IP area alone, and no other AGV area, was superior to the regular assembly line with a statistically significant difference in the following:

- better workability (5P, 1%)
- better feedback on reported defects (5P, 5%).

These items are potentially suspect in that they may have turned out to be statistically significant due predominantly to the leisurely work pace in the IP area, but practically, are not directly related with work pace to weaken this suspicion. On the other side of the equation, the door area taken alone was superior to the assembly line with a statistically significant difference in the following:

- fewer safety hazards (5P, 1%)
- better defect reportability (5P, 1%).

These items, proven in the less leisurely door area, deserve a high level of consideration.

To conclude the section, the most noteworthy finding of all is that AGVs are not any better than the assembly line at keeping stress in check. On the surface, the assembly line appears to stress out workers by continuously presenting them with a new work piece, irrespective of whether they are ready or not. In principle, AGVs do not force workers to take on a new work piece when they are not ready. It seems that stress has little to do with this dimension of the pacing mechanism.

4.9.6 Real News

Ignore all the quibbles I mentioned in the previous section regarding the AGVs for the moment. The survey data poses a much more fundamental question about them - or, to be precise, the assembly line they are supposed to replace. Although I have so far limited the reference point against which to compare the AGV areas to the assembly shop, as done by many researchers in the past, auto plants use the assembly line extensively in the other shops as well. When the AGV areas are compared against these other assembly lines, we are in for a big surprise.

Of the improvement credited to the AGV areas earlier, they in fact delivered no statistically significant difference at the 10% level when compared with the assembly line in the body shop and the paint shop in the following:

- overall satisfaction with Plant X (YN)
- job fulfillment (5P)
- frustration with work content (10L)

- frustration with work climate (10L)
- conviction that production takes precedence over safety at Plant X (5P)
- satisfaction with the training provided (YN)
- ergonomics (5P)
- overall supervision (5P: all three related questions).

The AGV areas actually fared worse than the assembly line in the body shop and the paint shop in:

- workability (5P, 10% and 1%).

The AGV areas also lack team spirit and kaizen attitudes as mentioned earlier. The AGV areas delivered improvement over the assembly line in the body shop in:

- fewer safety hazards (5P, 1%).

The AGV areas delivered improvement over the assembly line in the paint shop in:

- better feedback on reported defects (5P, 5%)
- better defect reportability (5P, 1%).

If mechanical pacing and repetitiveness are the main problems of the assembly line, one would be hard pressed to name any reasons why the body shop and the paint shop compare favorably with the assembly shop. The above result excludes the part of the body shop and the paint shop that does not use a conveyor belt. In what is included, these shops are not only mechanically paced and repetitive, much like the assembly shop, but also suffer from many disadvantages. What is going on?

The body shop is predominantly machine-paced. Many workers there stand sporadically in front of a conveyor line enclosed by a metal fence, and feed metal parts from a small opening of the fence every now and then when a work piece slides into the proper position. The parts loading takes about five seconds. In the next 30 seconds or so, they simply watch the machine emit sparks and smoke and weld the parts they fed until the conveyor line slides down with roaring sounds. When everything goes well, the work is literally so simple and repetitive that the workers, I presume, can be buried in nothingness – completely isolated in the landscape dominated by enormous machines. When something goes wrong, however, they can be injured quite badly. Finishing operations are an exception. They keep workers busy and upright all the time as the assembly shop does. But they include dirty operations, such as grinding, where noise and dust are so bad that the workers have to be wrapped up in special clothing and confined to a special booth. Isolation, long waiting, noise, and dirt are not usually found in the assembly shop.

The paint shop uses a conveyor line in such diverse operations as sealing, sanding, and spraying. These operations normally keep workers standing and busy as they work on every car that passes in front of them. Since they do not have to move around much, they work in close proximity with their neighbors. The paint shop requires that all workers wear a special uniform to prevent dust which has an adverse effect on the quality of paint. In the spray booths, some workers even wear a gas mask in addition to the heavier-duty uniform. The work in the paint shop calls for a certain degree of judgment because every car is different for the purpose of its operations. That may be a crucial reason why this shop is usually the preferred target destination for inter-shop transfers.

The survey data shows that the workers deem the body shop and the paint shop a better working environment than the assembly shop, even though these shops all use the assembly line, and that the AGV areas are at most on par with the body shop and the paint shop even though they represent improvement over the assembly shop. The culprit cannot be the assembly line, or its features such as mechanical pacing and repetitiveness. Nor can it be the conditions that exist in the body shop and the paint shop, such as short cycle time, isolation, idle time, noise, dirt, chemical odors, and special clothing requirements. I do not mean to argue that these obviously undesirable features are acceptable to auto workers. However, in order to understand the real problems in auto assembly, one needs to stop the confusion between the assembly *line* and the assembly *shop,* and ask what conditions prevail in the assembly shop that do not exist in the other shops which employ the assembly line.

The culprit of the assembly shop has two components. One is the ergonomics issue: heavy lifting, walking, bending, climbing, stretching, awkward body positions, and so on. The assembly shop has a disadvantage here because it involves: (a) numerous parts, some of which are quite heavy; and (b) extensive work inside the vehicle where space is limited.[15] The thing about these body movements is that none of them is particularly difficult if it is to be done just once. It is in the context of repetition that they become burdensome and even cause injuries. Prolonging the cycle of repetition probably helps, but the root cause is the bad ergonomics. A real solution here is to minimize heavy lifting, walking, and awkward body movements – things that are unrelated with the type of the process flow.

The other component of the culprit is defect reportability. The assembly shop, by its nature, constantly exposes problems: parts do not fit correctly, body shells are distorted, paint is imperfect, wrong parts have been installed, etc. Yet, the workers who detect these problems firsthand are not given the time to do something about them on their own. They need help. The thing about the assembly line, however, is that it disperses the workers on one dimension, a linear line, and thereby stretches the required attention zone of whoever is there to help, a supervisor or a team leader, in an artificially skewed manner. Therefore, it wastes the untapped ability to supervise – an ability that exists in the direction perpendicular to the assembly line and would have been available should all workers stay in a circle – and at the same time, leaves many neglected workers frustrated. The body shop mitigates this problem with an army of maintenance workers who wander around the shop floor. The paint shop does not have to deal with the problem because the defects workers find there – body or paint surface imperfection – are usually what they are supposed to correct in the course of their normal duties. Only the assembly shop here lacks a remedy unless one is positively provided.

[15] Jürgens, Malsch, and Dohse [5] make an important point. AGVs should be credited for improving the overall ergonomics of the main assembly line because they remove many jobs that would have been done inside the vehicle with awkward body movements.

Why did the AGV areas receive a better response than the rest of the assembly shop? The answer is self-evident with the understanding of what really bogs down the assembly shop and how the assembly line configuration augments it. The AGVs presented a good solution to the ergonomics problem because they are height-adjustable stationary platforms which eliminate walking and bending.[16] But almost any reasonable subassembly line would have delivered the same effect. The point is that the work piece is smaller and much more manageable than the entire car – thanks to modularization. The AGV areas also presented a good solution to the problem of defect reportability because they had a square shape and significantly improved supervisability.[17] Note that this positive feature, in itself, has no logical connection with the AGVs. It could have existed independent of the choice of the carrier if so desired. Furthermore, the solution did not have to be the square layout. The point is that the geographical area, which the supervisor must watch, is matched with his or her ability to actually supervise – an ability that can be enhanced with appropriate tools. Granted, the AGV areas did deliver some good news in Plant X. The credit, however ironical this may sound, does not belong to AGVs.

4.9.7 Conclusion

The assembly line revolutionized modern society. A case in point: a Ford Model T, priced at $13,861 in 1909, could be had for $2,179 in 1925 after its production volume grew more than 100 times (both prices are expressed in 1990 US. dollars). Not only did the assembly line make a car affordable to a large number of consumers, it made a large number of unskilled workers wealthy enough to afford a car of their own. Yet, the assembly line has become a negative symbol of modern society. Vast amounts of literature have quickly sprung up on the subject of inhumane working conditions under the assembly line, and unanimously blamed mechanical pacing and mind-numbing repetitiveness as the major culprits. AGVs are capable of doing away with these notorious features, and hence offer a promise to vastly improve the working conditions in auto assembly.

AGVs did accompany improvement in working conditions in the assembly shop at Plant X. The improvement was observed in such wide dimensions as job fulfillment, work content, safety-orientation, ergonomics, and defect reportability. Stress and boredom on the job were a notable exception, however. Contrary to popular belief, boredom was not a problem on the assembly line at Plant X, and

[16] An often-repeated comment in the interviews went like, "It is a good system and people like it. You stand still. You work while standing still."

[17] Curiously, many supervisors in the AGV areas made a comment to the effect that the assembly line is easier for them to "supervise" because the moving line would make sure that workers keep working and QC (quality control) check points would police their quality. Such a comment suggests that the supervisors, whether they like it or not, are engaged in often difficult, real supervision in the AGV areas by actively managing workers through dense communication.

did not leave much room for improvement to begin with. Stress was a problem, but AGVs were not necessarily the solution. AGVs also accompanied negative effects besides being an expensive investment proposition. They weakened team spirit and kaizen attitudes. AGVs may not be a prudent investment if these attributes are important to auto assembly.

The real problem of AGVs is not what they failed to accomplish. It is that most of the improvements did not materialize from the anticipated reasons. If mechanical pacing and repetitiveness were in fact the main culprits, working conditions must be at least as bad on the assembly line in the body shop and the paint shop as they are in the assembly shop. In reality, AGVs delivered little or no improvement over the assembly line in the body shop and the paint shop. It is quite likely that the problem of the assembly *shop* has long been confused with the problem of the assembly *line*. The former, I would argue, concerns two issues: ergonomics and defect reportability. These problems exist away from the assembly line, although they are surely enhanced by the line configuration. AGVs worked effectively in relation to the assembly shop because they improved ergonomics and defect reportability. Once this point is realized, however, there are other ways to reap the same benefit without resorting to expensive AGVs.

This story of AGVs is a great reminder of the difficulties associated with sorting out intertwined cause-effect relationships. In the face of severe labor problems in auto assembly, it seemed so obvious to convict the assembly line. Many have in fact done so. But the data from Plant X casts a serious doubt on this conviction. It may not be the assembly line that is wrong. Instead, the real culprit may well be the way in which the assembly line has been managed, or mismanaged, in the auto industry. Discussions on the humanization of auto assembly almost always begin with a presumption that the assembly line must go, and move on to a search for promising alternatives. However, it is possible that this presumption stands on shaky ground. At the very least, the experiment of Plant X with AGVs makes another close look at the assembly line deserving.

4.9.8 References

1 Berggren C (1992) Alternatives to Lean Production: Work Organization in the Swedish Auto Industry. ILR Press, Ithaca, NY

2 Beynon H (1973) Working for Ford. Allen Lane, London

3 Chinoy E (1955) Automobile Workers and the American Dream. Doubleday, New York

4 Gooding J (1970) Blue-Collar Blues on the Assembly Line. Fortune July 1970: 69-71, 112-113, 116-117

5 Jürgens U, Malsch T, Dohse K (1993) Breaking from Taylorism: Changing Forms of Work in the Automobile Industry. Cambridge University Press, Cambridge

6 Meyer III S (1981) The Five Dollar Day: Labor Management and Social Control in the Ford Motor Company, 1908-1921. State University of New York Press, Albany, NY

7 Tolliday S, Zeitlin J (eds) (1992) Between Fordism and Flexibility: The Automobile Industry and its Workers. Berg Publishers, Oxford

8 Walker C R, Guest R H (1952) The Man on the Assembly Line. Harvard University Press, Cambridge, MA

4.10 Organizational Change and Assembly Automation in the Dutch Automotive Industry

J. Benders · B. Dankbaar

4.10.1 Introduction

In the Netherlands, discussions about the limitations of traditional production concepts as well as the practical development of new production systems have been strongly influenced by the Dutch version of sociotechnical systems design (STSD). Japanese management concepts such as Lean Production (LP) tend to be compared with sociotechnical principles rather than with traditional Taylorist models as is the case in much of the international literature. This raises many questions. How do LP and Dutch STSD compare? Are they at odds, or might a combination of both be fruitful? What is the impact of these organization concepts on organizational design in practice?

This paper aims to at least partially answer these questions by discussing the developments at the two main Dutch players in the automotive industry, namely DAF Trucks and NedCar. Representatives of both companies claim to have been influenced by STSD as well as by LP in decision-making concerning their organizational designs. An important difference, however, is that the Japanese company Mitsubishi Motor Car became one of the owners of NedCar in 1991, whereas DAF Trucks has never had a Japanese stockholder. Thus, whereas the impact of LP is indirect at DAF Trucks, NedCar came under the direct influence of a Japanese company.

Before discussing these cases, a brief comparison of LP and Dutch STSD is presented in the next section. This provides the background for the discussion on both cases; at the same time, the main ideas of the Dutch version of STSD are presented. This is necessary for three reasons. In the first place, there are many different versions of STSD around [15]. Secondly, the term tends to be interpreted in many different ways. Thirdly, the Dutch version of STSD has its own particular characteristics which are different from the internationally better known versions of STSD. In the subsequent sections, the practical developments at both companies are discussed. Finally, on the basis of these cases, some conclusions are drawn concerning the relationship between LP and STSD.

4.10.2
Modern Sociotechnical Design and Lean Production

In this section, the Dutch version of sociotechnical systems design (STSD) and lean production (LP) are compared. As the readers of this book are assumed to be familiar with the concept of LP, only STSD is described explicitly. Internationally, the concept of sociotechnical design is probably far less ambiguous than that of Lean Production. It is generally acknowledged that its cradle stood in English coal mines around 1950. Much of the pioneering work took place in the United Kingdom by researchers of the London-based Tavistock Institute of Human Relations. In the following decades, the approach spread to Norway ("industrial democracy"), Sweden, the United States, ("participative design") and the Netherlands. Involving employees in the process of (re)designing their workplaces was a central point around which most sociotechnical projects evolved. Both the Swedish and Dutch approaches stress the importance of production organization. A line structure with short work cycles is rejected, as such a structure is seen to result in high system losses, among which line balancing problems are the most important. In Sweden, Volvo and Saab gained worldwide recognition for their innovative projects at several plants in the 1970s and 1980s. These were not restricted to the assembly of passenger cars, but included, and still include, many truck, bus and component plants as well [7,16].

In the Netherlands, a sociotechnical approach was developed that is in many ways different from the STSD approaches used in the United Kingdom, Australia and the United States (see [15] for an overview). The Dutch approach is known as *Moderne Sociotechniek* [9,35] and in the remainder of this paper the term *Modern Sociotechnical Design* or *MSD* is used. MSD is often characterized as an *expert approach*. Rather than focusing on employees' participation in the process of (re)design, its founding father De Sitter stresses that participation by itself does not take account of the unequal distribution of power and knowledge among the people in the redesign process. Powerful and knowledgeable people are likely to effectively resist changes which could undermine their positions. Participation will not change this state of affairs. In De Sitter's view, this problem must be evaded by following a design methodology which is offered by MSD. The methodology forces all participants to go through the same line of reasoning. Before going into details, it needs to be pointed out that it is the MSD rather than any other STSD version that influenced the developments at DAF and NedCar.

MSD involves the systematic (re)design of production organizations following a number of steps. The first step consists of grouping products into product families, which can be handled relatively independently from each other. The aim of this step is to reduce variety in production by the creation of parallel flows, thereby substantially simplifying the organization of production, and consequently the need for coordination and centralized production control. The next step is the segmentation of the individual production flows. The operations necessary to make the various product groups are clustered into segments, including preparatory and supportive tasks. The aim is to enable the segments to solve their own problems, thereby further reducing the need for centralized control. Small buffers may be introduced to decouple segments and give them a certain degree of autonomy. In the third step, so called *whole task groups* are formed which carry out all tasks occurring in a particular segment. Next, the control structure is designed bottom-up. That generally means that controlling tasks are placed as low as possible in the organization in order to encourage local control. Problems need to be dealt with locally as far as possible.

Before turning to a comparison between LP and MSD, a methodological consideration needs to be pointed out: it is not evident that it is actually possible to make a valid comparison. Unlike LP, MSD is a fairly worked-out and detailed design methodology. This is not the case with LP. The concept LP was presented to the world by the MIT study *The Machine that Changed the World* [39]. In spite of the fact that the book did not contain any reference to concepts that were not known before, it had an impact on management practice in the European automobile industry that can hardly be underestimated. The book catapulted Krafcik's [25] cleverly chosen term *Lean Production* to a desideratum for many managers. By counterposing the "devilish" mass producers against the "angels" in the form of lean producers, the easily accessible book had the ideal format to create a managerial fashion [1]. Its popularity had probably much to do with the attractive message it contained: Lean Production leads to superior performance and it is universally applicable. The timing of its publication proved to be lucky: the economic downturn, which started to set in when the book first appeared, led to an intensified search for improvement measures among European producers, many of whom claimed to be adopting "lean" practices (cf. several contributions in this book and Sandberg [31]). Nevertheless, LP is a rather ambiguous concept [23] which was constructed out of empirical data. This ambiguity is often resolved by taking LP to be identical with the Toyota Production System. This system, however, is the result of an evolutionary process which still continues [5,17,18]. Hence, any characterization concerns a specific point in time. Whereas these ambiguities may have been crucial for LP's viability in societal discourse and organizational practice [30], they have also given rise to many different interpretations of the concept [6, 30]. This makes any attempt to (re)construct LP's production logic vulnerable to debate. The following characterization of LP follows one that is conventional in academic circles, namely the Toyota Production System as it existed in the 1980s, but it has to be realized that this is not the only possible characterization [6,30].

Basic differences between LP and MSD concern the following characteristics:

1. Worker autonomy
2. View on buffers
3. Pacing of work
4. Use of standardized operating procedures (SOPs)
5. Length of cycle time
6. Form of teams.

Especially the different approaches to teamwork are a common source of misunderstanding. Although the MIT-study stresses the importance of teams and even calls them "the heart of the lean factory" and "a hallmark" of LP [39: 9 and back-flap], it does not discuss what teams in a lean production system look like. Jürgens [22] presented two quite different team concepts which he labeled respectively, *Japanese* and *German*. As Jürgens' portrayal of German teams fits well with the notion of *whole task groups* in Dutch MSD, his distinction can serve to highlight some basic differences (table 4.10.1.).

Table 4.10.1. Japanese versus German teams

Japanese	German
1. Simplification of tasks	1. Enriched jobs
2. Semi-skilled employees	2. Skilled employees
3. Replaceability	3. Polyvalency
4. Homogeneous composition	4. Heterogeneous composition
5. Line-paced work speed	5. Some control over time
6. Strict standardization of tasks	6. Some autonomy in execution of tasks
7. Hierarchical	7. Anti-hierarchical

Source: adapted from Jürgens [22: 29]

Although the question as to whether the (interrelated) differences in team concepts as mentioned in table 4.10.1. have to be ascribed to differences in the task environment (e.g. stable vs. dynamic output characteristics), national environment (Japan vs. Germany), or production concepts (lean vs. sociotechnical) needs to be elaborated, the frequent references to differences such as those displayed in table 4.10.1. leave little doubt about their existence.

The first item is crucial to the understanding of all other differences. This item concerns the difficulty of jobs and therewith the requisite qualification level of employees. Under lean (and mass) production, jobs are standardized and simplified as much as possible; this has consequences for the deployability of employees [4: 12-14]. Jobs can generally be considered as a bundle of tasks, and a certain period is required to master each individual task. The length of the total learning period for a job determines the ease of substitution between job holders. Job holders can only be substituted for each other as far as they master each other's tasks. A broad deployability of employees can be achieved in two different ways: by simplifying jobs or by providing training. In table 4.10.1., these different approaches to deployability are labeled *replaceability* and *polyvalency* respectively.

The "Japanese" workforce is homogeneous in the sense that there is little segmentation; this is possible thanks to the simplified work. The pace of work is controlled by the line speed. Standardization by using and enforcing standardized operating procedures (SOPs) is another control mechanism and essential to the renowned *kaizen* (continuous improvement) which is generally seen as a key element of lean production. Kaizen guru Imai is very explicit about the importance of SOPs to *kaizen*:

> „...management must first establish policies, rules, directives, and procedures for all major operations and then see to it that everybody follows SOP. If people are able to follow the standard but do not, management must introduce discipline. If people are *unable* to follow the standard, management must either provide training or review and revise the standard so that people can follow it"[21: 6].

Improvement is an ongoing process which never ends. Employees are strongly encouraged to provide suggestions for improving procedures so that higher standards can be achieved. In this way, employees are structurally involved in the design of their own work procedures which is supposed to have "morale-boosting benefits of positive employee participation" [21: 112]. Yet, *kaizen*'s influence on worker autonomy is contested. Womack et al. [39] see this system as providing a "creative tension" in which workers have many ways to address challenges and which offers possibilities to workers to improve their own working environment and is coupled to job security. In contrast, Delbridge, Turnbull and Wilkinson [14: 104] state:

> „collective autonomy is limited to task *design*, whereby workers are encouraged to make suggestions for "improvement" which must then be agreed firstly by the team and then by management. If the improvement is incorporated into the job and the task redesigned, it is effectively standardized, giving workers no effective control over subsequent task *execution*",

a view formulated succinctly by Conti and Warner [11:39]:

> „*The labour process [...] is contradictory,* with employees working four hours a month to make their work for the rest of the month even more Taylor-like".

This path of continuous improvement always requires close hierarchical supervision, as stressed by Imai (see quote above).

Apart from these differences with respect to shopfloor work, Lean Production needs to be credited for its broader approach of production issues. Unlike sociotechnical approaches, Lean Production also includes notions about product and process design, supplier relations and the marketing and distribution process. However, these areas are beyond the scope of this paper. Table 4.10.2. summarizes the arguments listed above.

Now that the main differences between both organization concepts have been discussed at a conceptual level, the cases of DAF Trucks and NedCar can be dealt with against this conceptual background. As stated in the introduction, MSD as well as LP have at some time influenced changes in the work organization at both companies, but as will be shown in the following, in quite different ways.

4.10.3 DAF Trucks

DAF Trucks is the only Dutch truck manufacturer. The company is a medium-sized player in the European market. In 1982, DAF started a program called *Quality of Working Life*, which was renamed as *Quality of Work and Organization* in 1986/87. This program was heavily influenced by sociotechnical ideas current at that time. The program's basic tenets were

1) production workers determine to a very important degree product quality and production costs;
2) the rising educational level in society at large and in the labor force have led to higher demands where the level of the work content is concerned;

Table 4.10.2. Key characteristics of lean production and modern sociotechnical design

Characteristic	Lean Production	Modern Sociotechnical Design
Goal	Organizational performance	Organizational performance; quality of working life
Organizational design	Partly ambiguous organization concept	Design methodology
Worker deployability	Interchangeable, based on experience	Polyvalent, based on education and training
Worker autonomy	Not recognized as issue	Important structural goal
Flow production	Yes	Yes
Lay-out	Line or U-form	Group structure
View on buffers	Have to be eliminated	Necessary in order to decouple
Line or machine-pacing	Yes	No
Cycle time	Short	Long
Standardization	SOPs emphasized and enforced	Limited, because it affects autonomy
Logistical concepts	Pull system: JIT, *kanban*	Not included in approach
Orientation to quality	Pervasive; extensive control procedures	Supportive, but no explicit attention
Orientation to improvement	Pervasive: *kaizen*; QCS	Supportive, but no explicit attention

Source: adapted from Aertsen and Benders [2] (Used with permission)

3) effective production automation requires independently functioning workers rather than "button pushers".

The existing organization was characterized by the separation of "thinking" and "doing", rigid mechanization, many hierarchical layers and functionally specialized departments. The program aimed at resolving the disfunctional effects of this organization by creating so-called "cells", i.e. teams of workers responsible for a segment of the production process, integrating direct and indirect tasks, including a decentralisation of control tasks and requiring the introduction of *DAF-kringen*, DAF's equivalent to quality circles.

Although the Quality of Work and Organization program was not uncontested within DAF, the company was open to and even stimulated innovations of all kinds, including this program. It was indeed "one large and ideal garden of experimentation", giving managers a relatively large degree of freedom in carrying out experiments. By 1986, cells were operational in a number of different departments. The firm's approach to sociotechnical re-organization was pragmatic. Sociotechnical principles were applied as far as the existing structural and political situation allowed. However, emphasis was placed on the quality of working life rather than on the redesign of organizational structures in order to achieve higher levels of flexibility and productivity.

When the program stagnated in 1987 due to technical reasons (logistics, the introduction of a new product, increase of production numbers, and a reorganization project) and a "lack of general support", a new department *Quality of Labor and Education* was created as part of manufacturing operations. This department was intended to give new impetus to the program. Probably the most elaborate project of this period consisted of a complete reorganization of the Motor Test Hall between 1987 and 1990 [37]. The project was typical of the sociotechnical approach at DAF. The changes involved not only a structural reorganization, but also an effort to change employees' attitudes and behavior. This and other projects, however, remained isolated pockets of change within the organization: "Many experiments were conducted, but some were turned back as well". There was resistance within the organization from those people whose jobs were threatened, i.e. especially from indirect personnel.

In a speech delivered in September 1988, the Board member who strongly backed the program acknowledged the existence of internal resistance, but was rather optimistic about the progress so far: the project "sold itself", and some 60 cells were already operational. When this top manager passed away unexpectedly in February 1989, the support for Quality of Work and Organization on the Board of Directors ended as well. Although there had hardly ever been open resistance to the program, the badly needed active support was often absent. "The project slowly died", according to one of the managers interviewed. Furthermore, "technicians viewed the program as a soft approach". Its advocates tried to revitalize the program by pointing to its economic advantages. While it was repeatedly stated that the experiments paid off handsomely, other sources mentioned the practically insurmountable difficulties in establishing the exact economic effects of reorganization projects [3].

In conclusion, DAF's Quality of Work and Organization program consisted in the end of a number of scattered experiments which were conducted during an eight year period and had varying degrees of success. It enjoyed the full support of one top manager, but was not uncontested within the firm. Its progress was dependent on the initiatives of individual middle and lower managers. The necessary full support of top management and an overall sociotechnical vision were missing. Favorable financial data could not change this situation.

Lean Production at DAF

The first signs of attention for Lean Production within DAF date back to the second half of 1990; a time when DAF started to experience financial difficulties. Lean Production drew the attention of the member of the Board who was responsible for manufacturing. Soon, top management embraced Lean Production: "Lean Production was the only idea that got the Board enthusiastic in ten years time", according to one of our contacts. This enthusiasm probably had much to do with DAF's financial difficulties which presumably would be resolved by enhancing performance through the introduction of Lean Production. As such, the coming implementation of Lean Production served as an indicator to banks that DAF was working seriously on its problems. DAF became increasingly dependent on the banks for a continuous supply of short-term loans.

One of the manufacturing manager's assistants was given the assignment of introducing Lean Production in the organization. He even hired professor Daniel Jones (one of the authors of the MIT study) to give a lecture to DAF's management early in 1991. Lean Production's main, and according to some respondents only, feature at DAF was "head count reduction". DAF had too many (eight) hierarchical layers and staff departments. The Quality of Work and Organization program had managed to reduce the number of layers, but with its *de facto* bottom-up approach, it failed to have an impact on the higher levels of the organization, where new layers and departments had been formed simultaneously with the implementation of sociotechnical ideas at DAF's lower levels. The head count reduction was pursued rigorously. The man in charge reportedly had as his motto: "I shall eliminate every job of which I do not understand the job description". During a public discussion in September 1992, the manufacturing manager announced that the first landmark had been reached in the beginning of 1992 and that DAF would be "lean" by the end of 1993. Official DAF documents contained similar statements.

An important difference with the sociotechnical approach was that Lean Production aimed to eliminate the lowest managerial level of "work masters" who had been leading figures in the sociotechnical approach. A second important difference between the two approaches concerned the time reserved for implementing them. Whereas the sociotechnical approach was characterized by a slow process of organizational change, Lean Production was to be implemented at high speed. Not surprisingly, the emphasis was on directly visible elements. "They had understood the lean side of Lean Production, but they failed to understand the flexible side of it". A salient illustration of this point: although the sales forecast had been adjusted downward twice in 1992, the production level was not cut back because that would make the factory's efficiency look bad.

Yet, there were also similarities between Lean Production and Modern Sociotechnical Design. As indicated above, LP and MSD have several elements in common, albeit in some cases at a nominal level only. Elements ascribed to LP and used within DAF include the emphasis on flow production, shopfloor teams and fewer hierarchical levels. However, the differences in team concepts were little understood. As a result, the work floor sometimes received conflicting messages about the measure of autonomy given to them. At DAF, many managers in the field of operations management regarded LP and MSD to an important extent as overlapping organization concepts, even on items such as the length of the cycle times and the use of standardization which in theory should have been discriminating factors [38]. Such perceived similarities, together with a lack of detailed knowledge about the exact contents of both approaches, even made it possible to carry out sociotechnically inspired projects under the label of Lean Production. Furthermore, the experience with change processes that had been built up with the sociotechnical program proved to be very useful. At the same time, projects were carried out that clearly fit better into the Lean Production philo-sophy, but these were isolated projects started by individual DAF employees without being incorporated into a larger "lean" design philosophy.

4.10.4 NedCar

In 1968, DAF opened a new plant to house its passenger car operations. The plant is located in Born in the south-eastern province of Limburg. This choice was influenced strongly by the Dutch government, as the coal mines in the region were in the process of being closed down hence contributing to mounting unemployment levels. In the newly established plant, many former miners were employed. Not much later, due to financial difficulties DAF wanted to withdraw from the passenger car market. In 1972, Volvo took a 33 percent stake in DAF's car operations which was increased to 75 percent in 1975. DAF withdrew completely, and the operation's name became *Volvo Car BV*. The bad economic circumstances after the 1973 oil crisis necessitated substantial state subsidies. After an upsurge in the late 1970s, a second crisis in 1979/80 induced Volvo to sell part of its shares to the Dutch state which became a 70-percent shareholder in 1981. Largely thanks to state financing, the development of the new Volvo 400 series was made possible, and helped by a favorable economic environment, production and profits developed positively during the second half of the decade. In the beginning of the 1990s, however, the pendulum swung back to "recession", and as the Dutch government was neither willing nor able to provide additional support, a new partner was sought and found. In the fall of 1991, an agreement was reached between the Dutch government, Volvo and the Japanese manufacturer Mitsubishi to share ownership on an equal basis. The three owners committed themselves to invest 2.1 billion guilders (700 million each). The Dutch state was to sell its shares to its two commercial partners by 1998. The new name *NedCar*, an abbreviation for "Netherlands Car BV", symbolizes the transition. A Mitsubishi and a Volvo variant based on the same platform were to be built on a new production line. Total production volume was scheduled to be 200,000 cars, i.e. 100,000 of each model. In the spring of 1995, Mitsubishi's Carisma model was taken into production; Volvo's S40 (sedan) and V40 (hatchback) models followed in the autumn. In the following, organizational changes before and after Mitsubishi joined are discussed.

4.10.4.1 Organizational Changes in the Pre-Mitsubishi Period

In the first half of the 1980s, when the plant was still called Volvo Car, several change projects were undertaken. The so-called "New Style" program of management combined Japanese-inspired quality programs with sociotechnical restructuring [12,13]. On top of this, some major investments were undertaken in further mechanization. The sometimes conflicting and incoherent series of change projects failed to meet the high expectations raised at the program's launch and led to widespread disillusionment and resistance in the workforce.

A major investment project involved the introduction of Automated Guided Vehicles (AGV) in the final assembly line for the 400 series [10]. The AGVs were expected to help handle the increasing number of variants and options, which had led to increasing differences in the length of assembly times. The AGVs permit the assembly of cars in parallel flows, so that these differences can be coped with. In additon, the AGVs were used to position the body in various ways, thus improving ergonomic conditions and reducing the risk of errors. Unlike conveyors, AGVs can be used easily in different lay-outs, i.e. they allow a great degree of freedom for future changes. An elaborate cost-benefit analysis was made, confronting AGVs and an overhead conveyor. The higher costs of AGVs were expected to be off-set by even higher savings. The impact on the quality of working life, among other factors, was taken into account as well. Ultimately, AGVs were implemented almost throughout the assembly process (1986). A later evaluation, however, showed that they scored positively with regard to the quality of working life and flexibility, but cost savings had not been achieved.

In 1987, the management team of the Dutch plant presented the Board of Directors with a report on production policy. The report argued that a lot of attention had been given in the recent past to products, production processes and human resources, but not to the structures of work that tie these things together. A new production policy would have to be developed that would define targets for which the individual departments could be held accountable. The plant management's initiative coincided with comparable actions by the Board of Directors which had decided to introduce a flatter structure for the whole organization.

After extensive debate, the plant management initiative resulted in the introduction of so-called production units in the plant during the course of 1989. This involved a considerable change in the composition of the management team for the plant. Before the introduction of the unit organization, the management team consisted of the plant manager, the production manager and the managers of functional departments, such as quality assurance, manufacturing engineering, materials management and personnel. The heads of the individual production departments (press shop, body shop, paint shop and assembly lines) all reported to the production manager. In the new setup, the management team consisted of the managers of four production units (the four main production areas just mentioned) and the manager of a unit called *location affairs* who was responsible for various matters related to the plant as a whole. As far as possible, all of the functional departments were divided up over the four production units, which became almost self-sufficient in this respect. Every production unit received its own personnel, engineering, quality and finance/controlling functions. Maintenance had already been reporting to the production manager and was now further decentralized to the units. Some plant-level personnel and organization functions were maintained as a kind of staff function to the plant manager.

The managers of the production units obviously had to be more than simply production managers. Whereas the former heads of the production departments were responsible only for production, the new unit managers were responsible for the support functions as well. Depending on the unit, between 30 and 40 people were employed on maintenance, engineering and quality alone. Additionally, there were those people involved in personnel, organization, management information and control. The unit managers had to shift responsibility for production further down the line. In fact, the unit managers had a head of manufacturing reporting to them whose job it was to oversee production. From the very beginning, a further decentralization of responsibilities was envisaged. The proposal for the unit structure had included a proposal for further decentralization of responsibilities towards so-called task groups on the shop floor. The new unit managers were charged with creating such task groups. For this particular responsibility, they could rely on a so-called "mentor", usually a former supervisor with extensive experience.

4.10.4.2
Lean Production at NedCar

After the signing of the agreements in the fall of 1991, the newly named NedCar company entered a difficult period of reorganization and restructuring. A transformation plan had been agreed between the partners for the period 1992-1998 in which the plant would have to become "one of the most productive in Europe": it was announced that the plant's annual production capacity would be doubled (to 200,000 cars) and this was to be accompanied by a fifty percent reduction of the workforce. As it happened, the transition to the new lines for the new model(s) had to be made during a period of declining demand in most European markets; this affected the sales and hence production of the Volvo 400 series. As early as the fall of 1991, a first reorganization was announced which involved downsizing the workforce by 412 employees over the course of 1992. Furthermore, several parts of the company were sold, the Research and Development Center was made independent and corporate headquarters moved to the plant at Born. As a result of such measures, NedCar's total workforce numbered 6,109 by the end of 1992, compared to 9,276 two years earlier. Under pressure of further market contractions, additional cost-cutting measures were undertaken in 1993 and 1994, so that by the end of 1994, the workforce had come down to approximately 4,200.

In the following, three interrelated topics are discussed subsequently, namely the new technical equipment, organizational changes and the introduction of task groups. In addition, Exhibit 1 sketches some of the views of NedCar's Japanese vice-president Mr. Norio Takehara as background information.

A. The Technical Equipment
As was to be expected from its strong engineering reputation, Mitsubishi took the lead in developing the new models and production line. Investments include two Hitachi Zosen transfer presses in the press shop, and a completely automated warehouse for pressed parts between the press shop and the body shop. Remarkably, the warehouse is to some extent "unproven technology" and as such quite exceptional in combination with a new product. Normally, like the other Japanese manufacturers, Mitsubishi is very reluctant to introduce too many new features at the same time in a new production system.
In final assembly, a conveyor system has replaced the AGVs. Work is carried out in short cycle times and is line-paced. The new line has small buffers of two cars between line segments in order to handle unforeseen disturbances. Overall, the new assembly line is a quite traditional, continuous line with three separate automated islands (Exhibit 2). The degree of automation, as assessed by the extent to which the workforce would have to be enlarged if the automated stations did not exist, is approximately 10 per cent. For the moment, the three islands are

Exhibit 1: Views of NedCar's Japanese vice-president

Mr. Norio Takehara is vice-president of NedCar on behalf of Mitsubishi and responsible for production engineering. In a lecture held in February 1993, he expressed some of his views on NedCar's future. He mentioned three driving forces behind the high productivity of Japanese producers:

1) rationalization through peoples' participation, especially by the active participa tion and contribution of blue collar workers;
2) process-driven design;
3) a flexible production system, parts of which are a multi-skilled workforce and flexible equipment.

Such a system, in his view, cannot be attained without "teamwork with willingness", or a "positive contribution by all people working in the company". He expressed a clear awareness of the limitations of the Dutch setting for applying Mitsubishi's ways of working:

"When implementing the Japanese Lean Production Concept in Europe, a careful choice needs to be made between what can be realised in Japan, with only Japanese employees and what can be realised in the Netherlands with Dutch employees. It is our objective to select those Lean Production Methods which will work in Europe and add them to the already outstanding European ones, to create a Lean Production System suitable for our company in Holland" [36].

Yet, he also stressed that some changes needed to be made, among which:

- the introduction of *kaizen*;
- target costs reductions;
- the replacement of the AGVs in final assembly by a conveyor;
- the reduction of buffers to an absolute minimum, as buffers conceal problems.

---

organizationally treated as one task group. In the future, it may be possible that adjoining task groups also become responsible for the automated islands. The work of the operators in these islands, however, is obviously quite different from manual assembly work and requires considerable technical training. There is one operator per island per shift. A total of 16 operators have received training for the islands. Several maintenance engineers (working for Production Engineering) have followed the same training and can be called for if the need arises. The operator is in each case responsible for the transfer from the last automatic station in the island to the moving assembly line. In front of each island, a worker of the preceding task group checks if the cars being transferred into the island are ready for the automatic operations.

Niepce [29: 35-40] points to the fact that all decisions concerning technical matters were made by Mitsubishi; this also applies to the production targets. One of the lines in Mitsubishi's Mizushima plant in Japan served as a model for the new line in Born. Yet, within these constraints, decisions concerning organizational matters are largely left to NedCar's management, provided, of course, that sufficient progress is made in reaching the production targets. Nevertheless, Mitsubishi's influence on organizational matters may be stronger than appears at a first glance. In the first place, the equipment has consequences for, and sometimes even embodies organizational matters. In the second place, in the period August 1993 - July 1994, some 600 NedCar staff stayed for an average of about one month in Japan or the so-called *P-trial project* which involved building, installing, and testing the new equipment [27: 13]. In addition, some 200 Japanese

Exhibit 2: Assembly automation in NedCar's new assembly line

Island 1:	6 robots / 5 stations: body measuring, rear seat supply, clutch fluid supply, brake fluid supply (pre vacuum), brake fluid supply (filling), body primer application (rear, left quarter), body primer application (front, right quarter)
Island 2:	6 robots / 4 stations: body measuring, third side window glass assembling (right and left; only Volvo), window glass assembling (rear), window glass assembling (front), front seat supply (right), front seat supply (left)
Island 3:	12 robots / 6 stations: body measuring, assembly tailgate strip, battery supply, radiator fluid supply, tire assembling (right rear), tire assembling (left rear), wheel nut supply (right), wheel nut supply (left), tire assembling (left front), tire assembling (right front), spare tire supply, washer fluid supply, power steering oil supply

engineers came to the Netherlands during the installation phase. Throughout this project, much knowledge on Mitsubishi's working methods must have been transferred as well.

B. Organizational Changes

The proponents of the work structures introduced at Volvo Car since 1987 experienced great difficulties when explaining them to Mitsubishi's representatives. These work structures included the creation of four production units for each of the main processes in the plant and the organization of work in so-called *task groups*. Production units and task groups have a certain measure of autonomy which is the source of their responsibility. Their autonomy increased by assigning responsibility and decision latitude in areas such as financial and personnel management to the production units. This decentralization could go quite far. For instance, finance and controlling tasks carried out by the production units included multi-year planning and budgeting, periodic reporting, task analysis, management information and controlling. In the Japanese approaches, this particular notion of autonomy is not present. Consequently, Japanese manufacturers have shown little interest in relocating indirect and staff tasks to production units. Hence, it does not come as a surprise that the Mitsubishi managers found little value in the idea of production units as developed at the Volvo plant. They preferred to return to an organization structured along functional lines.

The Japanese preferences were reinforced by economic pressures. After considerable debate in the plant management team, the production units were partly dismantled in the spring of 1993. The responsibility for personnel, finance, controlling and main engineering functions were removed from the production units and returned to functional departments. In the course of the process, considerable streamlining took place and it was possible to reduce the total number of workers. To a large extent, logistics had always been a centralized function and continued to be so. What remained within the units were some quality control and quality engineering functions, engineering tasks concerned with changes in the existing process (adaptations of purchased parts, changes in the product, adaptations of the layout), and maintenance. In the course of 1994, these were further reduced. The units now only have a small support group for typical industrial engineering tasks (time and motion studies, personnel calculations) and continuous improvement projects. As a result of these changes, the unit managers have little staff left and have become more traditional production managers. This also allowed for the elimination of the function of head manufacturing in the units. NedCar now has an flat and functionally structured organization.

C. Task Groups

First pilots projects for the design of task groups were started in the middle of 1990. Identification of "complete tasks" for the task groups was carried out by line management supported by the plant staff group for organizational development. Taking layout and equipment as given for the time being, complete tasks included not just an identifiable set of production tasks, but also quality assurance, materials management and what are called "regulatory" tasks. Once these tasks had been identified, manning levels and (on-the-job) training requirements were established. Between 1990 and 1993, task groups were gradually introduced in all the production units, and a total of 140 groups is now operational.

The discussion about the role of functional departments and the value of semi-autonomous production units has hardly affected the introduction of task groups. The policy of forming task groups was continued and expanded, although the creation of teams requires a minimal level of continuity for the purposes of forming and developing the task groups; this was disrupted by the almost continuous downsizing process described above.

The differences in approach between LP and MSD disappeared from the surface and the task groups were presented as implementations of the Japanese system; the firm's former training and development manager saw Lean Production as

> „a social system, in which responsibilities are brought to the lower levels of the organization, and in which employees work in teams, in which one should be able to take over each other's tasks and actually does so. Employees in such a system are responsible for as many aspects as possible, including continuous improvement of product and process" [32].

Each task group has a set of tasks consisting of approximately 10 production tasks and 12 indirect and regulatory tasks such as quality control, materials supply, maintenance of tools, detailing production schedules and work assignments, maintaining standard operating procedure forms, production statistics, planning of holidays, advisory role in the hiring of new personnel. The ideal of all workers being able to carry out all tasks (polyvalence) and rotate among different jobs may not be reached due to worker's personal constraints. The reward system is now based on workers' competencies, i.e. the amount and level of tasks they master. As a result of the training measures, there was an upward influence on the average wage of production workers as they acquired additional skills.

An increasingly important aspect of task group activities concerns continuous improvement. Both individual suggestions (which are individually rewarded at a level related to the savings they help to create) and group suggestions are encouraged. A special event in the improvement process is an annual "Challenge Day", first held in November 1993. On this day, a special effort is made to achieve superior performance. Another activity is a convention concerning group improvements which was held for the first time in the summer of 1994. From the participating groups, one is chosen to represent NedCar at the annual Circle International Convention of Mitsubishi in Japan.

In a survey held among 900 workers who had been involved in task groups for at least 2 months, it was found that acceptance and support of task groups increased according to the length of time people had worked in such groups. Management continued to view task groups as the main mechanism to enhance motivation and gather support for productivity increases.

4.10.4.3
Future Developments at NedCar

By definition, any prediction about future developments is to some extent speculative. Unforeseeable circumstances, such as economic up- and downturns may change organizational policies drastically, even on a short term basis. Yet a few observations concerning the strategic positions and motives of NedCar's shareholders may help to assess the direction of future developments.

Obviously, Mitsubishi is of crucial importance for the plant's future. The cooperation between Volvo and Mitsubishi came about because Volvo was not able or willing to afford the plant and related product development costs. The success of the Volvo S40 and V40 will be crucial for the continuation of the Swedish engagement. If the new car is not successful, Volvo will probably feel forced to go back to the larger luxury models, or it will have to find a partner with a complete line of cars. In both cases, the need for a plant in the Netherlands will be questionable (unless of course the new partner is Mitsubishi). For the moment, signs are pointing in the opposite direction due to the economic recovery which set in in 1994 and the introduction of the new models. Mid 1995, some 1,500 temporary staff were employed at NedCar to cope with increased demand. Both partners have even publicly expressed an interest in expanding the capacity of the plant to 300,000 units. In December 1995, Volvo announced plans to increase production capacity due to positive sales of the new Volvo model. A year earlier, Mitsubishi had announced its plans to invest an additional Dfl. 700 million to produce another model. Whereas it remains to be seen whether or not such announcements materialize, for the time being, NedCar's future seems to look brighter than it has looked for quite some time.

Other strategic considerations are important too. The Born plant is Mitsubishi's only manufacturing plant within the European Union. Not only can - and will - protectionist measures be circumvented in this way, the rising yen and the sustained economic problems in Japan also add to the need for a European presence. For Volvo, the Born plant offers a unique opportunity to learn at first hand about Japanese working methods, not only in manufacturing, but also in product and process development and suppliers relations. Mitsubishi clearly played first fiddle in setting up the new production line, whereas Volvo refrained from active involvement. In technical matters, the Japanese dominated, whereas in social and organizational matters they are conscious of the restrictions posed by the Dutch environment and leave most decisions to the Dutch. Nevertheless, the impact of Mitsubishi's working methods is increasingly visible, although the process of transferring Mitsubishi's knowledge seems to be highly pragmatic.

It should be recognized that the learning process may be mutual. The Dutch experience may prove valuable for Mitsubishi in case of future investment in European countries. In addition, since the labor market shortages around 1990, Japanese car manufacturers seem to realize the importance of the quality of working life [5,8,18,19,33,34]. Asked in an interview about "promising developments" in Mitsubishi's relationship with Volvo, Mitsubishi president Nakamura [26: 4] answered:

> „Volvo's strengths are its environmental technology and the thought that goes into the working environment of its employees. Volvo excels in these areas. I think that both sides will derive great gains and make significant product improvement through the partnership."

Thus, the lessons learned in Born may be useful for Mitsubishi's home base. Volvo's experiences elsewhere, for instance in its well performing Gent plant [20], play a role as well. In conclusion, given both partners' interest in their joint venture, it seems unlikely that either one will withdraw for the time being.

4.10.5 Discussion

In the first section, some questions were raised about the relationship between Modern Sociotechnical Design (MSD) and Lean Production (LP). Rather than embarking on an elaborate comparison of both organization concepts at a conceptual level, in this concluding section, some observations are made based on the cases of DAF and NedCar. Both cases show that the interpretation of Lean Production by managers plays an important role in the implementation of "lean" practices. The concept of Lean Production as given by Womack c.s. [39] left considerable room for interpretation. When organizational changes take place under the banner of "lean production", it is necessary to fill this interpretation room. In the Dutch, and probably also German and Swedish practice, this often happens by using originally sociotechnical notions [6]; this is not surprising given the overlap between both concepts, at least at a superficial level.

Confronting DAF with NedCar, the difference in applying "lean" techniques is striking. Whereas both companies experimented, albeit with varying degrees of success, with Japanese-inspired production techniques, their current ways of organizing vary widely. DAF rode the wave of lean production's popularity, but apparently without really knowing what the concept entailed. In contrast, at NedCar, the Japanese influence is pervasive, and the company seems to be steadily moving towards more resemblance with Japanese car factories. Where DAF was dependent upon indirect sources about Japanese car manufacturing, NedCar had access to first-hand knowledge after Mitsubishi joined. The constant contacts and exchanges between Japanese and Dutch employees at different hierarchical levels played a vital role. Not only was the technical design of the new production line made by Mitsubishi in Japan, several hundred NedCar employees even stayed in Japan for some weeks to get acquainted with and exposed to Mitsu-bishi's ways of working. Although the Japanese vice-president clearly expressed a sensitivity for limitations in transferring Mitsubishi's way of working to the Netherlands and the Japanese seem hesitant to get involved in non-technical matters, they do appear to intervene when, in their view, there is insufficient progress in such matters. The re-functionalization of the organizational structure may be an example. At NedCar, an interesting learning process seems to be taking place. In a continuous trial-and-error process, indigenously Japanese ways of working are gradually being introduced in a Dutch car assembly plant which was formerly influenced by (Dutch) sociotechnical and Swedish ideas. The arguably most interesting example is the philosophy behind the task groups which was originally based on MSD, but is now influenced by insights from Mitsubishi, for instance with respect to continuous improvement.

Another possible reason for the differences between DAF and NedCar lies in the different products: NedCar concentrates on producing large volumes of fairly identical products, whereas DAF has a larger range of products (trucks) which are produced in smaller quantities. As a consequence, NedCar's production situation lends itself better for applying Japanese management practices than that of DAF [41].

Work organization in the NedCar plant appears to become a hybrid of various different approaches. It is too soon to speak of a unique hybrid concept. One must distinguish here between the conceptual level on the one hand, and the practical or empirical level on the other. Organization concepts can be described in their pure forms. In this respect, Wood [40: 582-583] pointed out that just-in-time management is a model, even in Japan. But if such pure concepts are used to analyze empirical situations, almost every real organization turns out to be a hybrid of various concepts and traditions. In the case of NedCar, this problem is compounded by the fact that it is unclear to what extent Mitsubishi's production system resembles Toyota's production system - which is generally considered to be the archetypical "lean" producer.

Not every hybrid organization therefore represents a new "hybrid concept". Concepts can be used by academics to shed light on the empirical reality, and by managers to give direction to organizational changes. Yet, the discussion above shows that there are at least two problems in dealing with such organization concepts. First, organization concepts contain a certain degree of vagueness [30] which makes it impossible to establish their exact content, and which leaves room for interpretation [6,24]. Second, organization concepts can have similarities which increases the difficulty of making clear distinctions between them. Both problems make the boundaries between organization concepts rather hazy. Academics may define their own precise and discriminating categories to describe reality, but as soon as these categories are used by practitioners, the problems that academics wanted to evade inevitably surface once again [cf. 28].

What is clear from the above discussion is that the relationship between organization concepts and their manifestation in practice is complicated, and our understanding of it needs improvement. Naturally, the globe will continue revolving when academics try to single out such an academic issue. Undisturbed by it, managers will continuously be affected by the launch of new or only allegedly new organization concepts. Within politically feasible boundaries, they will act on the basis of their understanding of the organization concepts as the DAF case clearly shows. Whether or not the actions within an organization resemble the initial concepts is then less important than studying the role of organization concepts in the contingent [30: 62] decision-making processes where the outcome is uncertain. In other words: there is more to organizations than organization concepts. The DAF and NedCar cases can also be seen as examples where the application of organization concepts is enabled and constrained by both structural and coincidental factors. Structural factors include institutional and cultural factors in the national environment as well as the output characteristics. In a certain sense, these set the stage on which the play is performed. Yet, as organization concepts do not always, and perhaps can never, provide a sufficiently clear script for the play, the actors can take a certain freedom to move at will. This freedom is constrained within the confines of the stage (concrete production setting) and other actors' moves (organization members). Purely coincidental events, such as the passing away of DAF's top manager supporting MSD, can greatly affect the course of the play. Often, the importance of events can only be assessed retrospectively, as the outcome depends on complex interaction patterns between moves which can also only be established *ex post*. If, in the long run, this contingent outcome is economically successful, a company's way of organizing may, and is even likely to, become a model for other companies. When described in general terms, it can ultimately be awarded the status of an organization concept, as has happened in the past in the case of Ford and Toyota. Thus, organization concepts are both input and output of hybridization processes.

4.10.6 References

1. Abrahamson E (1996) Management Fashion. Academy of Management Review Vol 21 No 1: 254-285
2. Aertsen F, Benders J (1993) Tricks or Trucks; Ten Years of Organizational Renewal at DAF? Research Memorandum 627, Faculty of Economics, Tilburg University
3. Bell J H J (1990) Profiteren van de profits? Unpublished Master's thesis, Tilburg University
4. Benders J (1993) Optional Options: Work Design and Manufacturing Automation. Avebury, Aldershot
5. Benders J (1996) Leaving Lean? Recent Changes in the Production Organization of some Japanese Car Plants. Economic and Industrial Democracy Vol 17 No 1: 9-38
6. Benders J, Bijsterveld M van (1995) Leaning on Lean; The Processing of a Management Fad. Unpublished paper, Nijmegen Business School
7. Berggren C (1993) The Volvo Experience; Alternatives to Lean Production in the Swedish Auto Industry. Macmillan, Basingstoke\London
8. Berggren C (1995) Japan as Number Two: Competitive Problems and the Future of Alliance Capitalism after the Burst of the Bubble Boom. Work, Employment & Society Vol 9 No 1: 53-95
9. Bijsterveld M van, Huijgen F (1995) Modern Sociotechnology: Exploring the Frontiers. in: Benders J, de Haan J, Bennett D (eds) The Symbiosis of Work and Technology. London\Bristol PA, Taylor and Francis: 25-45

10. Boom H van den (1992) Keuzeproces carriers bij eindmontage Volvo BV. in: Frambach R T, Nijssen E J (eds) Technologie en Strategisch Management. LEMMA, Utrecht: 219-237
11. Conti R F, Warner M (1993) Taylorism, new technology and just-in-time systems in Japanese manufacturing. New Technology, Work, and Employment Vol 8 No 1: 31-42
12. Dankbaar B, Diepen B van (1990) Vernieuwing en herstrukturering bij Volvo Car BV. MERIT, Maastricht
13. Dankbaar B, Diepen B van (1991) Voordat Mitsubishi kwam... Vernieuwing en herstructurering bij Volvo Car BV. Tijdschrift voor Arbeidsvraagstukken Vol 7 No 3: 43-55
14. Delbridge R, Turnbull P, Wilkinson B (1992) Pushing back the frontiers: management control and work intensification under JIT/TQM factory regimes. New Technology, Work and Employment Vol 7 No 2: 97-106
15. Eijnatten F van (1993) The Paradigm that Changed the Work Place. Assen\Stockholm, van Gorcum\Arbetslivscentrum
16. Engström T, Medbo L (1994) Intra-Group Work Patterns in Final Assembly of Motor Vehicles. International Journal of Operations and Production Management Vol 14 No 1: 101-113
17. Fujimoto T (1994) Reinterpreting the Resource-Capability View of the Firm: A Case of the Development-Production System of the Japanese Auto Makers. Discussion paper 94-F-20, Faculty of Economics, University of Tokyo
18. Fujimoto T (1996) An Evolutionary Process of Toyota's Final Assembly Operations - The Role of Ex-post Dynamic Capabilities. Discussion paper 96-F-2, Faculty of Economics, University of Tokyo
19. Grönning T (1995) Recent developments at Toyota Motor Corporation: The emergence of "Neo-Toyotism"? in: Sandberg Å (ed) Enriching Production; Perspectives on Volvo's Uddevalla plant as an alternative to lean production, Avebury, Aldershot: 405-425
20. Huys R, Hootegem G van (1995) Volvo-Gent: a Japanese transplant in Belgium or beyond? in: Sandberg Å (ed) Enriching Production; Perspectives on Volvo's Uddevalla plant as an alternative to lean production. Avebury, Aldershot: 231-248
21. Imai M (1986) KAIZEN - The key to Japan's competitive success. New York, McGraw-Hill
22. Jürgens U (1992) Lean Production in Japan: Mythos und Realität. in: IAT/IGM/IAO/HBS (eds) Lean Production; Schlanke Produktion, Hans-Böckler-Stiftung, Düsseldorf: 25-34
23. Kieser A (1993) Die "Zweite Revolution in der Autoindustrie" - Eine vergleichende Analyse und ihre Schwächen. in: Meyer-Krahmer F (ed) Innovationsökonomie und Technologiepolitik. Forschungsansätze und politische Konsequenzen, Physica-Verlag, Heidelberg: 103-134
24. Kieser A (1996) Moden & Mythen des Organisierens. Die Betriebswirtschaft Vol 56 No 1: 21-39
25. Krafcik J F (1988) Triumph of the Lean Production System. Sloan Management Review Vol 30 No 1: 41-52
26. Nakamoto M (1995) Partnerships: A Driving Force for Sales Growth. The JAMA Forum Vol 13 No 4: 3-6
27. NedCar (1994) Personeelsjaarverslag 1993. NedCar, s.l. [Born]
28. Nederveen Pieterse J (1994) Globalization as Hybridization. International Sociology Vol 9 No 2: 161-184
29. Niepce W (1994) Lean Production Versus Sociotechnical Systems at NedCar. Unpublished master's thesis, State University Groningen
30. Ortmann G (1995) Formen der Produktion; Organisation und Rekursivität. Westdeutscher Verlag, Opladen
31. Sandberg Å (ed) (1995) Enriching Production; Perspectives on Volvo's Uddevalla plant as an alternative to lean production. Avebury, Aldershot
32. Schramade P (1992) Learning for lean & lean for learning? Opleiding & Ontwikkeling Vol 5 No 7/8: 3-7
33. Sey A-P (1994) Soziale Aspekte in den gegenwärtigen Modifizierungen von Produktionskonzepten in der japanischen Automobilindustrie; Japan auf dem Weg zur "arbeiterfreundlichen Fabrik"? Unpublished Master's thesis, Free University of Berlin

34. Shimizu K (1995) Humanization of the production system and work at Toyota Motor Co and Toyota Motor Kyushu. in: Sandberg Å (ed) Enriching Production; Perspectives on Volvo's Uddevalla plant as an alternative to lean production, Avebury, Aldershot: 383-403
35. Sitter L U de (1989) Modern Sociotechnology. Internal paper, KOERS, Den Bosch
36. Takehara N (1993) NedCar's Way of Thinking. Proceedings Conference on Lean Manufacturing and Automation, University of Nijmegen, February 1-2
37. Verlaar A J J M, Buyse J J (1990) Integrale organisatievernieuwing bij DAF; Rol P&O bij veranderingsproject in Motoren Testhal. Human resource management III.11-(1-20)
38. Vloet M (1993) Lean Production versus Moderne Sociotechniek; Een vergelijkende studie tussen de arbeidsorganisatie van Lean Production en Moderne Sociotechniek. Unpublished master's thesis, University of Nijmegen
39. Womack, J P, Jones D T, Roos D (1990) The Machine that Changed the World. Rawson Associates, New York
40. Wood S (1991) Japanization and\or Toyotaism? Work, Employment & Society Vol 5 No 4: 567-600
41. Young S M (1992) A Framework for Successful Adoption and Performance of Japanese Manufacturing Practices in the United States. Academy of Management Review Vol 17 No - 4: 677-700

4.11 Recycling and Disassembly – Legal Burden or Strategic Opportunity?

G. Seliger · C. Hentschel · A. Kriwet

4.11.1 Legal Framework in Germany

The recycling of technical consumer products after usage is a subject of rising importance. The lack of natural resources, the necessity to save energy and the reduced permission for landfill areas or waste incineration plants, reflected in the steeply rising cost for waste disposal, have increased the awareness that components and/or materials of used products must be regained and reused.

In Germany, recycling is mainly stipulated by law. In 1986, the German government proposed several measures to reduce waste. This led to a relatively good system for simple, one-material products, e.g. glass bottles and paper. It was not until 1991 that the German Ministry for Environmental Protection submitted a legislation to be enacted on January 1st, 1994, which obliges producers and distributors of complex electronic products to take back their products after use, and

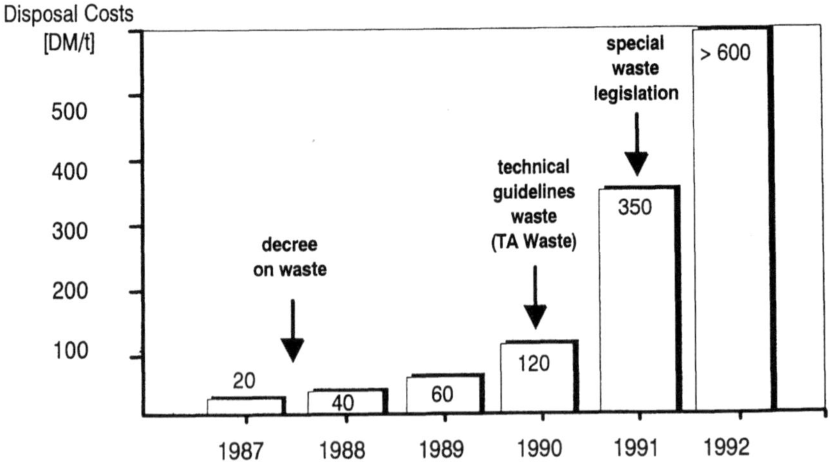

Fig. 4.11.1. The rise of disposal costs in Germany. Source: Prognos

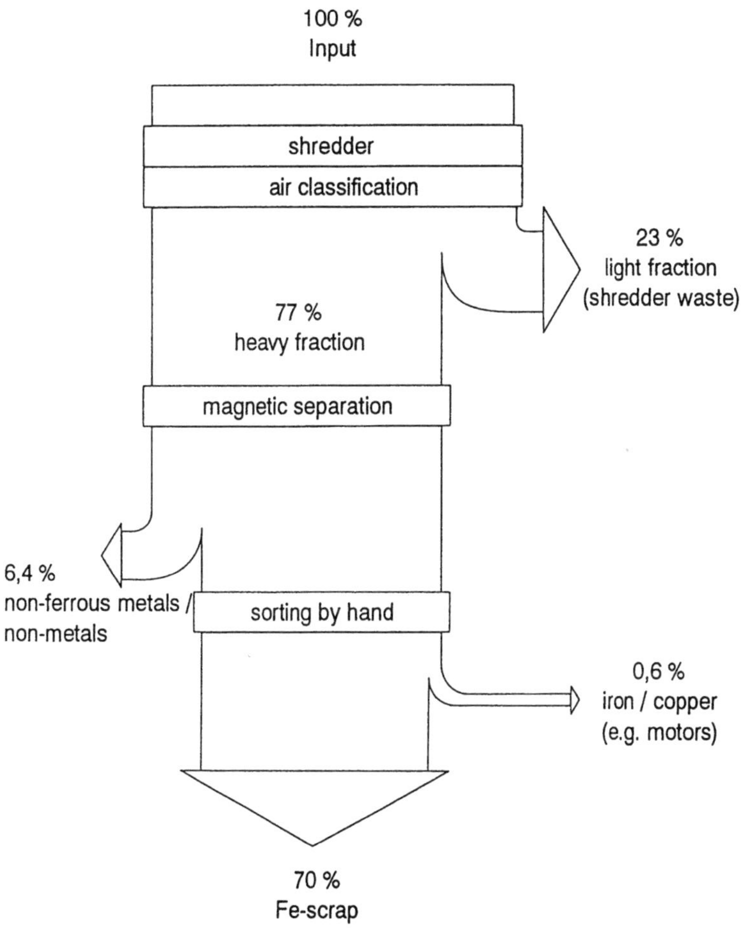

Fig. 4.11.2. Material flow of used car reprocessing. Source: Bilitewski [16]

to have these products recycled. The scope of the term *complex electronic products* comprises, amongst other things, electronic toys, watches, laboratory equipment, household appliances, lighting equipment and computers. Another law, dealing with cars, is also under preparation. It is astonishing that the idea to give producers and consumers joint responsibility for the recycling of products was formulated as early as July 1975 in the EC Directive 75/442/EEC.

Complex consumer products, however, still require a sophisticated recycling system in order to recover valuable materials and sort out harmful substances. Since landfill capacities are limited, prices have already exploded, e.g. from about DM 20.00 per ton in 1987 to DM 1000.00 per ton in 1993 for the shredding of residues from used cars (fig. 4.11.1.). German legislation is becoming more stringent and European legislators are quickly picking up the pace.

Waste can be defined as redundant goods, by-products or residues that have no value and must be disposed at a cost. Once an adequate recycling technique is found, materials and components can be reused, and what was formerly just waste becomes a valuable and well-paid-for resource. Different technologies, e.g. shredding, metallurgy and energy recycling, which are already applied in order to reduce the volume of waste, hardly consider the variety of materials in complex products (fig. 4.11.2.).

4.11.2
Scrap from Automobiles

4.11.2.1

Dimension of the Problem

Currently in Germany 2.6 million, in Western Europe about 14 million and worldwide about 34 million old passenger vehicles are scrapped every year (fig. 4.11.3. and 4.11.4.) [15].

Due to their relatively high content of metals, 90 % of used cars are routinely processed in shredders. Sometimes they are cannibalised for serviceable electric motors and easily recoverable non-ferrous metals, and the rest is all too often left to rot in the open air. Automobile scrap is complex and self-contaminating, characteristics which can adversely affect recovery values [3].

4.11.2.2
Current Recycling Technology

The approx. 5,000 used-car-recyclers in Germany seperate reuseable parts and materials, e.g. motors, electronic and chassis parts, and sell them to private consumers. The partly dismantled cars are brought to shredder firms, where they are seperated into the fractions steel scrap, shredder scrap and shredder fluff. When non-ferrous metals are not separated from steel, the latter must be classified as low-grade material. Therefore, the shredder scrap is separeted further into non-ferrous metals.

This kind of car recycling is also found in other West-European countries, as well as in the US and in Japan (fig. 4.11.5.). Shredders for processing scrap from automobiles were primarily used in the USA in the 1960's. In Germany, this process has been used since 1970. To date, about 45 shredders are in use in Germany, in western Europe about 240.

Due to high processing temperatures in the shredder and insufficient separating technologies, plastics, elastomers and other non-metallic materials are difficult to recover from shredder scrap (fig. 4.11.6.) [9]. These fractions have to be dumped on landfills. Their volume is stated to be too high. To date, the volume

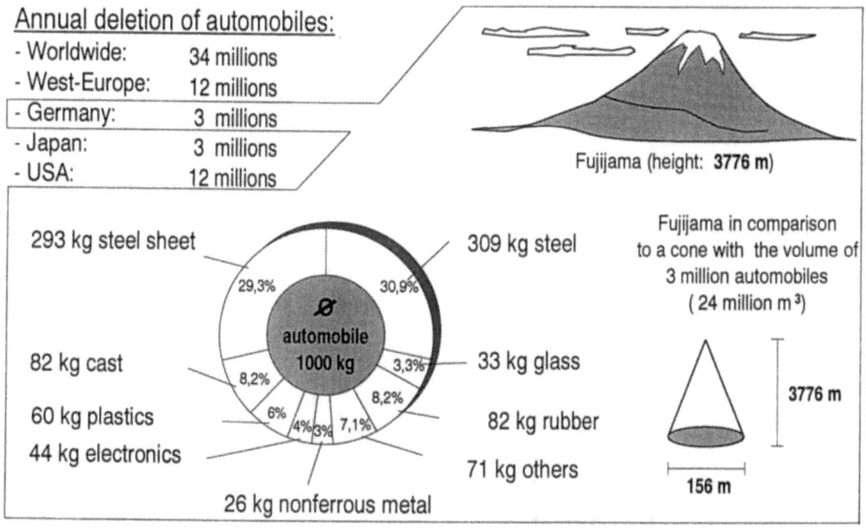

Fig. 4.11.3. Scrap from automobiles. Source: Automobil-Industrie 3/90

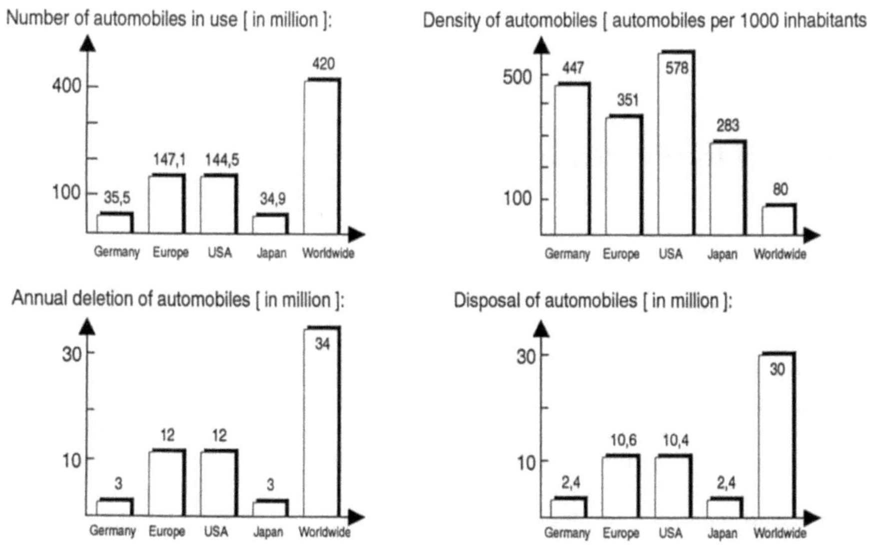

Fig. 4.11.4. Number, density, deletion and disposal of automobiles in 1990. Source: Müllmagazin 1/93

from shredding used cars is estimated to be 2.2 million tons. Worldwide, about 6 million tons of shredder scrap are not recyclable. Alternative recycling methods should make an attempt to separate pure materials from used products.

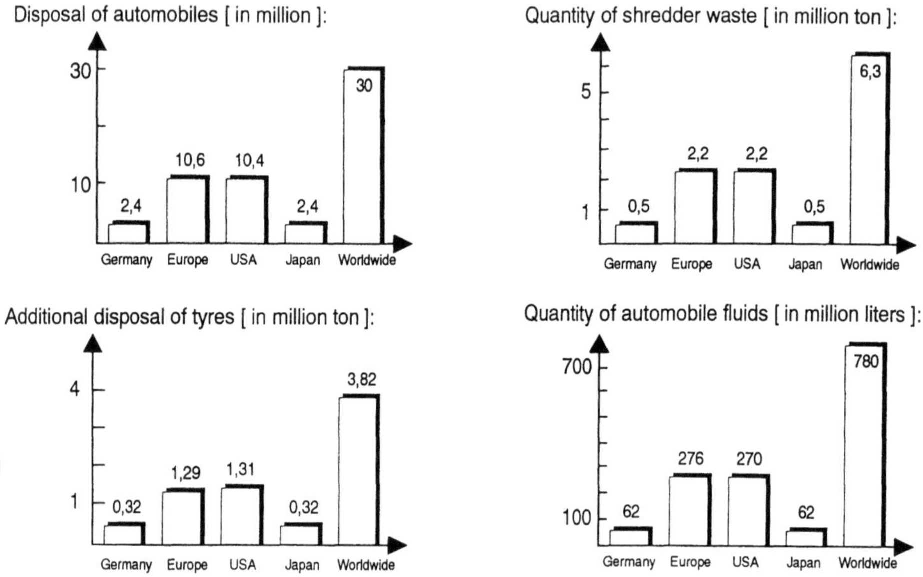

Fig. 4.11.5. Residues from automobile scrap in 1990. Source: Müllmagazin 1/93

4.11.3
Ways out of the Problem: Disassembly

4.11.3.1
Why Disassembly?

In order to recover a valuable material or isolate a toxic substance from scrap, shredding must always be followed by a separation process [7], i.e. wind sifting or sink-floating. These additional processes are inefficient if the shredder scrap contains materials difficult to separate. Additionally, if harmful substances in the scrap contaminate the shredder scrap as a whole, it must be dumped on special landfills. Used, complex products must hence be pre-processed in order to reduce their complexity and facilitate further recycling. Disassembly is considered an adequate form of preparation. Furthermore, only disassembly allows the recovery of reusable components and parts, which is not possible with any other preparatory step for recycling (fig. 4.11.7.).

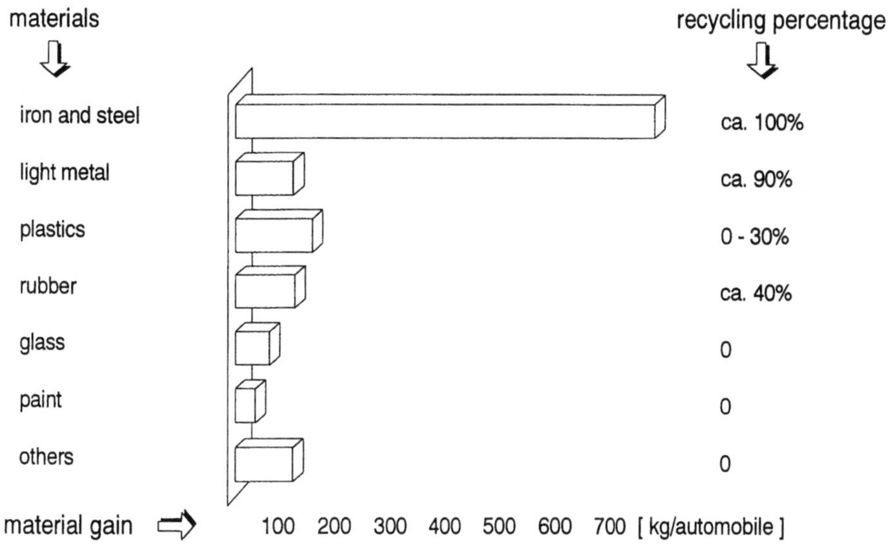

Fig. 4.11.6. Recycling of automobile materials in Germany 1990. Source: Automobil-Industrie 4/91

4.11.3.2
Disassembly Is not the Opposite of Assembly

Whilst the complete assembly process has to be carried out with utmost precision, the disassembly process does not necessarily have to be performed right down to the last screw and may not require the same accuracy. The aim of any assembly process is to ensure a product's functionality by bringing all the components of a product into a defined position and orientation. The assembly planner knows the number of parts, their quality and the materials for the final product, which may also have been designed for ease of assembly.

As today's products are usually not designed for ease of disassembly, one is confronted with inseparable components, a lack of characteristics distinguishing one material from another and lost manufacturing information. In addition, used products are subject to many imponderables; unexpected changes in product components and joining elements are unavoidable. Today, getting one's hands on all the information required for product disassembly is illusory. Unlike assembly, disassembly planners are confronted with the difficulty of developing one-of-a-kind processes on the basis of incomplete product information.

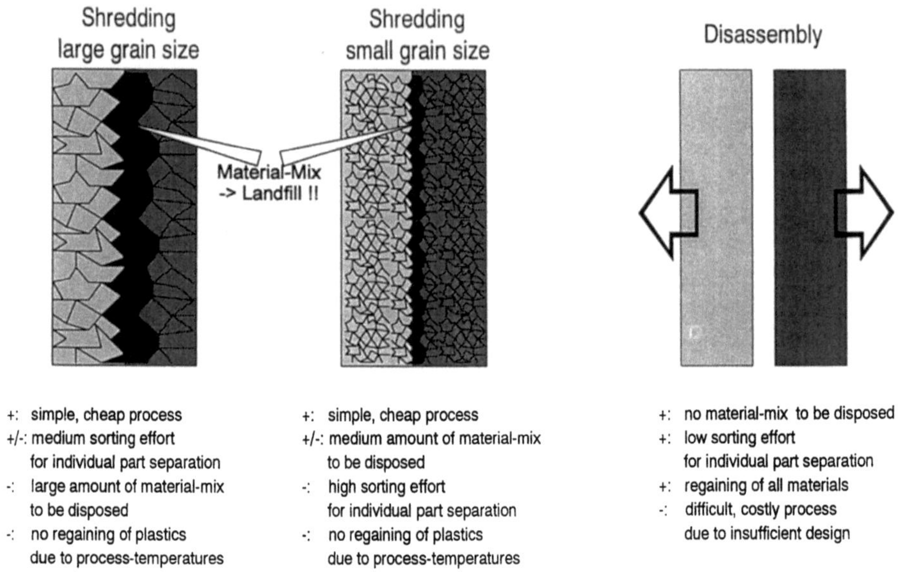

Fig. 4.11.7. Disassembly versus shredding

4.11.4 Disassembly Planning

4.11.4.1 Product Analysis and Evaluation

Disassembly process planners can not yet proceed from known process parameters. They require support by systematic planning methods. IWF Berlin is working on a structured planning method which consists of a product analysis phase, the evaluation of alternative disassembly sequences and the determination of disassembly tools and equipment (fig. 4.11.8.) [4,8].

In the product analysis phase, materials and parts which are considered to be valuable or reusable are defined; this provides information for optimum disassembly depth, the point after which further disassembly is no longer worth the effort. Substances which are harmful for the environment and require isolation have to be identified. If a recycling technology for a part is already available, as is the case for cables and printed circuit boards, their dismantling is useful for further recycling [19].

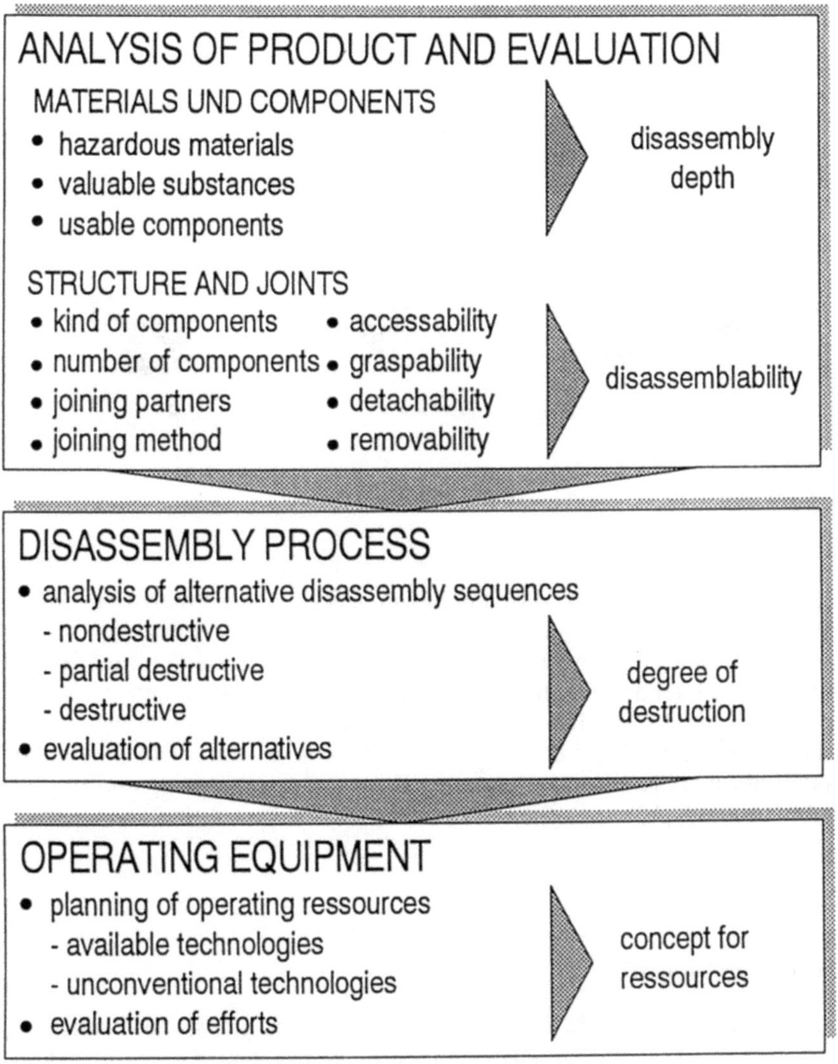

Fig. 4.11.8. Steps of disassembly planning

4.11.4.2
Disassembly Sequences

In the second step, product component attachment technology, the components' hierarchy and former assembly sequence are analysed with regard to the compon-

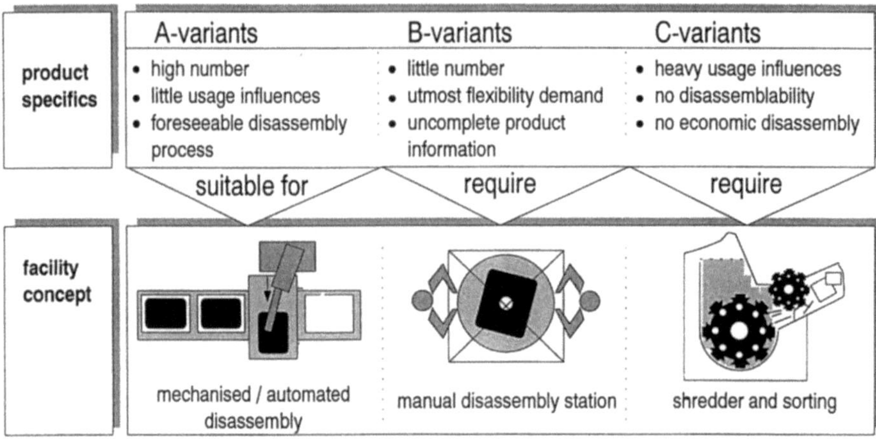

Fig. 4.11.9. Holistic facility concept for disassembly and recycling

ents' separability and possible dismantling sequences. Furthermore, the imponderables of the dismantling process have to be taken into account. [5] A *usage mode and effects analysis* generates possible deviations of a used product from its original state. The intended function of a component already gives hints to possible changes to its original state. If a product is intended to be used in humid environments, the planner probably has to consider corrosion of parts. Other deviations may result from manipulation and damage. This requires utmost flexibility of the disassembly process. [6].

4.11.4.3
Disassembly Equipment

The above planning steps deliver data and influencing factors for the decision as to whether destructive, non-destructive or partly destructive disassembly should be considered. Non-destructive dismantling seeks to reverse the assembly process in order to diminish the risk of destroying valuable components. Examples are unscrewing screws and unsnapping snap connections. In the case of inseparable attachments, whether intended, e.g. bonding and welding, or caused by corrosion and wear, destructive dismantling methods gain in importance [13]. Either attachment components or surrounding components are destroyed in order to recover a valuable part. Accessibility of components and joining elements can be increased by altering the product's position. The result is a process plan, which provides information on disassembly steps, dismantling tools, fixtures, times, safety measures for workers and qualification needs [11].

An advanced step in the planning process deals with automated disassembly (fig. 4.11.9.). For the above-mentioned reasons, a flexible disassembly process

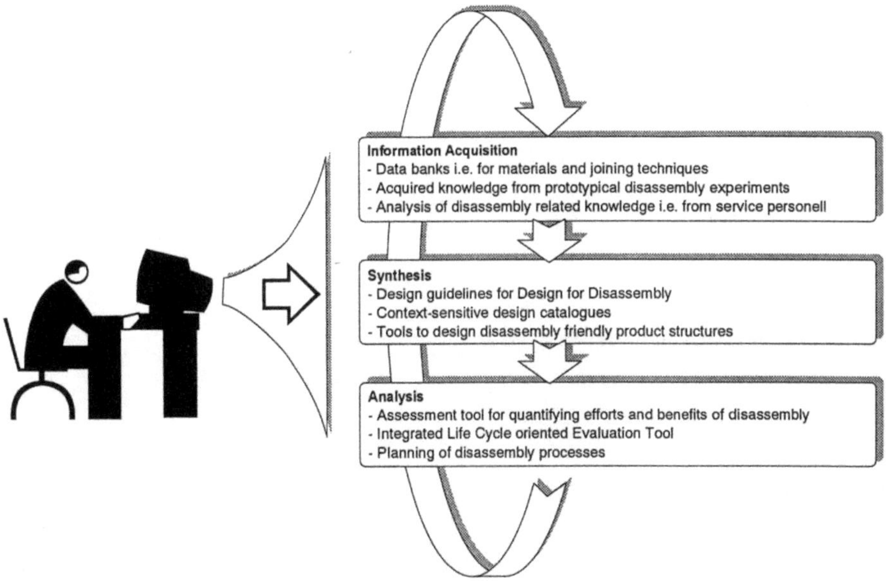

Fig. 4.11.10. Tools to support the designer

requires a higher degree of flexibility than an automated assembly process. On the one hand, flexible automation gives rise to completely new demands on process parameters, not yet sufficiently known even for assembly processes. On the other hand, automation offers the opportunity to keep workers away from dirty, and possibly dangerous, workplaces. A cost-to-benefit evaluation, parallel to these planning steps, suggests termination criteria for disassembly.

4.11.4.4 Evaluation

In current research carried out by the authors at the Technical University of Berlin, an assessment methodology to support product design for the "end-of-life" phase is being developed. It is based on the assessment of feasible options for disassembling a product and applying recycling processes to its components and subassemblies. The methodology balances the future effort to be invested in recycling processes with the future benefits from a reduction in disposal charges and sales of recovered materials, components and subassemblies. It identifies the optimal recycling strategy for a given product with respect to multiple economical and environmental objectives by the designer. The assessment supports the designer by allowing him to compare design alternatives and identify weak spots in the design. The methodology is illustrated by assessing alternative future scen-arios for the recycling of a washing machine subassembly [8,17]

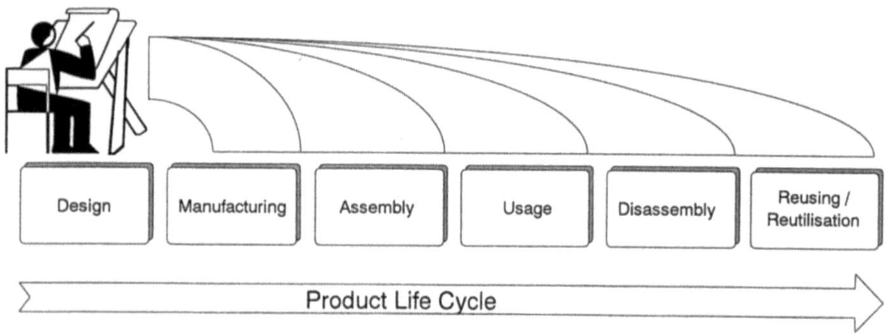

Fig. 4.11.11. Enlarged optimization scope of design engineers

4.11.5 Product Design

4.11.5.1 Optimization Scope

During the design process, different and sometimes conflicting goals have to be achieved. Technical designs, optimized for ease of manufacturing, are often difficult to assemble, whilst lightweight materials which save energy during the product life create problems during the recycling process. It is difficult for the designer to evaluate the importance of the conflicting goals in order to choose the optimal solution. What is hence needed are methods and tools which allow a reconciliation between the advantages and disadvantages of alternative designs, especially in view of the subsequent life cycle phases in order to find a global optimum (fig. 4.11.10.).

4.11.5.2 Design Strategy

The basis for Life Cycle Design is the cooperation of experts from all stages of the product's Life Cycle, right from the beginning of the design process (fig. 4.11.11.).

The implementation of Life Cycle Design strategies will boost efforts focussing on concurrent engineering. The role of the designer is changing along this path: he used to be a single combat fighter, who received a specific task description as an input, processed it and delivered the required design data. In the future, he must become the moderator of a design team involved in all the stages of design, from the specification of basic product requirements to the final touch of the product's outer appearance.

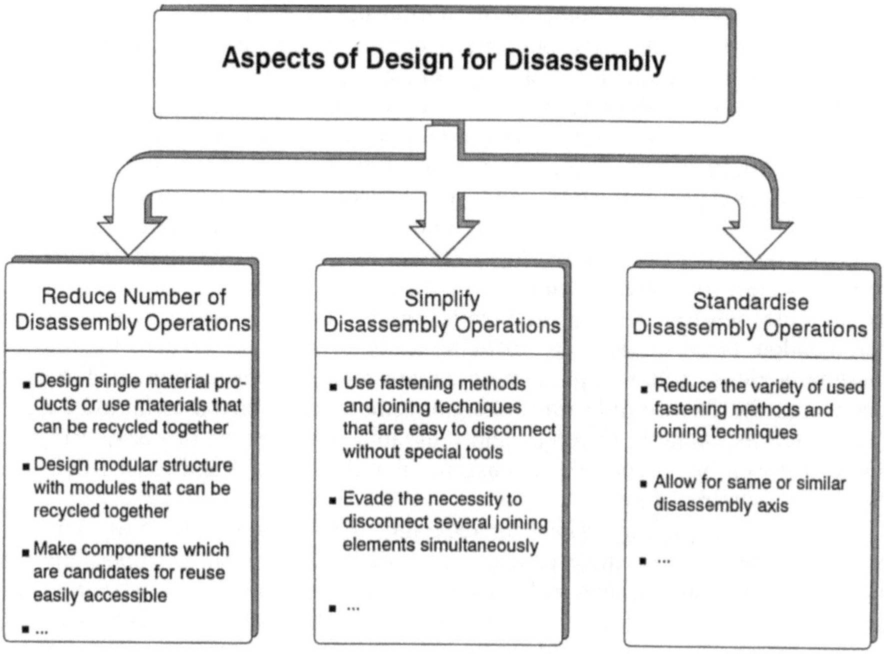

Fig. 4.11.12. Aspects of design for disassembly

4.11.5.3
Design Rules and Guidelines

Design For Recycling is still subject to research [14]. A specific problem lies in the complexity of the problem, because most design decisions are influenced by many factors and because they also influence many downstream decisions [2]. The aim of research must be to provide the designer with a set of guidelines which should be simple, easy to apply and easy to evaluate. The current state of knowledge, however, is a more or less unstructured collection of specific rules which are hard to apply to realistic design problems. Forming groups of rules related to the same design aspects indicates a possibility for structuring the rules presented here.

The first priority for recycling is to extend the life span of products by recycling during usage. It aims at ensuring that the product can fulfill its function for a longer period, using minimal resources. Significant to the Design For Recycling rules is the ease of disassembly (fig. 4.11.12.), since the disassembly process allows the regaining of intact components and materials of higher purity than the alternative shredding process.

Design for disassembly can also be an advantage for other stages of the product life cycle, i.e. for convenient packaging and transport during the distribution phase or for repair and maintenance during the usage phase.

This paper focusses on disassembly during the recycling phase. Disassembly should not be regarded as the opposite operation of assembly, because the time horizon of the operation, the task requirements and the product conditions are different. However, while taking a system approach, the design-for-assembly and the design-for-recycling activities should be done concurrently in order to achieve a global optimum of the system. However, conflicts cannot always be avoided. For example, design-for-assembly rules require reducing part counts by combining functions [1]. This reduces assembly times, but also complicates disassembly efforts and recycling operations when different materials are involved. As another example, design-for-assembly rules recommend snap or press-fit mounting whenever possible, whilst - from a disassembly point of view - metal inserts are not recommended because they are difficult to remove by a worker. Design-for-disassembly rules include e.g. forming subassemblies from harmful materials in order to cut the disassembly time, designing the easiest access to harmful materials, as well as valuable materials and reuseable components, or considering the optimal disassembly sequence, i.e. ensuring that the harmful and reused components are disassembled first.

An issue of major importance for recycling is the usage of the proper materials. If recycling of the whole product or product components is impossible, the combination of materials has to be separated and regained. Rules regarding this aspects are, for instance [16], to select environmentally compatible and recyclable materials for components, to reduce the volume of plastic and composite materials used, because most recycled plastic materials today find their way into less demanding applications, and to avoid secondary finishing operations such as painting, plating, coating and so forth. In order to ensure convenient sorting, similar materials and colors should be used for a given part or assembly. Dissimilar materials must be identified and separated, often a labour-intensive operation. If different materials must be used, the parts should be identified. In order to permit simple shredding, non-shreddable materials should be avoided, e.g. concrete blocks as a counterweight in a washing machine.

Besides the design of the product for ease of recycling, another important factor is also to make sure that the product is actually fed back to the recycling process by the last user. This can be assisted by designing the product in such a way that it can be transported easily after usage, i.e. by allowing pre-disassembly and by developing a simple and efficient system support approach which will encourage the consumers to start the recycling process. At the same time, this approach must also be cost-effective.

All the given rules affect the efficiency and effectiveness of the recycling process. They must be integrated into the conceptual design stage and evaluated simultaneously to the manufacturing, maintainability, reliability and other design rules of a product.

4.11.5.4
Other Design Targets

Today's experiences gained in the field of disassembly can not be considered as sufficient. Only the systematic collection, processing and exploitation of know-how about the disassembly of products which comes back from today's usage market, will lead to the improvement of future product generations. Additionally, design for recycling and disassembly is not the only target to be fulfilled. Other goals, such as design for function, security, ergonomic requirements, production and assembly simplicity, quality fulfilment and cost requirements can dominate or even contradict the requirements for design for ease of disassembly.

An example in which design for ease of disassembly is not consistent with design for ease of assembly should illustrate this. As an example, consider a complex product, designed for ease of disassembly by integrating predetermined breaking points into parts. Predetermined breaking points enhance the product's accessability to components but, at the same time, reduce the ability of the parts to withstand forces during the assembly process.

In most cases, a design for ease of disassembly also facilitates the assembly process, but a design for ease of assembly does not always mean easier disassembly. The following is an example in which ease of assembly is not consistent with ease of disassembly. Bonding is considered to facilitate the assembly of parts. Bonded parts, however, have proved to be very difficult for both, disassembly and recycling. Adhesive tapes, for example, are very difficult to separate from parts because they are non-rigid and difficult to manipulate. Unless separated from the part, the glue often causes harmful emissions when parts have to be processed for material recycling.

4.11.6 Conclusion

Legislation on resource saving and environment protection is hitherto regarded as a strategic disadvantage for manufacturing enterprises. On the contrary, this is a rather promising challenge that can only be met by the development of innovative disassembly and recycling processes for products already designed and by the analysis of the entire life cycle of future products from a more holistic point of view. Innovative enterprises could pioneer a new image for ecological thinking as a strong competitive factor.

4.11.7 References

1 Boothroyd G, Alting L (1992) Design for Assembly and Disassembly. Annals of the CIRP Vol 41 No 2
2 Garbe E, Salomon H (1989) Recyclinggerechtes Konstruieren - Erfordernis moderner Produktgestaltung. VDI-Z 4: 79-83
3 Henstock M E (1988) Design for Recyclability. Institute of Metals, London
4 Hentschel C (1993) The Greening of Products and Production - A New Challenge for Engineers. In: Advances in Production Managment Systems (B-13). Pappás I A, Tatsiopoulos I P (eds), Elsevier Science Publishers B V (North Holland), IFIP

5 Hentschel C, Seliger G, Zussman E (1994) Recycling Process Planning for Discarded Complex Products: A Predictive and Reactive Approach. In: Proceedings of the 2nd International Seminar in Life Cycle Engineering CIRP RECY '94. Meisenbach Verlag, Bamberg: 195-209
6 Hentschel C, Seliger G, Zussman E (1995) Grouping of Used Products for Cellular Recycling Systems. Annals of the CIRP, Hallwag Science Publishers, Bern Stuttgart Vol 44 No 1: 11-14
7 Käufer H, Stricker U, Windelen-Hoyer U (1992) Möglichkeiten und Bewertungskriterien für Heraustrennen und Recycling von Kunststoffen aus komplexen Altprodukten. Müllverbrennung und Reststoffverwertung. Berlin
8 Kriwet A (1995) Bewertungsmethodik für die recyclinggerechte Produktgestaltung. Dissertation, Carl Hanser Verlag, München
9 Riller P (1991) Wege zur recyclingfreundlichen Konstruktion von Elektrohausgeräten. VDI-Berichte 906, Recycling, VDI-Verlag, Düsseldorf
10 Seliger G, Hentschel C (1994) Disassembly Process Planning to Support the Recyclability of Used Technical Products. In: Proceedings of the Vision EUREKA Conference, Industrial Opportunities in Waste Management, Lillehammer/Norway, June 13-16. EUREKA Büro, Köln
11 Seliger G, Hentschel C, Wagner M (1995) Disassembly Factories for Recovery of Resources in Product and Material Cycles. In: Proceedings of the International Conference PROLAMAT on Life Cycle Modelling for Innovative Products and Processes, IFIP Working Group 5.3 Life Cycle Modelling for Innovative Products and Processes. Great Britain. Chapman & Hall: 56-67
12 Seliger G, Krüger St, Neu St (1992) Rechnerunterstützte Bereitstellung von Montageprozeßwissen für die Konstruktion. VDI-Berichte 999, Montage und Demontage, VDI-Verlag, Düsseldorf
13 Seliger G, Schmidtmann R, Hentschel C (1995) Wirtschaftliches Recycling von Bildröhren durch Demontage. Zeitschrift für wirtschaftlichen Fabrikbetrieb (ZwF) Vol 90 No 6: 295-297
14 Seliger G, Zussman E, Kriwet A (1993) Integration of Recycling Considerations into Product Design - A System Approach. Paper presented at the NATO ARW, Italy, June 6-11
15 Schmidt J (1993) Im Schrittempo - In zahlreichen Pilotprojekten werden Erfahrungen mit der Altautoverwertung gesammelt. Müllmagazin No 1: 34-42
16 Thomé-Kozmiensky K J (1992) Materialrecycling durch Abfallaufbereitung. EF-Verlag für Energie und Umwelttechnik GmbH
17 Zussman E, Kriwet A, Seliger G (1994) Disassembly oriented Assessment Methodology to support Design for Recycling. In: Annals of the CIRP Vol 43 No 1, Hallwag Publishers Ltd, Bern Stuttgart

5
Conclusions and Outlook

U. Jürgens · T. Fujimoto · K. Shimokawa

In this book we have tried to identify the trends and options of car manufacturers with regard to assembly automation and work organization. Through various contributions from researchers and practitioners, it has become clear that there are various solutions to the problems they face in this area. There does not seem to be one best way.

The first section of this concluding chapter deals with the lesson for factory automation to be learnt from the Japanese experience in developing the Just-in-Time production system: factory automation has to correspond with the human resource and personnel development requirements which are essential to lean production systems. In the second section we would like to highlight some of the design choices with regard to assembly automation drawing from the contributions to the book and finally, in the outlook we would like to show some of the alternative directions assembly factories might take in the future.

5.1
Lessons to be Learnt from the Japanese Style of Production and their Application to Factory Automation

5.1.1
Diversity of Strategies Between East and West in the 1980s

In the past, the development of automation technology provided the major impetus for changes to the production system and to the work organization of automotive factories in all major auto-producing countries. The dramatic advances in automation have been especially apparent in areas such as casting, machining, body assembly, painting. The final assembly line, however, has been the greatest bottleneck towards automation. One of the common concerns of world auto makers has been how to develop the automation of this stage of the work process, which is highly labor intensive and thus creates many problems in the organization of work. The final assembly line requires a higher level of skillfulness and individual judgment, and has thus been considered unsuitable for automation for a long time.

In the period from the early 1980s to 1988, when assembly automation boomed worldwide, Japanese car manufacturers tended to be conservative in introducing automation into their assembly lines. In contrast, European auto makers such as Volkswagen and Fiat tried more aggressively to develop assembly automation, and as a result the automation ratio among European makers was generally higher than among their Japanese counterparts. Among US auto manufacturers there was a contrast between those more eager to push ahead with assembly automation and the more conservative ones. For example, GM's Hamtramcks or Linden plants started with higher automation levels at an early stage, whereas Ford's Chicago and Atlanta plants tended to be more conservative and took a step-by-step approach while waiting to enhance the development of skill levels among their shop-floor workers.

In more recent years, as the assembly automation boom took hold among Japanese car makers as a response to the labor shortage, the contrast between Western auto makers eager to introduce assembly automation and the more reluctant Japanese makers has diminished. During this process, however, the question of what kind of production system is best suited has emerged as the most critical issue with regard to the further increase of assembly automation. At the beginning of the 1990s the view expressed by the authors of the "Machine that Changed the World" had a major impact on Western automakers. According to this view there are two main paradigms of production systems: high-volume mass production, which has traditionally been dominant in the US and in Europe, and the Lean Production system, which was developed mainly in Japan. There are dramatic differences in the way production systems operate between the two extreme paradigms. One way is based on a system of subdividing working tasks and creating monotonous jobs, while the other attaches more importance to workers' multi-skillfulness, to broader job classification and to the improvement of team work through quality circle activities and Kaizen. Thus, introducing higher assembly automation could either be guided by the attempt to preserve the high-volume mass production principle or by the aim to establish lean production principles.

The difficulty of discerning strategies for automation is compounded by the very diverse approaches followed by manufacturers nationally and internationally. Even within one country, manufacturing strategies for assembly automation differ among individual auto makers. For example, many auto manufacturers have introduced robots for installing batteries, tires, instrument panels, bumpers, window glass, and so on. These processes are relatively amenable to automation and have a more obvious rationale from the viewpoint of eliminating heavy work and mistakes. But with respect to some operations, such as assembling the hood and the trunk or installing the engine, car makers are hesitant to introduce automation because sequence continuity requires higher precision. One major reason for the delay of automation in the assembly line in Japan was the fact that the reliability of high-tech devices such as robots, sensors and artificial intelligence (AI) was regarded as inadequate.

Furthermore, even if these devices become more reliable, it would be impossible to operate flexible assembly line and production systems without the ability to respond quickly to abnormal trouble situations and changes in the assembly line process in the same way as in the case of an assembly line manned by multi-skilled workers. In contrast to Japanese plants, Fiat's Cassino plant, for instance, tried to keep its operation ratio up by means of vehicle buffers flanking the automated zone. Japanese plants, in general, keep a lot less buffers, and therefore have to ensure a high level of reliability and maintenance capability for automated equipment, making multi-skillfulness all the more important. This raises the question of the relationship between the type of automation and the skillfulness and judgment capability required from shop-floor workers.

5.1.2
Factory Automation and the Just-in-Time System

The success of the just-in-time production system (JIT) has had a global impact, and the JIT revolution has spread worldwide. With everybody participating in quality control and other work place activities so as to improve the existing process, the JIT system inspired such issues as team work, problem solving and humanization of work, for example job enrichment.

In recent years automation and total system rationalization have advanced with the use of various high-tech machinery and information systems, and the challenge has been how to combine the JIT way of thinking with these innovations. As a result, a high-tech JIT system making use of local area networking (LAN) and the *electric sign board (denshi kanban)* system has emerged. More-over, a total JIT system will grow out of the moves to speed up the development process, which is itself linked to the development of the CAD/CAM network. Thus, the process of automation may lead to a situation where just a handful of production engineers will be able to operate the whole production process. This apparently would be in contradiction to the traditional practice of fostering a multi-skilled work force

When discussing this vision of the unmanned production line, it is important to note, however, that there are basic conceptual differences between seemingly similar approaches when it comes to the question of just what sort of automation to aim for and how to integrate multi-skilled workers into the automated work environment.

Automation in the US and in Europe has been based on the high-volume mass production principle pioneered by Ford: reducing task variety, as far as possible, to achieve a sequence of unchanging tasks carried out by workers trained for one role only. The idea was to improve the efficiency of large lot production which relies on inspections to ensure product quality. Consequently, the US and European automated systems traditionally lacked the flexibility required to adapt to design changes or to produce a wide variety of products in small quantities. High-volume production of a small range of products can be done very efficiently and product quality can be maintained if inspections are carried out properly, but any changes in product design or in the production system require special equipment and the mobilization of groups of specialists, and all this costs a lot in terms of time and money. A necessary condition for the continuation of this high-volume, low-variety system with regard to further automation would be a sharp reduction of the work unskilled force as opposed to the work force with multiple skills. There has therefore been a general tendency to eliminate multi-skilled workers from the shop floor in the Western system. More often, specialists such as production engineers and industrial engineering technicians will be responsible for designing and operating automated systems.

In contrast, the emerging Japanese concept of automation is one of flexible production. Thanks to the multiple skills and judgment ability cultivated on the shop floor, this system can cope with short product lead times and consequently with a wide variety of products produced in varying quantities. The reason for this is that the workers on the Japanese final assembly lines have, to a high degree, developed a variety of abilities, enabling them to deal with the constantly changing demands of the flexible JIT production system. These are abilities which would be very difficult to transfer to a group of robots. In other words, any automated system introduced in Japan must be an extremely flexible one, one which will carry on the tradition of JIT production and which can fully absorb the range of technical skills and the power of judgment cultivated in the work force.

As systems become more highly automated, it will become necessary to transfer to computers the skills and subtle judgment abilities which enable highly trained workers to handle all aspects of the production process from machine layout to the fast and efficient change-over of jigs and tools. It is, however, inevitable that breakdowns and abnormalities of one sort or other will occur in such a highly integrated automatic system. The ability to deal with such problems quickly, or rather to anticipate and prevent them, is only to be found in expert workers with many years of experience. A computer programmer can input given data to control certain aspects of the automated process such as the tool position, cutting order, conveyor speed, and so on. But it is too much to ask of a programmer to judge just which tools, speeds and positions are most appropriate.

It is commonly said that it takes between ten and twenty years of experience to become an expert at jobs such as welding, die-making and machining. Such experience cannot be disregarded when designing and operating an automated production line. The task of "teaching" a specific action to a robot should properly be entrusted to a worker with such experience. As mentioned earlier, the move towards unmanned automated production lines has been taking place at a startling pace in recent years. As automation proceeds, the accuracy of production processes and the standards of reliability requirements get higher, but no matter how automated the line may ever become, it is still dependent for its operation on technically skilled personnel, on the sophisticated software they design, and in particular on the knowledge and judgment which comes from personal experience and training. No robot, however advanced, can replace this experience, which means it will be impossible to transform all processes to become completely automatic and unmanned. We will rather see a situation where production lines appear to be totally unmanned at first glance, but where in fact highly skilled workers are present to operate the system and to play a back-up role.

The JIT system, which depends on the participation not only of production engineers but also of foremen and skilled workers for the design and implementation of new production processes and work standards, shows the importance of abandoning the elitist management system of the European and the American high-volume manufacturers. MIT's International Motor Vehicle Program called this sort of JIT production system "lean" and indicated that the "lean revolution" is spreading worldwide. The JIT system has been adopted worldwide thanks to the trials it has gone through in the production facilities of Japanese companies, where its strength has gradually become clear. Adapted to constantly flexible production of a varying range of products, the JIT system itself has no fixed form because it is the result of the combined effort of production engineers, foremen and multi-skilled workers all constantly working together to locate and solve problems. The danger of pursuing automation is that, beguiled by the expanding capabilities of sensor-/AI-equipped robots taking the place of humans, CIM computer systems and all the other technology which will make a *total system* increasingly possible, the stampede towards unmanned automation may go ahead without due regard being paid to shop-floor improvement activities.

Nowadays, of course, a high level of automation cannot be achieved simply by involving large numbers of people in shop-floor improvement activities. The development of today's automation up to higher levels demands that workers who are already skilled have their knowledge extended further to cover an even wider range of skills. In this case, the "knowledgeable" worker needs, in addition to his multi-skill know-how, a knowledge of micro-computer software to be able to transfer his own abilities into the form of a computer program or data base. This ability will come about through close cooperation with system engineers. As a recent example of this, Honda invited skilled workers from its mass production plants to undergo training at its Tochigi plant, where it produces the sophisticated NSX sports car. After their training, these workers returned to their plants and played a central role in the implementation of new highly technical automated production lines.

An issue of vital importance when considering the development of high-level automation is that of preventive maintenance. "Knowledgeable" skilled workers are essential for the anticipation and effective prevention of malfunctions in the process. Specially trained workers in many Japanese industries now hold the key to high-level automation, which is doing away with the so-called "3K" jobs. Here then we are not talking of automation simply for the sake of cutting the work force but automation to create a more human working environment employing highly trained people.

It may be the ultimate direction of assembly automation to move towards totally unmanned lines such as the assembly lines for cameras, VTRs, semiconductors etc. Yet in considering unmanned lines there are several different images. One is the image of eliminating direct workers by introducing a completely unattended line, the other is a line backed up by shop-floor workers and engineers who have the human skills and judgment capabilities, the appropriate software as well as the maintenance know-how. The risk of the first image would be to exclude the multi-skilled workers who had maintained the line up to now and to weaken their growth potential. Thus the unmanned line should not be an objective in and of itself. Knowledgeable skilled workers educated in the use of micro-computer software, with the support of expert systems and with machine-maintenance capabilities should remain the master of the unmanned line.

Looking back at the history of today's factory management, the enormous contribution made by the Taylor and Ford systems hardly needs to be emphasized once more. This was the introduction of the principle of systematic management, based on research into methods of motion and time studies for various activities, which brought about great improvements in efficiency and led the way towards a revolutionary *total system* style of synchronized mass production. However, a flaw lurked in Taylor's system in that management was placed exclusively in the hands of an expert elite. Ford too, after experiments in line synchronization in which foremen, skilled workers and other staff from a variety of areas participated by making suggestions and carrying out improvement activities, came up with a total system, but then reverted right back to Taylor's elitism. The Japanese-style production system has avoided these shortcomings. If, as automation progresses further, the lessons learnt from the Japanese experience are implemented in factories worldwide, then we will perhaps be able to talk of a human revolution in manufacturing industry.

5.2 Design Choices for Assembly Systems

In the following section we want to highlight some of the design choices drawn from the contributions of this book specifically with regard to assembly automation. One of the striking facts that we identify from the case studies in this book is the variety of the approaches that the auto companies have chosen. Such a variety can be found not only between the Japanese and European groups, but also within each regional group. There is a wide range of alternative assembly systems from high-tech automation to neo-craft concepts, even among the Japanese makers.

(1) Layout: Long Assembly Line, Short Line, or Single-Station Booth
As for plant layout for assembly lines, though, there seems to be two approaches: long line approach versus short line approach.

(a) The long line approach apparently assumes that assembly automation requires longer main lines and more work stations (typically around 200). Flexibility of layout for future automation is emphasized in this case.

(b) The short line approach, by contrast, tries to shorten the length of the main line (typically less than 150 effective stations) by introducing modular vehicle designs and expanding sub-assembly lines accordingly. The main idea in the latter case seems to be to make the main assembly line simple.

Japanese auto assembly lines have usually been long yet, though, relatively compact partly because of multi-skilled workers, each of whom can conduct multiple jobs in a limited space and time. State-of-the-art robot assembly systems are not as space-efficient as these human workers, however. As a result, introducing robot systems into existing assembly lines tended to make them significantly longer, creating a space problem. Some companies did patchwork such as appending extra work stations and expanding the building, but more fundamental changes would be required to automate assembly further. When building new assembly factories, many of the Japanese makers tried to incorporate this space problem into their plant design by expanding the building areas.

There is a radically different layout of assembly processes, though: the Volvo-Uddevalla-type stationary assembly system, which eliminated assembly lines altogether.

(2) Carrying the Bodies: Continuous, Shuttle, or AGVs
Conventional assembly lines have used continuous conveyors (overhead hangers and floor-based) since the era of Fordism. Unlike human workers, though, robots are not good at tracking motions of moving conveyors.

There are several ways of dealing with this problem. (a) Stationary intelligent robots automatically track the continuous motion of the conveyor while assembling parts. This method requires sophisticated sensors and robots, and is likely to be expensive. (b) Synchronized robots move with the conveyor, so that relative speed between them is zero. The robots and the conveyors may be mechanically linked, or they may be synchronized by remote sensing devices. This type of synchronized robot assembly, which tends to consume space, was found occasionally when one isolated robot station existed in the middle of a manual assembly line. (c) Shuttle transfer mechanisms, which are common in automated welding lines, are disconnected from the continuous conveyor lines of the manual area. Several robotized stations form the *automated area* in this case. This type has been found most often in Japan, including the earlier versions of assembly automation. (d) Automatically Guided Vehicles (AGVs), which were popular particular among European manufacturers until recently: Some of the latest assembly lines use rail-guided AGVs to carry the bodies (tape-guided AGVs are used for parts feeding, but not on the body side). Unlike the conveyors, AGVs can change speed and pitch semi-independently among the bodies; they stop for robot assembly. This is the most flexible system, but it is also rather expensive and has other draw-backs as discussed in various chapters in the book.

(3) Manual Areas and Robot Areas: Separated or Mixed
It is said today that assembly automation over 50% would be difficult to justify, and that direct assembly workers will remain in the lines in the long run. If we cannot eliminate direct assembly work altogether, should robots work closely with assembly workers, or should they be totally isolated from manual assembly operations? There seems to be two approaches.

(a) Separation: this approach isolates manual work stations from the robotized stations. One reason for this is to avoid potential alienation and isolation of workers in the automated assembly lines. When robots and direct assembly works are mixed, the teams of workers may be disjointed by the robots intervening between the human links. Some companies put several robotized zones together in one area, so that manual work stations and robotized stations are designed to be totally separate. Another company envisions an overhaul of product designs and process layout which will concentrate all the robotized stations in the upstream area of the assembly lines. In any case, the basic idea is to decouple the two zones so that manual stations and robot stations can be separately designed and optimized. "Human-friendly" concepts for assembly organizations may be implemented for the latter zone.

(b) Mixed: In this case, more than one island of automation exists in the middle of manual assembly areas. In some companies, this is simply a result of space constraints on plant layout; other companies may pursue this idea on purpose, so that direct workers can interact with robots. The basic question here is who should handle the "residual work" that robot assembly creates. The separation idea seems to assume that robot operators handle such residual operations, or that a separate group of workers does the residual jobs. The latter case would lead to a classic problem of alienation by automation. The *mixed approach*, however, assumes that direct assembly workers in the manual assembly stations handle minor maintenance and monitoring of the robot stations next to their areas. Thus, the mixed approach may be accompanied by a further job enlargement program in which direct assembly workers are trained to handle a part of robot operations.

(4) Job Design and Training: Maintenance, Robot Operators, and "Residual" Workers

Assembly automation poses a challenge to the traditional system of multi-(on-the-job-) skilling. Thus, job classification of the Japanese final assembly lines has been relatively simple, having two main categories: direct or semi-direct workers and maintenance workers. However, introduction of assembly robots may complicate this job structure. That is, introduction of robot automation may create two "classes" of workers: operators doing teaching, monitoring, minor maintenance etc., and direct workers doing residual activities that robots do not do. This means that assembly automation may create three basic job categories: maintenance, robot operators, and direct workers. The first two types of workers control the robots, while the latter job may be controlled by robots.

Robot operators are recognized as fairly skilled workers. In one company, a robot operator's skill level is regarded as equivalent to that of team leaders in manual assembly areas. These operators have been formally trained through off the job training programs and pilot plant operations at this company.

A potential danger of the emerging work organization described above is that it is difficult to make the residual job meaningful to the workers doing just this. Although "humanization" of assembly jobs may progress on the manual assembly area, it would be more difficult to solve the problem of work alienation in the automated zone. This is particularly the case when the robotized zone is clearly separated from the manual zone, as this may mean isolation of the "residual" workers on the automation side. The problem would be solved if the residual work for automation could be eliminated altogether, but this would be technically and economically difficult for the time being.

One way to avoid the work alienation of residual workers might be to let the robot operators do the peripheral jobs through job enlargement and job rotation. Another possibility may be to combine direct assembly jobs and the residual jobs in a "mixed" assembly line where robots and manual assembly workers are located adjacently to each other.

(5) Automating Main Line or Automating Sub-assembly Lines

When modular product designs are adopted and sub-assembly lines are expanded in many of the latest assembly plants, a question may arise as to where to automate first. In most of the automated assembly plants, automation of the main lines (trim, chassis, final, etc.) has preceded automation of sub-assembly lines.

Generally speaking, it may be easier to automate sub-assembly operations, as they tend to handle smaller, lighter and simpler objects. In addition, the operations become more easy to automate as they can be organized as a stationary process, and thus, problems related to the line flow do not have to be taken into account here. From the point of view of improvements in work conditions, though, higher priority may be placed on certain main-line jobs that handle heavy components.

(6) Parts Alignment: Mechanical versus Visual Sensing
Dimensional accuracy is a bottleneck for automating many of the assembly operations. Alignment of parts (robot hands) and bodies (jigs) is a key to such accuracy. The question is how to assure the body-to-parts alignment. There are two approaches adopted by different auto makers.

(a) The first approach is to rely on mechanical accuracy of jigs and components while using relatively simple robots and open-loop controls. The motions of the robots and clamps are not automatically adjusted, although defects in bolting may be automatically detected and manually reworked at the downstream stations. If the rework ratio can be kept within a certain limit, the mechanical alignment is relatively inexpensive.

(b) The second, and technologically more advanced approach relies on visual sensing mechanisms (e.g. video cameras) and closed-loop controls. Before a component is bolted into a hole on the body by a robot, for example, another (or the same) robot locates the hole and sends signals to the assembly robot, which adjusts its motion to the location. Video cameras, laser and other visual sensing devices are used for measurement. The visual approach is consistent with the *do it right the first time* principle of Total Quality Control, but it is usually more expensive than the mechanical method. Also, the visual method adds complexity to the control system of assembly operations, which may deteriorate the productivity of the plant through frequent breakdown. The choice between the mechanical and visual methods thus seems to be subtle. The best solution may be a certain combination of both approaches.

(7) Mixed Model Production: Sub-assembly Lines, Bypass Stations, Variable Pitch
Most of the Japanese assembly lines are mixed-model lines in that the same assembly line can handle a random sequence of multiple body types and even multiple platforms. A potential problem of such lines is how to handle models of different product content (e.g. large cars and small cars). A significant idle time will occur if the line speed is adjusted to the most time-consuming highest labor content version. The conventional Just-in-Time factories have solved this problem by a combination of continuous conveyor lines, which can absorb variation of cycle time, and levelization (Heijunka) of model mix.

When models of very different product content are assembled by a robotized assembly line, though, the above approach may not be enough. (a) One approach to this problem is to absorb the inter-model difference in sub-assembly lines, so that the main line can handle any models without much adjustment. (b) The second way is to use bypass stations in the main assembly line, a system which is used only for models of high labor content. (c) The third and most sophisticated way is to use AGVs which can automatically adjust pitches (i.e., intervals between bodies) according to the product content of each product. We found these three methods in some of the latest assembly plants in Japan.

It should be noted, however, that the Udevalla-type assembly operation, by definition, is free from the *line balancing* problem, as there is in principle only one station per *line*. The system is extremely capable of absorbing the differences in product content across products.

(8) Sequential Parts Delivery

Another potential problem in automating mixed-model assembly is how to handle the complexity that is added by the variety of the model mix. An assembly task normally includes identifying the model type and picking the right component for it. Eliminating these preparatory activities would facilitate robotization of mixed-model assembly lines. An approach to this problem is the sequential delivery system, in which components are supplied to the assembly station in a sequence that exactly matches the sequence of body variations. In this way, the robots can concentrate on direct assembly jobs. This system is particularly effective when the component is specific to vehicle variations (e.g. seats and instrument panels).

There is an increasing number of parts that are shipped in sequence by outside suppliers, or are sequenced inside the plant. So far, however, the number of parts to which sequential delivery systems can be applied is limited due to lead-time requirements. That is, information on final sequence of painted bodies is released as late as the beginning of the final assembly itself, which makes delivery lead time for sequential delivery very short (typically a few hours). For example, seat makers can supply seats sequentially only when the seat factory is located within an hour or two by truck. The lead time allowed would be much longer if the sequence information could be released when painting starts, but this confuses the system, as the sequence usually changes due to the needs for paint touch-up. Thus, there is a trade-off on how early the sequence information is released to the suppliers.

(9) Design for Automation

It is generally said that design for automation (DFA) is more advanced among some of the European volume producers (e.g. VW, FIAT) than among the Japanese makers. Unlike the Japanese consumer electronics makers, who aggressively pursued the concept of design for automation (e.g. reduction of parts number, stack-up design, etc.), the Japanese auto makers were not so active in DFA in the past. Even after the auto makers accelerated their efforts to automate assembly operations in order to respond to the problem of labor shortage in the late 1980s, the companies were not as interested in DFA as some of the Europeans. If the Japanese auto companies were to increase the degree of assembly automation beyond the current frontier of about 20% (based on the definitions by the auto makers), they would have to make the designs more friendly to assembly equipment, reduce product variety, reduce parts variety, and reduce the number of parts for each product. Some of the Japanese have already started to move in this direction. With DFA efforts, according to some Japanese engineers, it is technologically possible for the assembly automation ratio to eventually go up to about 50%. Further automation of assembly may, however, require breakthroughs in product technologies.

There is a concern that design for automation, when pursued too much, might deteriorate product marketability. When the degree of assembly automation reaches a certain level, the companies may face a subtle trade-off between marketability and manufacturability for automation.

(10) How Far to Automate

As the auto companies increase the degree of assembly automation, they may face the fundamental question of how far to automate in the first place. There may still be a belief among production engineers that the higher the automation ratio the better, and that the ultimate goal is 100% automation with total elimination of human labor. Other engineers, though, say that total elimination of human jobs is unrealistic, and that the degree of automation should be carefully controlled in order to optimize the human-machine interface. In other words, there may be a certain optimal level of automation, beyond which motivation of workers for process improvement is suppressed, work teams are disjointed, attractiveness of the production areas to potential workers decreases, and total manufacturing cost increases.

The decision on how far to automate also applies to the design of automation systems at each work station. Some Japanese companies, for example, install front seats into the cabin by robots but bolt them manually; some others go ahead and have the seat bolted to the body by robots. The former group argue that their decision is based on the consideration of the costs and benefits of assembly automation. The cumulative effect of such microscopic decisions for robotization might create a significant difference on optimal automation ratios among the companies.

As of the mid 1980s, virtually all the Japanese auto companies were still emphasizing the principles of low-cost automation. The upper limit of investment for a robot that eliminates one person per shift of human labor was about 5 to 10 million yen across the companies, which was very conservative. As of 1991, though, they split into two groups: one still emphasizing low-cost automation, and the other raising the upper limits for automation investment. The latter group argued that, in order to cope with the long-term labor shortage, they had to make the production lines more attractive by automating certain processes even though this meant cost increase. The investment for assembly automation was also accelerated by the favorable financial conditions of the late 1980s.

Now that the labor shortage problem has been alleviated, ironically, by the recession, and that financial conditions are less favorable, some of the companies are emphasizing cost factors again. Although it is hard to distinguish a long-term trend toward more liberal investment on assembly automation and short-term cycles of macro economies and profit performance of the firms, the decisions on how much to pay for assembly automation will become a strategic issue for each of the auto companies.

The above list of potential choices which auto firms may make is by no means complete. The cases presented in this book also indicate a variety of approaches in designing, automating and organizing assembly processes. To sum up, the book has demonstrated that both convergence and diversity existed in the auto firms' efforts towards transforming assembly. There may have been a tendency towards convergence at the "philosophy" level, but at the concrete implementation level, a wide variety can still be observed - and thus a wealth of opportunities for mutual learning.

5.3 Outlook

In the early 1990s, we experienced a peculiar asynchronism regarding the direction of design for assembly systems: whilst Western manufacturers re-focused on the classic issues of balancing and scheduling assembly systems for better efficiency and quality, not least by introducing team work, Japanese manufacturers were thinking of how to improve the quality of work in assembly systems. At the same time, a consensus had been reached among car manufacturers with regard to the basic principles of a Just-in-Time process organization. Thus, team work and the problem-solving approach - from bottom up - which both contradicted previous principles of mass production in the West, - has by now been introduced almost everywhere. In this way, Western manufacturers have improved their capabilities in the field of manual labor, whilst Japanese manufacturers have recognized the limits of automation.

One may wonder now, if the questions raised in this book may perhaps have lost some of their relevance since the wave of investments in new plants and the refurbishment of existing plants seemed to have ceased by the mid 1990s and it may take some time before the next wave of modernization takes place. The dynamics of restructuring in the automotive industry are unbroken, however, and due to new configurations and company strategies, the current solutions may become obsolete faster than expected. It seems to us that with a view to the future, a further differentiation of types of assembly systems corresponding with different strategic contexts could evolve. In this context, we see four concepts which could be of relevance for the assembly systems of the future:

(1) The *re-standardized serial flow system* which seeks the classic economies of scale advantages by reducing model mix variation in order to minimize balance losses or quality problems. Such a system would correspond to attempts towards parts commonization and standard option packaging, as well as to the platform approach which has become part of the new competitive strategy by many car manufacturers. Standard option cars could be manufactured on *runner lines* and optimized by extensive effort to standardize the process. A constitutes the common *underbody* for different model lines. This approach which allows the standardization of bodyshop operations calls to mind the separation of chassis and assembly plants which had historically already existed in the auto industry. Certainly, a centralization of the chassis production at one location, with deliveries to different assembly locations as in the Ford system of the 1920s, will not be feasible any longer due to just-in-time considerations and environmental concerns. Nevertheless, standardization of the body shop processes in different plants would offer scale economies in purchasing process equipment and certain other advantages. In any case, under these conditions, a high degree of automation seems most feasible and appropriate.

(2) The *flexible model mix and model change system* which is capable of dealing with a high degree of variation in terms of the production program and model change activities. This type of system would correspond to the platform ap-

proach, i.e. it has the flexibility to produce the different "hats" on a common platform. Technology and work organization have to allow for high model mix flexibility as well as for the frequent intermittent new product innovation. Automation will have to assist rather than substitute human labor in coping with the challenges of such an environment.

(3) The *supplier consortium system*, this is where the car manufacturer delegates much of the manufacturing activities to suppliers and limits himself to coordinating and auditing the process. The suppliers build their modules near to the main line and deliver or even install them here, taking full responsibility for the process and work organization. In such a system, assembly work would, to a large extent, lose its central role in the process of car manufacturing. Installing the modules into the car would be the task of fitting experts employed by the suppliers; car assembly is hence the result of a supplier team cooperation. They would be assisted by mechanical devices; a high degree of automation, however, does not seem to be appropriate as this would possibly interfere with the clear lines of responsibility.

(4) The *team-centered cellular assembly system*, this is where self-directed teams are responsible for assembling the complete car. This assembly concept corresponds to a decentralized, near-the-customer approach, where manufacturing becomes part of the marketing strategy. Customers or dealers could be closely involved in this process. The Uddevalla system of assigning responsibility for the assembly of the complete car to teams is an example of this. The process layout of parallel assembly cells offers little economic incentive for process automation, processes would have to be "human-centered" under these conditions.

Each of these concepts requires a different type of assembly plant or line. They can, of course, be complementary to each other within the framework of the overall company strategy. The implications regarding assembly automation and work organization differ drastically in each case, however. The corridor of strategic choices seems to be broadening with the consequence that questions of assembly automation and work organization remain a topic on the agenda.

6
The Authors of the Book

Jos Benders
Dr., received his doctorate from the Nijmegen University Business School. He received an MBA from Tilburg University, and a Ph.D. from Nijmegen University. His research interests include technical systems and work design, the transfer of Japanese management practices, and management trends and fashions.

Christian Berggren
Prof. Dr., studied sociology and management at the Royal Institute of Technology in Stockholm. He is a senior researcher at the National Worklife Institute in Stockholm and a professor of industrial management at Linköping University. He is currently involved in a study of competitive strategies and personnel policies in four major Japanese industries after the burst of the Bubble Boom. His major publications include *The Volvo Trajectory. Work Organization in the Swedish Auto Industry*.

Arnaldo Camuffo
is an associate professor of Human Resource Management at the Department of Business Economics and Management at the Venice University, Ca' Foscari. He holds a Ph.D. in management from Venice University and a Master of Science in management from the Sloan School of Management of the Massachusetts Institute of Technology. He is a research associate at the International Motor Vehicle Program (IMVP) of the M.I.T. and is the Italian representative of the TIM division of the Academy of Management. He has published books and articles both in Italy and abroad.

Ben Dankbaar
Prof. Dr., is a professor of Business Administration at the Business School of the University of Nijmegen. He studied social sciences and economics at Amsterdam University and received a Ph.D. from Limburg University. His research interests include organizational change, technology management and socio-technical design of organizations. He is a specialist on the automobile industry.

Frédéric Decoster
entered the Renault Car Corporation as an ergonomist in 1982; his first studies were related to the human factors in connection with plant automation; he has written a book describing socio-technical planning for the design of new plants (ANACT, 1989). As a member of the Vehicle Engineering Department, he is now working on industrial organization studies.

Kajsa Ellegård
Dr., associate professor, Department of Human and Economic Geography, School of Economics and Commercial Law, Gothenburg University. She works in the tradition of time-geography and her empirical studies are related to the automobile industry and to the everyday living conditions of individuals in households. Her first study concerned the development of new work organization principles in the process of automation in carbody shops within Volvo - followed by an actor-oriented study during the planning of Volvo Uddevallaverken. She holds a research position in Time Geography at the Swedish Council for Research in Humanities and Social Sciences.

Peter Enderle
Dipl. Ing. (graduated engineer), former Board Member at Adam Opel AG and Executive Director for Powertrain and Chassis Components GM Europe. He studied mechanical engineering at the Darmstadt Technical University and joined Opel in 1962. He has held various management positions, including Plant Manager at Opel's Bochum Plant and was deeply involved in Opel's creation of the most modern and lean manufacturing factory in Eisenach. He also participated in Harvard's International Senior Management Program.

Michel Freyssenet
Dr., sociologist, is a research director at the French National Research Center (CNRS). He is a co-director of the GERPISA (Permanent Group for the Study of the Automobile Industry and its Employees) and the International Program *The Emergence of the New Industrial Models*. His latest article was published by the review *Sociologie du travail* (no. 3, September 1995): *Can 'Reflexive Production' be an Alternative to 'Mass Production' and 'Lean Production'?*.

Takahiro Fujimoto
is an associate professor at Tokyo University, Department of Economics, since 1990. After receiving a BA in economics from Tokyo University in 1979, he joined the Mitsubishi Research Institute where he participated in various projects for the automobile industry. He received his doctorate from Harvard Business School in 1989 with his thesis *Organizations for Effective Product Development*. His major publications include Product Development Performance, which was co-authored by Professor Kim Clark of Harvard.

Claudia Hentschel
Dipl. Ing. (graduate engineer), studied industrial engineering, focusing on mechanical engineering at the Berlin Technical University and the Ecole Nationale des Ponts et Chaussées, Paris. Since 1991, she has been a research assistant at the department of assembly technology, specializing in the planning and control of assembly and disassembly systems. Since January 1995, she has also been a management assistant of the special research program 281 *Disassembly Factories*.

Seiji Honda
is a manager of Body Section No. 2 at the Nissan Motor Corporation Oppama plant. He joined Nissan as a body production engineer in 1978. He has carried out various studies related to body assembly engineering for various types of vehicles.

Liang-Han Hsieh
Dr.-Ing. (doctorate in engineering), studied mechanical engineering at the National Taiwan University and Berlin Technical University. He worked for three years as an engineer at the Taiwan Power Company and several years as a researcher at the Institute for Machine Tools and Manufacturing Technology (IWF) at Berlin Technical University and the Fraunhofer Institute for Production Systems and Design Technology (IPK-Berlin). Since 1990, he has been head of the assembly department at IWF as Chief Engineer.

Ulrich Jürgens
Dr., is a senior researcher at the Social Science Research Center Berlin (WZB) and *Privatdozent* (external professor) at Berlin Free University. He has directed internationally comparative projects in the fields of company strategies, industrial relations and work organization with a focus on the automobile industry. His recent major publications include *Breaking away from Taylorism - Changing Forms of Work in the Automobile Industry* co-authored by Thomas Malsch and Knuth Dohse.

Hiroshi Kinutani
joined Mazda as an assembly engineer in 1967 after graduating from Hiroshima University with a Master's degree in mechanical engineering. Having played a key role in Mazda's CKD projects in Burma and Colombia he became Deputy General Manager of the Trim and Final Assembly Engineering Department in 1987 where he was responsible for the modular assembly project at Mazda's state-of-the-art Hofu Plant. He is currently the General Manager of the Thai Joint Venture Project Office, preparing for the launch of a new pick-up truck plant in Thailand.

Yutaka Kodama
is a senior manager of Production Engineering Department No. 3 at the Nissan Motor Corporation. He joined Nissan as a production engineer in 1970. Over the past two years, he has participated in the new assembly plant project.

Ansgar Kriwet
Dr.-Ing. (doctorate in engineering), studied mechanical engineering at Aachen University. From 1989 to 1995, he was a research assistant at the department of assembly technology, specializing in the development of evaluation methods for design for ease of recycling - this became the subject of his doctoral thesis, published in 1995. In March 1995, he left the department of assembly technology in order to take on new tasks in industry.

Martin Kuhlmann
Dipl. Soz. (graduated sociologist), is a research assistant at the Department of Sociology and is also involved in research work at SOFI (Soziologisches Forschungsinstitut) at Göttingen University. He is a co-author of the „Trendreport Rationalisierung“ [*Trend Report Rationalization*] (1994) and is currently working on work organization, quality of work and team-based work.

John Paul MacDuffie
is an assistant professor in the Management Department at the Wharton School of Business, University of Pennsylvania. He received his BA degree from Harvard University and his Ph.D. degree from the Sloan School of Management at M.I.T. (Massachusetts Institute of Technology). His research explores the relationship of technology, production systems, and human resource policies in manufacturing settings. With Frits Pil, he has recently completed the second survey for the International Assembly Plant Study, from a sample of 88 automotive assembly plants representing 20 companies and 21 countries, under the sponsorship of the International Motor Vehicle Program (IMVP) at M.I.T.

Yasuhito Matsudaira
is Assistant General Manager of the Technical Development & Administration Dept., Assembly & Design for the Manufacturing Engineering Division at the Toyota Motor Corporation. Over the past 4 years, he has participated in new concept assembly line projects. He holds a Master's degree in Electrical Engineering from Nagoya University.

Takashi Matsuo
is a graduate student in Business Management at the Tokyo University Graduate School of Economics. He received his Master's degree (u. A.) from Tokyo University in 1995. His recent research focuses on the evolution of technology and organization in the Toyota Motor Corporation.

Yutaka Mishima
joined Shin Mitsubishi Heavy-Industries Ltd., Nagoya Motor Vehicles Works in 1960. He became General Manager of Production Engineering Department in 1986. He was Chief Executive Engineer in 1989, and in 1990, he became Deputy Works General Manager of Nagoya Motor Vehicles Works. In 1991, he was appointed Director and Deputy Corporate General Manager of the Office of Passenger Car Production Engineering & Control. He is now Corporate General Manager of the Office of Production Engineering of Mitsubishi Motors Corporation. Over the years, he has been conducting various studies related to production engineering for passenger cars.

Kazuhiro Mishina
is an associate professor at the Japan Advanced Institute of Science and Technology (JAIST). He received a BA and an MA in commerce from Hitotsubashi University in 1982 and 1984 respectively. After receiving a Ph.D. in Business Economics from Harvard University in 1989, Mishina taught *Technology and Op-*

erations Management in the MBA program at Harvard Business School as an assistant professor until he joined JAIST in 1995.

Tatsuo Naitoh
is a manager of the Simultaneous Engineering Center at the Nissan Motor Corporation. He joined Nissan as a control systems engineer in 1974. He has been carrying out a study related to new welding and assembling machine control systems.

Atsushi Niimi
graduated from Nagoya University and became involved in Assembly & Design for the Manufacturing Engineering Division at the Toyota Motor Corporation. He has gained experience as Executive Coordinator in Toyota Motor Manufacturing in Kentucky, USA. He is currently General Manager of Toyota's Production Control Division.

Frits K. Pil
is an assistant professor at the Katz School of Business and a research scientist at the Learning Research and Development Center at the University of Pittsburgh. After receiving a BA in economics from Harvard University in 1990, he received a Master's degree in corporate strategy and human resources from Wharton Business School in 1992. In 1996, he received a Ph.D. from Wharton Business school for his thesis *Understanding the International and Temporal Diffusion of High-Involvement HR and Work Practices*. His research and publications have focused on understanding organizational learning and change with a particular emphasis on HR and work practices, and productions systems more generally.

Thomas M. Schmahls
Dipl.-Ing. (graduated engineer), studied mechanical engineering at Berlin and Brunswick Technical University. He specialized on plant layout and assembly technology. Since 1992, he has been partner in the consulting company Thomas M. Schmahls + Partner.

Michael Schumann
is a professor of Sociology and Director of SOFI (Soziologisches Forschungsinstitut) at Göttingen University. He is a co-author of several studies on technical and organizational change and working life questions, e.g. „Industriearbeit und Arbeiterbewußtsein" [*Industrial Work and Worker Consciousness*] (1970), „Ende der Arbeitsteilung?" [*The End of the Division of Labor?*] (1984), „Trendreport Rationalisierung" [*Trend Report Rationalization*] (1994).

Günther Seliger
Prof. Dr.-Ing. (doctorate in engineering), studied economics and technical engineering at the Berlin Technical University. He received his Ph.D. from the Berlin Technical University in 1983. From 1983 to 1988, he was head of the *Planning Technology* department at the Fraunhofer Institute for Production Systems and Design Technology (IPK) in Berlin. Since 1988, he has been a professor of as-

sembly technology at the Berlin Technical University and head of the *Assembly Technology* department at the Institute for Machine Tools and Manufacturing Technology (IWF).

Koichi Shimokawa
is a professor at Hosei University in Tokyo, Department of Business Administration. His research fields cover American and Japanese business history, marketing history and industrial research in the automobile industry. He participated in M.I.T.'s first and second round of the International Motor Vehicle Program. His most recent publication concerns itself with *The Japanese Automobile Industry - a Business History.*

Roland Springer
Dr., is a senior manager for Work Organization and Labor Politics at Mercedes-Benz in Stuttgart. In the 1980s, he worked as a researcher on technical and organizational change.

Katsuhiko Tanase
joined Honda Motor Co., Ltd. in 1964 and was assigned to the production engineering department. He was head of production engineering in the body section of the Honda Engineering Co., Ltd. at the time when this article was written.

Joseph Tidd
Dr., is a director of the Executive MBA Program and head of the Management Innovation Group at The Management School, Imperial College of Science, Technology & Medicine, University of London, UK. Before this, he was a research affiliate in the International Motor Vehicle Program (IMVP) at the Massachusetts Institute of Technology in the US, and technology policy adviser to the Confederation of British Industry (CBI). His current research interests are in the relationships between technology development and market dynamics in complex industries, such as automobiles and telecommunications.

Giuseppe Volpato
is a professor of Management and Director of the Department of Business Economics and Management at Venice University, Ca'Foscari. He is currently involved in the ICDP within the IMVP of the M.I.T. and the GERPISA of the Université d'Evry Val d'Essonne. He has carried out extensive research in the automobile industry all over the world, participating and giving presentations at a number of conferences. He has published books, essays in books and articles in Italy, the UK and France.

Daniel E. Whitney
is a senior research scientist at the Massachusetts Institute of Technology. After teaching at M.I.T. from 1968 to 1974, he spent 19 years at Charles Stark Draper Laboratory, doing research and industrial applications in robotics, mechanical assembly, and CAD of assembly processes. At M.I.T., his research has focused on product development, CAD, and worldwide surveys of industrial productivity. He

is the author of over 80 technical papers and co-author of *Concurrent Design of Products and Processes*.

Bernd Wilhelm
Dr.-Ing. (doctorate in engineering), technical director at Volkswagen Brussels S.A, Belgium. He studied technical engineering at Brunswick University. He joined Volkswagen in 1973 as an engineer for Production Planning. In the 1980s, he was responsible for the Production Planning Department *Paint and Assembly*. Highlights: launching the automated car assembly *Halle 54* and the installation of their flexible variation as the second generation of automated car assembly systems. Afterwards: more then four years head of VW Production Strategy Department world-wide, General Production Manager of the Wolfsburg assembly area and special tasks in Europe and China.

Koichi Yamamoto
is an engineer at the Machine and Tool Engineering Division at the Nissan Motor Co., Ltd. He joined Nissan as a software system engineer in 1982. He was a visiting scholar at Stanford Robotics Laboratory between 1991 and 1992. His recent work has focused on robot offline programming.

MIX
Papier aus verantwortungsvollen Quellen
Paper from responsible sources
FSC® C105338

If you have any concerns about our products,
you can contact us on
ProductSafety@springernature.com

In case Publisher is established outside the EU,
the EU authorized representative is:
Springer Nature Customer Service Center GmbH
Europaplatz 3, 69115 Heidelberg, Germany

Printed by Libri Plureos GmbH
in Hamburg, Germany